Environmental Science

Series editors: R. Allan · U. Förstner · W. Salomons

Springer

Berlin
Heidelberg
New York
Barcelona
Budapest
Hong Kong
London
Milan
Paris
Santa Clara
Singapore
Tokyo

Walter Geller · Helmut Klapper · Wim Salomons (Eds.)

Acidic Mining Lakes

Acid Mine Drainage, Limnology and Reclamation

With 155 Figures and 55 Tables

Springer

Prof. Dr. Walter Geller
Prof. Dr. Helmut Klapper
UFZ Centre for Environmental Research Leipzig-Halle plc
Dept. Inland Water Research
Am Biederitzer Busch 12
D-39114 Magdeburg
Germany

Prof. Dr. Wim Salomons
GKSS Research Centre
Max-Planck-Strasse
D-21502 Geesthacht
Germany

ISSN 1431-6250
ISBN-13:978-3-642-71956-1

Library of Congress Cataloging-in-Publication Data
Acidic mining lakes : acid mine drainage, limnology, and reclamation / Walter Geller, Helmut Klapper, Wim Salomons, eds. p. cm. – (Environmental science) Includes bibliographical references and index.
ISBN-13:978-3-642-71956-1 e-ISBN-13:978-3-642-71954-7
DOI: 10.1007/978-3-642-71954-7 1. Limnology – Congresses. 2. Acid pollution of rivers, lakes, etc. – Congresses. 3. Acid mine drainage – Environmental aspects – Congresses. I. Geller, Walter. II. Klapper, Helmut. III. Salomons, W. (Willem), 1945-Series Entry:
GB1605.A28 1998 628.1/6832 21

Typesetting: Fotosatz-Service Köhler OHG, Würzburg
Coverdesign: Struve & Partner, Heidelberg

SPIN: 10509909 32/3020-5 4 3 2 1 0 – Printed on acid-free paper

Preface

Lignite was the most important source of primary energy in the former German Democratic Republic (GDR), reaching a share of 90% in 1988 with an output of 310 million tons of coal. This worldwide maximum of per-capita consumption of brown coal containing high amounts of sulfur, and out-dated technical equipment of power plants effected two kinds of acidification. The first is acid rain and long-distance acidification of soils and waters in sensitive areas via atmospheric transport. The second is more direct, affecting the near environment of the lignite open-cast mining sites due to geogenic acidification of the aerated overburden and groundwater from pyrite oxidation. Thus, the acid input into ground and surface waters has severe effects, resulting in mining lakes with pH values between 2 and 3.

Initially, the lignite mining was planned over long time scales. After shifting from lignite to other sources of primary energy after 1990, these plans became obsolete. The antedate for closing down the mines and filling the empty basins resulted in consecutive problems, which are now concentrated into one decade, while at least half a century was initially scheduled.

Facing the now-apparent problems of ground and surface waters threatened by sulfuric acid, limnologists of concerned institutions, such as the UFZ Centre for Environmental Research, organized the collection of scientific data and knowledge and practical experiences of how to handle these acidification problems. One part of these activities was an International Symposium on the "Abatement of Geogenic Acidification in Mining Lakes" held in September 1995 at the UFZ Department of Inland Waters Research in Magdeburg, Germany, where 26 invited presentations were discussed by 60 participants from the UK, Canada, the USA and Germany. The results of the meeting are presented in this volume which, hopefully, may be seen as an actual baseline for acid mining lakes. Since the symposium had a function to connect ideas and, thereby, to initiate new approaches, we are looking forward to a second symposium to bring together the large body of new results emerging from currently running work.

Walter Geller

April 1998, Magdeburg, Germany

Contents

Part 1
Introduction

Part 2
Limnological Case Studies on Acid Lakes

List of Contributors

Beisker, W.
GSF-Forschungszentrum, AG Durchflußzytometrie, Postfach 1129, D-85758, Oberschleißheim, Germany

Bender, J.
Clark Atlanta Univ., Georgia 30314, Atlanta, USA

Davison, W.
Lancaster Univ., Institute Environmental & Biol. Sciences, Lancaster, UK

Deneke, R.
Brandenburgische Technische Universität Cottbus, LS Gewässerschutz, Seestr. 45, D-15526 Bad Saarow, Germany

Dodds-Smith, M.
Knight Piésold, 35/41 Station Road, Kent TN23 1PP, Ashford, UK

Evangelou, V.P. (Bill)
Univ. of Kentucky, Agronomy Dept., KY 40546-0091, Lexington, USA

Fischer, R.
TU Dresden, Inst. für Wasserchemie & Wassertechnol., Mommsenstr. 13, D-01062 Dresden, Germany

Flather, D.H.
Rescan Environmental Services, Vancouver, Canada

Friese, K.
UFZ-Umweltforschungszentrum, Gewässerforschung, Am Gouvernementsberg 1, D-39104 Magdeburg, Germany

Geller, W.
UFZ-Umweltforschungszentrum, Gewässerforschung, Am Biederitzer Busch 12, D-39114 Magdeburg, Germany

George, D.G.
Institute for Freshwater Ecology, Windermere Laboratory, Ambleside, Cumbria LA22 0LP, UK

Glässer, W.
UFZ-Umweltforschungszentrum, Hydrogelogie, Hallesche Str. 44, D-06246, Bad Lauchstädt, Germany

Guderitz, T.
IDUS GmbH, Biol. Analyt. Environm. Lab., Dresdner Str. 43, D-01458, Ottendorf-Okrilla, Germany

Gusek, J.
Knight Piésold, Denver, USA

Hedin, R.S.
U.S. Bureau of Mines, P.O. Box 18070, PA 15236-0070, Pittsburgh, USA

Hupfer, M.
IGB-Insitut für Gewässerökologie & Binnenfischerei, Müggelseedamm 310, D-12587 Berlin, Germany

Johnson, D.B.
University of Wales, School of Biol. Sciences, LL57 2UW, Bangor, UK

Kalin, M.
Boojum Research Ltd., 468 Queen Street East, M5A 1T7, Toronto, Canada

Katzur, J.
Forschungsinstitut für Bergbaufolgelandschaften, Brauhausweg 2, D-03238 Finsterwalde, Germany

Klapper, H.
UFZ-Umweltforschungszentrum, Gewässerforschung, Am Biederitzer Busch 12, D-39114 Magdeburg, Germany

Kleinmann, R.L.P.
US Dept of Energy, P.O. Box 10940, PA 15236-0940, Pittsburgh, USA

Kwong, Y.T.J.
Klohn-Crippen Consultants Ltd., Suite 114, 6815-8th Street N.E., T2E 7H7, Calgary, Canada

Lamb, H.M.
Knight Piésold, 35/41 Station Road, Kent TN23 1PP, Ashford, UK

Lawrence, J.R.
National Hydrology Research Institute, 11 Innovation Boulevard, S7N 3H5, Saskatoon, Canada

Liebner, F.
Forschungsinstitut für Bergbaufolgelandschaften, Brauhausweg 2, D-03238, Finsterwalde, Germany

McNee, J.J.
Rescan Environmental Services, Vancouver, Canada

Mueller, B.
Univ. of British Columbia, Dept. Earth and Ocean Sciences, V6T 1Z4 Vancouver, Canada

Mull, R.
Universität Hannover, Institut für Wasserversorgung, Appelstr. 9a, D-30167 Hannover, Germany

Nairn, R.W.
Ohio State University, Columbus, OH, USA

Neu, T.
UFZ-Umweltforschungszentrum, Gewässerforschung, Am Biederitzer Busch 12, D-39114 Magdeburg, Germany

Nixdorf, B.
Brandenburgische Technische Universität Cottbus, LS Gewässerschutz, Seestr. 45, D-15526 Bad Saarow, Germany

Pedersen, T.F.
Univ. of British Columbia, Dept. Earth and Ocean Sciences, V6T 1Z4 Vancouver, Canada

Peiffer, S.
Universität Bayreuth, Limnologische Forschungsstation,
D-95440 Bayreuth, Germany

Peine, A.
Universität Bayreuth, Limnologische Forschungsstation,
D-95440 Bayreuth, Germany

Pelletier, C.A.
Rescan Environmental Services, Vancouver, Canada

Phillips, P.
Clark Atlanta University, 30314 Georgia, Atlanta, USA

Pietsch, W.
Brandenburgische Technische Univ. Cottbus, LS Bodenschutz & Rekultivierung, Postfach 101344, D-03013 Cottbus, Germany

Prein, A.
Prof. Mull & Partner GmbH, Osteriede 5, D-30827 Garbsen, Germany

Reißig, H.
Arltstr. 10, D-01189 Dresden, Germany

Salomons, W.
GKSS Research Centre, Max-Planck-Strasse, D-21502 Geesthacht, Germany

Schäfer, H.
GSF-Forschungszentrum, AG Durchflußzytometrie, Postfach 1129,
D-85758 Oberschleißheim, Germany

Schreck, P.
UFZ-Umweltforschungszentrum, Hydrogeologie, Hallesche Str. 44,
D-06246 Bad Lauchstädt, Germany

Scharf, B.
UFZ-Umweltforschungszentrum, Gewässerforschung, Am Biederitzer
Busch 12, D-39114 Magdeburg, Germany

Schimmele, M.
UFZ-Umweltforschungszentrum, Gewässerforschung, Am Biederitzer
Busch 12, D-39114 Magdeburg, Germany

Schultze, M.
UFZ-Umweltforschungszentrum, Gewässerforschung, Am Biederitzer Busch 12, D-39114 Magdeburg, Germany

Steinberg, C.E.W.
IGB-Institut für Gewässerökologie & Binnenfischerei, Müggelseedamm 310, D-12587 Berlin, Germany

Tittel, J.
UFZ-Umweltforschungszentrum, Gewässerforschung, Am Biederitzer Busch 12, D-39114 Magdeburg, Germany

Wendt-Potthoff, K.
UFZ-Umweltforschungszentrum, Gewässerforschung, Am Biederitzer Busch 12, D-39114 Magdeburg, Germany

Wisotzky, F.
Ruhr-Univ. Bochum, Fak. Geowiss., LS Geologie III, Universitätsstr. 150, D-44801 Bochum, Germany

Wollmann, K.
Brandenburgerische Technische Universität Cottbus, LS Gewässerschutz, Seestr. 45, D-15526 Bad Saarow, Germany

Part 1
Introduction

1 Natural and Anthropogenic Sulfuric Acidification of Lakes

W. Geller, H. Klapper and M. Schultze

UFZ Centre for Environmental Research, Leipzig Halle, Department of Inland Water Research, Am Biederitzer Busch 12, 39114 Magdeburg, Germany

1.1 Introduction

Since the early 1960s, acid rain has affected surface waters in regions with soils low in carbonates. The acidification of whole lake districts resulted from industrial emissions distributed through the atmosphere on a regional scale. A large body of literature exists about this acidification of surface waters by acid rain, its ecological consequences, and countermeasures to restore landscape and waters (e.g., Drablos and Tollan 1980; Fleischer 1993; Steinberg and Wright 1994; Stumm 1995). The hydrochemistry and the ecological consequences were extensively described (e.g., Dillon et al. 1984), leaving the impression that all acidic lakes are anthropogenic and artificial waters that principally need to be restored, for instance by liming measures (Henrikson and Brodin 1995).

Strong acids may be introduced into surface waters by two further pathways, at least. Sulfuric acid in lakes and rivers can originate from volcanic sources and from mining activities. Here, the transport is linked to the flow of water and, therefore, is locally restricted. Both the natural process and that induced by human activities can lead to nearly identical results: strongly acidic lakes with pH values below 4, often accompanied by high concentrations of dissolved iron, and in some cases by further metals.

1.2 Rain-Acidified Softwater Lakes

Due to the pH of acid rain, only softwaters with their low buffering capacity are affected. One characteristic of rain-acidified softwaters is the increased content of dissolved aluminum, which has two effects: (1) it is toxic for the biota; (2) the aluminum has a beneficial effect as its ionic and hydroxide species are a buffering system keeping the pH range between 5.5

and 4.5. Therefore these rain-acidified waters might be generally described as 'sulfur-acidic/aluminum-buffered' systems.

1.3 Strongly Acidic Lakes in Volcanic Areas

There has been no recent review on acidic crater lakes or other volcanic waters, although some early investigations exist (Ueno 1934, 1958; Yoshimura 1934; Ohle 1936; Perwolf 1944, cited in Zhadin and Gerd 1963; Hutchinson 1957). The publications are restricted to descriptions of the volcanic lakes of New Zealand (Timperley 1987) and to singular case studies (e.g., Ueno 1934; Yoshimura 1934; Ivanov and Karavaiko 1966; Satake and Saijo 1974). Most of the early papers were published in Japanese

Table 1.1. Acidic crater lakes and their chemical characteristics

Lake	PH	Sulfate (g/m^3)	Fe (g/m^3)	Location
Kawah Idjen	0.7			Indonesia
Yugama	0.9	1.656	163	J
Katanuma	1.4–1.8	1.003	5.8	J
Rainbow North	1.8	1.530		NZ
Ixpaco	2.3			Guatemala
Rotokawa	2.3	408		NZ
Echo	2.4	380		NZ
Rainbow South	2.5	162		NZ
Okama	2.7–2.9	421	10.8	J
Ngakoro	2.8	369		NZ
Kipyashch	2.8	238		Kurile Islands
Goryach	3	210		Kurile Islands
Akadoronuma	3	3.476	80	J
Rotowhero	3.1	205		NZ
Emerald	3.2	48		NZ
Osoresau	3.5			J
Akanuma	3.8	649	9.7	J
Onuma-ike	4.1	94	0.1	J
Opal	4.2	42.2		NZ
Misuma-ike	4.3	69	0.1	J
Fodo-ike	4.5		0.13	J
Rokkannon-miike	5.1		Tr	J
Blue	5.2	0.72		NZ
Byakushi-ike	5.5		Tr	J

Former Soviet Union and Japan (J) after Zhadin and Gerd 1963; Ivanov and Karavaiko 1966; Satake and Saijo 1974; NZ, New Zealand after Timperley 1987; other countries after Hutchinson 1957; Tr, trace.

or Russian, reflecting the geographical distribution of crater lakes following the geological ridge of the Japanese and Kurile Islands to Kamchatca. Additionally, there are a few reports from other regions such as Indonesia or Guatemala (Hutchinson 1957). Data on chemical characteristics of these acidic volcanic lakes are listed in Table 1.1.

1.4 Acidic Mining Lakes

The acidification of dumps and waters due to acidic mine drainage (AMD) is a common problem in many mining districts (e.g., Fischer et al. 1987; Alpers and Blowes 1994; US Department of the Interior 1994). Considerable quantities of the deposit soils in the dumps contain iron sulfides, pyrite, and marcasite, which are oxidized when the layers are removed and mixed, and, thereby, aerated (Wisotzky, Chap. 11, and Evangelou, Chap. 10, this Vol.). In a first step, oxidation of FeS_2 results in dissolved acidic products. The acidic components, H_2SO_4, and iron species can be washed out by leaching rainwater and/or by the fast-rising groundwater table of closed-down lignite pits. Thus, many mining lakes are expected to become strongly acidic with high contents of iron after filling of the open-cast basin.

1.5 The State of Mining Lakes in East Germany

Within the coal-mining districts of the central European countries Germany, Poland, Czechia and Slovakia, the districts of East Germany covered the most important European reserves of lignite (Fig. 1.1). After the unification of Germany, the close-down of lignite mines has been drastically accelerated (von Bismarck 1993), leaving a basic activity of about 30 % in 1993, as compared with a maximum in 1988 (Table 1.2).

The open-cast mining lakes of East Germany show the full scale from very acidic to neutral. This spectrum might reflect an aging process from young waters to old lakes (Pietsch 1979 and Chap. 9, this Vol.). The duration of this process is presumed to be relative and lake-specific. In the Lusatian district, some small lakes are well protected against wind and, therefore, have a high stratification stability as a prerequisite for the observed biological neutralization in the hypolimnion. In contrast, some large lakes are very acidic, with no major changes for about four decades. These well-mixed, large and deep, acidic mining lakes obviously need the allochthonous alkalinity of riverwater to become circumneutral. For example, the water of the main basin of Senftenberger See is neutral because it has

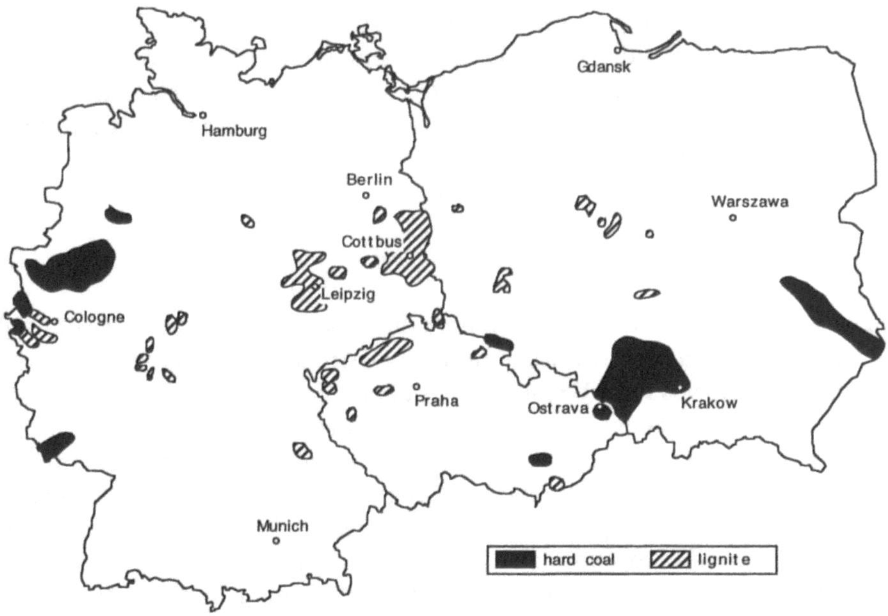

Fig. 1.1. Coal mining districts in the countries of central Europe, Germany, Poland, Czechia, and Slovakia. (Redrawn from Pätz et al. 1989)

Table 1.2. Output of brown coal and number of active surface mines in the two lignite mining districts of East Germany (former GDR) between 1963 and 1993. (After Möbs and Maul 1994)

	Central German district		Lusatian district	
	Output of lignite in 10^6 tons/year	Number of mines	Output of lignite in 10^6 tons/year	Number of mines
1963	145.5	–	108.7	17
1985	115.4	20	196.8	17
1988	109.8	20	200.5	17
1989	105.6	20	195.2	17
1990	80.9	19	168.0	16
1991	50.9	11	116.8	12
1992	36.3	9	93.1	9
1993	28.5	5	87.5	9
Plan 2000	< 20	3	70	5

Table 1.3. List of 19 open-cast mining lakes in East Germany; pH values and concentrations of total iron and aluminum (μmol/l) as investigated in July 1993

Mining lake	pH	Fe	Al	Fe:Al
Schlabendorf	2.69	3040	760	4.0
Skado	2.7	2690	550	4.9
Restloch 78	2.85	555	111	5.0
Sedlitz	2.9	700	471	1.5
Restloch 117	2.94	304	79	3.9
Halbend. See	3.0	358	73	3.9
Koschen	3.08	251	48	5.2
Waldsee	3.26	17.9	8.5	2.1
Roter See	3.3	172	148	1.2
Cospuden	3.4	491	742	0.7
Auenhain [a]	3.9	35.8	196	0.2
Felixsee	3.95	5.6	146	0.04
Rösa	4.0	12	52	0.2
Blauer See	6.8	8.2	11	0.8
Markkleeberg	7.5	21.5	8.0	2.7
Grüner See	7.6	3.4	3.7	0.9
Schlabend. North Basin	8.28	1.6	1.5	1.1
Hufeisensee	8.4	1.8	1.1	1.6
Hörlitz	9.05	19.7	33.7	0.6

[a] September 1992.

been exchanged. In a side basin where the riverine throughflow is lacking, the water is still acidic (Klapper et al., Chap. 22, this Vol.). A list of a few, recently investigated lakes and their chemical characteristics are presented in Table 1.3.

1.6 Ranges of pH in Acidic Lakes

Usually the pH of circumneutral lakes is stabilized by the CO_2–bicarbonate–carbonate-buffering system, which dominates the ionic composition of neutral waters in the range from softwater to hardwater lakes independently of different ionic strengths. With addition of acid the buffering capacity of the carbonate system is largely lost beyond pH 6, resulting in a rapid decrease in the pH value with further addition of acid. In softwater lakes and rivers acidified by acid rain, the pH range was found again to be stabilized between pH 4.5 and 5.5. This is caused by the aluminum species that are dissolved by the acidic rainwater. The intermediate transition range between pH 6.0 and 5.5, however, where the respective buffer is lacking, can

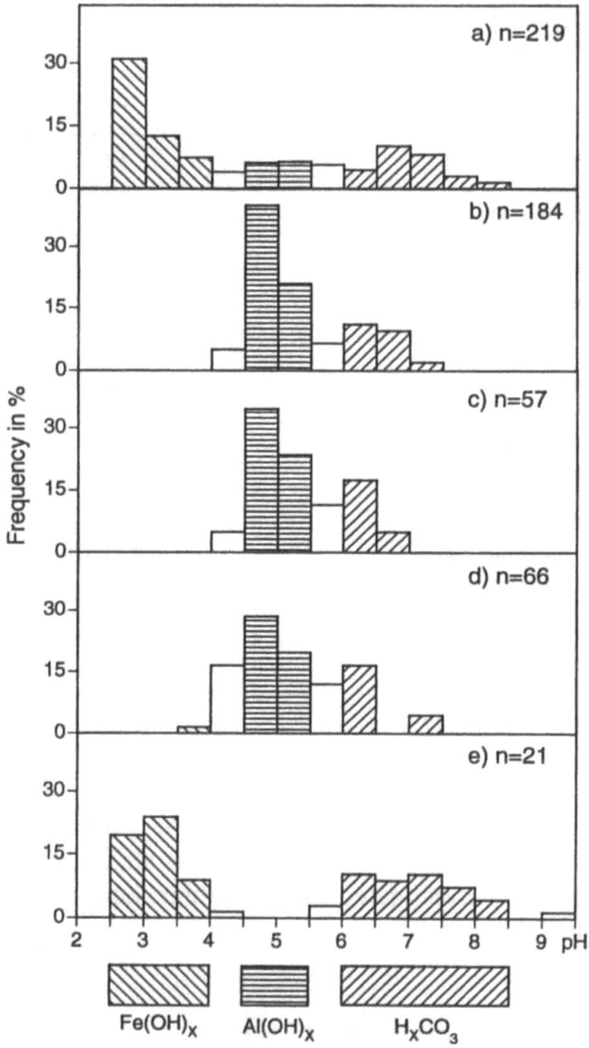

Fig. 1.2. Frequency distributions of lake pH in the East-German lignite districts: **a** 219 lakes (after Pietsch 1979) and **e** 21 lakes (from an investigation of the authors of this chapter); and **b** 184 lakes in the USA; **c** 57 lakes in rain-acidified areas in Norway; and **d** 66 lakes in Canada. (After Harvey 1980)

be found only episodically, i.e., after snow melt. In many acid-rain-affected districts of softwater lakes in Europe and North America (e.g., Harvey 1980), a bimodal frequency distribution of lake pH is observed. A comparison with strongly acidic mining lakes in Germany (Table 1.3) showed a third maximum between pH 2 and 4 (Fig. 1.2).

The solubility of metals in water increases with lower pH (Stumm and Morgan 1981), leading to elevated concentrations of Al and Fe. In the strongly acidic mining lakes, therefore, concentrations of Fe often reach amounts of 1 kg Fe/m^3. The different species of ferric hydroxides and the ional Fe(III)-form show a buffering capacity which is comparable with that of Al and of the carbonate system. This Fe-buffering system stabilizes the pH values between 2 and 4 in surface waters. In soils with more complex conditions the efficacy of the Fe buffer is assigned to the range of pH 2.4–3.0 (Ulrich 1981).

A comparison of Table 1.3 and Fig. 1.2 shows that the three different buffering systems are found in different lake types, e.g., in neutral and acidified softwater lakes, and in mining lakes. Presumably, this also holds for volcanic lakes from which only a limited data set is available (pH and sulfate; Table 1.1). With mining lakes, the three types of buffers occur within the same districts. In the Lusatian and the central German mining areas, there are (1) many neutral lakes; (2) a few hardwater lakes with higher concentrations of Al than of Fe (these lakes show pH values near to those of rain-acidified softwater lakes); and (3) many strongly acidic lakes where Fe is dominating. The molar ratio of Al and Fe apparently determines the dominating buffering system in types (2) and (3). The Al:Fe ratio, however, may be negligible in the neutral lakes because of the low levels of Al and Fe.

The authors propose a chemical typology of lakes based on the ionic concentrations in softwater and hardwater lakes (Fig. 1.3). This figure shows three ranges of pH for both softwater and hardwater lakes, depend-

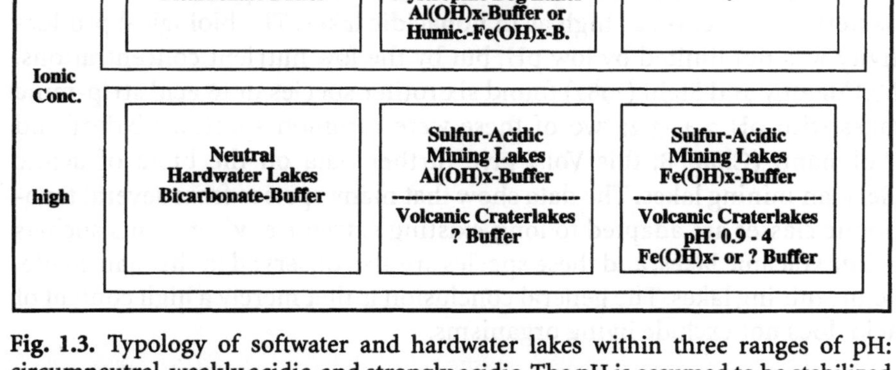

Fig. 1.3. Typology of softwater and hardwater lakes within three ranges of pH: circumneutral, weakly acidic, and strongly acidic. The pH is assumed to be stabilized by three different buffering systems that may be modified by organic substances in dystrophic lakes

ing on the given buffering systems: pH 6–8 (carbon), pH 4.5–5.5 (aluminum), and pH 2–4 (iron). In weakly acidic, humic bog lakes the buffering system can be modified by interactions between the dissolved metals and humic substances. No softwater lake with strongly acidic characteristics has been described until now, and the question is open whether or not such lakes exist. On the whole, however, the scheme covers all types of sulfur-acidic lakes and dystrophic modifications.

1.7 Acidic Mining Lakes: Extreme Habitats and Environmental Problems

1.7.1 Biota of Strongly Acidic Lakes

Strongly acidic mining lakes seem to be highly artificial, extreme habitats without natural counterparts, and one would expect extremely poor communities in such lakes owing to the data from acidified softwater lakes (Dillon et al. 1984). However, investigation of the biota of volcanically acidic crater lakes revealed the presence of a multitude of species which are well adapted to the extreme but natural environment. Such lakes show communities composed of few species, but these may occur in high numbers, as usual in extreme habitats. From 22 lakes in Yellowstone National Park, USA, Brock (1973) reported that blue-green algae were missing above pH 4, but he found 15 eukaryotic algal species in the pH range 1.9–4.0. In the Japanese crater lake Katanuma with the extreme pH of 1.8, Yamamoto et al. (1997) found a densely green population of *Chlamydomonas acidophila*.

In acidic strip-mine lakes of pH 3, Gyure et al. (1987) investigated the activity of algae and bacteria. They found rates of photosynthesis and bacterial production as high as in nonacidic lakes. The biological productivity was not limited by low pH, but by the low nutrient concentrations. McConathy and Stahl (1982) found six rotifer species in 16 acid strip-mine lakes with pH 2.4–3.2; two of these were common species. Nixdorf and Wollmann (Chap. 8, this Vol.) give further data on the biota of acidic German mining lakes. The data show that many species from several taxonomic classes are adapted to long-existing extreme environments such as acidic volcanic lakes, and these species are also observed in the man-made, acidic mining lakes. The general conclusion is that merely a high content of acid does not exclude living organisms.

1.7.2
Acidic Mining Lakes as an Environmental Problem: A Scaling Approach

Any demand for the restoration of multitudes of differently sized acid lakes has to face important technical and economical consequences. Hemond (1994) presented a ranking scheme discriminating anthropogenically altered lakes which have to be restored to a more 'natural' state and naturally acidic lakes which have to be protected. Following this approach, it is a crucial point whether the acidic waters appear as natural ones or as anthropogenic and artificial waters. The acidic mining lakes are 'natural' in regard to the source of acidity, which is pyrite, and the formation of acid by a natural weathering process. The lake basins, however, are man-made, and by moving and, thereby, aerating the pyrite-containing overburden, the weathering process is anthropogenically induced.

This example shows that the above criterion alone of being 'natural' or 'man-made' is an insufficient basis for political decisions. Therefore, additional and more appropriate criteria to judge the importance of this environmental problem are required. A promising approach is scaling the problem in time and in spatial dimensions.

1.7.2.1
Time Scales of Acidification

In a group report, Emmet et al. (1994) summarized the sources, pathways, and the resulting time scales of acidification of surface waters. In Fig. 1.4, the pool of sulfuric acid with its different sources and the time scale of sinks is shown. This figure, considering only the acidification processes via the element sulfur, is based on the scheme in Emmett et al. (1994). The authors added volcanic sources of acid and the anthropogenically induced weathering process of sulfidic minerals such as pyrite in mining dumps. Emmett et al. discriminated two categories of response time and time of maintained acidic conditions: episodic and chronic acidification. The authors added an intermediate time scale of multiannual to decennial acidification. This time scale appears to be more appropriate to judge processes observed over a few decades, such as acid rain and the strong acidification coming from mining dumps and tailings.

1.7.2.2
Spatial Scales of Acidification from Mining Sources

Black and brown coal is found in all continents (Pätz et al. 1989). In some areas, the open-cast mines are dry (Australia), but in most temperate

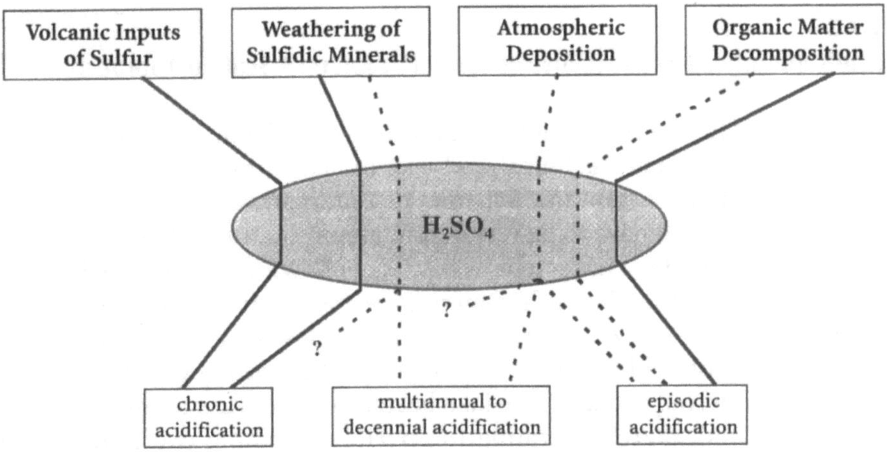

Fig. 1.4. Pool and sources of sulfur acidity and effects on surface waters, ranked by duration of observed episodic, intermediate, and chronic effects. The natural processes of acid formation or pathways are presented by a *solid line*; anthropogenically caused processes and pathways are shown by a *broken line*. (Based on a scheme in Emmett et al. 1994)

regions the rest holes will be flooded after closing the mining activities. The spatial extension of the brown-coal mines in East Germany enlarges local problems to an environmental problem of regional scale (BMU 1994). Besides the described central European areas, large areas in North America east of the Rocky Mountains, in western and central Siberia, and also in China are facing the same problem of acidic mining lakes. The spatial extensions of these districts urge the scientific community to look for simple and cheap approaches to restore these acidic lakes or to find ways of avoiding the acidification process. The question of restoring a lake or protecting a given area is one of the items addressed in the group discussion chapter (Chap. 25, this Vol.).

1.8 Conclusions

Many mining lakes are strongly acidic and are considered to present a major environmental problem. Some volcanic lakes, too, are very acidic, being the natural counterparts of mining lakes. Many aquatic species are able to live in these anthropogenic and natural extreme habitats, showing that merely a high content of acid does not suppress living organisms. However, faced with some hundred open-cast lakes in one mining district, it is not acceptable to leave the expected 50% which are or will become acidic untreated. We may maintain some of these peculiar sulfur-acidic lakes as subjects for comparative scientific research on extreme habitats, but the majority of these artificial lakes ought to become accessible for public use. The aim of this book is to specify approaches and concepts with respect to this goal which are scientifically well based and technically and economically applicable.

References

Alpers CN, Blowes DW (eds) (1994) Environmental geochemistry of sulfide oxidation. American Chemical Society, Washington, DC

von Bismarck F (1993) Changes of German lignite mining: just the effects of reunification? In: OECD/IEA (ed) The clean and efficient use of coal and lignite: its role in energy, environment and life. Conf Proc, Hong Kong, pp 667–673

BMU (ed) (1994) Umweltpolitik. Erster Bericht der Regierung der Bundesrepublik Deutschland. Bundesministerium für Umwelt, Naturschutz und Reaktorsicherheit, Bonn

Brock TD (1973) Lower pH limit for the existence of blue-green algae: evolutionary and ecological implications. Science 179:480–483

Dillon PJ, Yan ND, Harvey HH (1984) Acidic deposition: effects on aquatic ecosystems. CRC Crit Rev Environ Control 13:167–194

Drablos D, Tollan A (eds) (1980) Ecological impact of acid precipitation. Proc Int Conf, SNSF Project, 11–14 March, Sandefjord, Norway

Emmett D, Charles D, Feger KH, Harriman R, Hemond HF, Hultberg H, Leßmann D, Ovalle A, Van Mietgroet H, Zoettl HW (1994) Group report: can we differentiate between natural and anthropogenic acidification? In: Steinberg CEW, Wright RF (eds) Acidification of freshwater ecosystems. Wiley, Chichester, pp 117–140

Fischer R, Reißig H, Peukert D, Hummel J (1987) Untersuchungen zur Beeinflussung der Markasitverwitterung. Neue Bergbautechnik 17:60–64

Fleischer S (ed) (1993) Acidification of surface waters in Sweden – effects and counteraction measures. AMBIO 12(5):257–337

Gyure RA, Konopka A, Brooks, A, Doemel W (1987) Algal and bacterial activities in acdidic (pH 3) strip mine lakes. Appl Environ Microbiol 53:2069–2076

Harvey HH (1980) Widespread and diverse changes in the biota of North American lakes and rivers coincident with acidification. In: Drablos D, Tollan A (eds) Ecological impact of acid precipitation. Proc Int Conf, SNSF Project, 11–14 March, Sandefjord, Norway

Hemond HF (1994) Role of organic acids in acidification of fresh waters. In: Steinberg CEW, Wright RF (eds) Acidification of freshwater ecosystems. Wiley, Chichester, pp 103–115

Henrikson L, Brodin YW (eds) (1995) Liming of acidified waters. Springer, Berlin Heidelberg New York

Hutchinson GE (1957) A treatise on limnology, vol 1. Geography, physics, and chemistry. Wiley, New York

Ivanov MV, Karavaiko GI (1966) The role of microorganisms in the sulphur cycle in crater lakes of the Golovin caldera. Z Allg Mikrobiol 6:10–22

McConathy JR, Stahl JB (1982) Rotifera in the plankton and among filamentous algal clumps in 16 acid strip-mine lakes. Transact Ill Acad Sci 75:85–90

Möbs H, Maul C (1994) Sanierung der Braunkohlegebiete in Mitteldeutschland und in der Lausitz. Wasserwirtschaft Wassertechnik 1994/3:12–18
Ohle W (1936) Der schwefelsaure Tonteich bei Reinbek. Monographie eines idiotrophen Gewässers. Arch Hydrobiol 30:604–662
Pätz H, Rascher J, Seifert A (1989) Kohle – ein Kapitel aus dem Tagebuch der Erde, 2. Aufl. Teubner Verlag, Leipzig
Perwolf JW (1944) The acidic lakes of Japan. Priroda 1,2 (in Russian)
Pietsch W (1979) Zur hydrochemischen Situation der Tagebauseen des Lausitzer Braunkohlen-Revieres. Arch Naturschutz Landschaftsforsch Berl 19:97–115
Satake K, Saijo Y (1974) Carbon content and metablic activity of microorganisms in some acid lakes in Japan. Limnol Oceanogr 19:331–338
Steinberg CEW, Wright RF (eds) (1994) Acidification of freshwater ecosystems – implications for the future. Environmental science reports 14. Wiley, Chichester
Stumm W (ed) (1995) Chemical processes in lakes. Wiley, New York
Stumm W, Morgan JJ (1981) Aquatic chemistry, 2nd edn. Wiley, New York
Timperley MH (1987) Regional influences on water chemistry. In: Viner AB (ed) Inland waters of New Zealand. DSIR Bull 241, Wellington, pp 97–111
Ulrich B (1981) Die Rolle der Wälder für die Wassergüte unter dem Einfluß des sauren Niederschlags. Agrarspektrum 1:212–231
Ueno M (1934) Acid water lakes in north Shinano. Arch Hydrobiol 17:297–302
Ueno M (1958) The disharmonic lakes of Japan. Verh Int Ver Limnol 13:217–226
US Department of the Interior (1994) International Land Reclamation and Mine Drainage Conference and 3rd Int Conf on the Abatement of acidic drainage, vols 1–4: mine drainage. Proc Conf, Pittsburgh 1994, Bureau of Mines Spec Publ SP 06 A/B/C/D–94
Yamamoto Y, Tatsuzawa H, Wada M (1997) Effect of environmental conditions on the composition of lipids and fatty acids of *Chlamydomonas* isolated from an acid lake. Proc Int Assoc Limnol 26 (in press)
Yoshimura S (1934) The most acidic waters of the world: Katanuma by Naguro hot spring. Science 4:498–499
Zhadin VI, Gerd SV (1963) Fauna and flora of the rivers, lakes and reservoirs of the U.S.S.R. Translated from Russian and published by Israel Program Scientific Translations, Jerusalem

2 Regional Geology of the Lignite Mining Districts in Eastern Germany

P. Schreck and W. Glässer

UFZ Centre for Environmental Research, Department of Hydrogeology, Hallesche Str. 44, 06246 Bad Lauchstädt, Germany

2.1 Introduction

This introduction to the regional geology of the two lignite mining districts (the central German district and the Lusatian district) in eastern Germany is a condensed synopsis to provide basic information to those unfamiliar with the region's geological conditions. It is intended to highlight the geological background and the composition of the lignite-bearing Tertiary sequences. A number of facts have been simplified in order to illustrate the characteristics of the two mining districts.

2.2 Lignite Mining Districts in Germany

Lignite has been mined in Germany since the seventeenth century. In the past, even underground mining was used to obtain lignite as heating fuel; nowadays only open-pit mining is carried out. Germany contains a total of eight lignite mining districts. The three most important, from west to east, are the Lower Rhine district (Cologne, Aachen) in western Germany, and in eastern Germany, the central German district (Saxony, Saxony-Anhalt), and the Lusatian district (mainly in Brandenburg) (Fig. 2.1). All subsequent characteristics and descriptions refer to the two latter mining districts in eastern Germany.

In former state of East Germany, lignite was the only domestic fossil energy resource of economic relevance. About 50 % of the coal was used as a raw material for the chemical and petrochemical industries, and the other 50 % for power generation and heating. Until 1989, lignite mining continuously increased, reaching a peak of 330 million tons/year. Current production is down to about 100 million tons/year, used for industrial power generation and domestic heating. The annual output target for the end of the century is a steady rate of 70 – 90 million tons (BMU 1994).

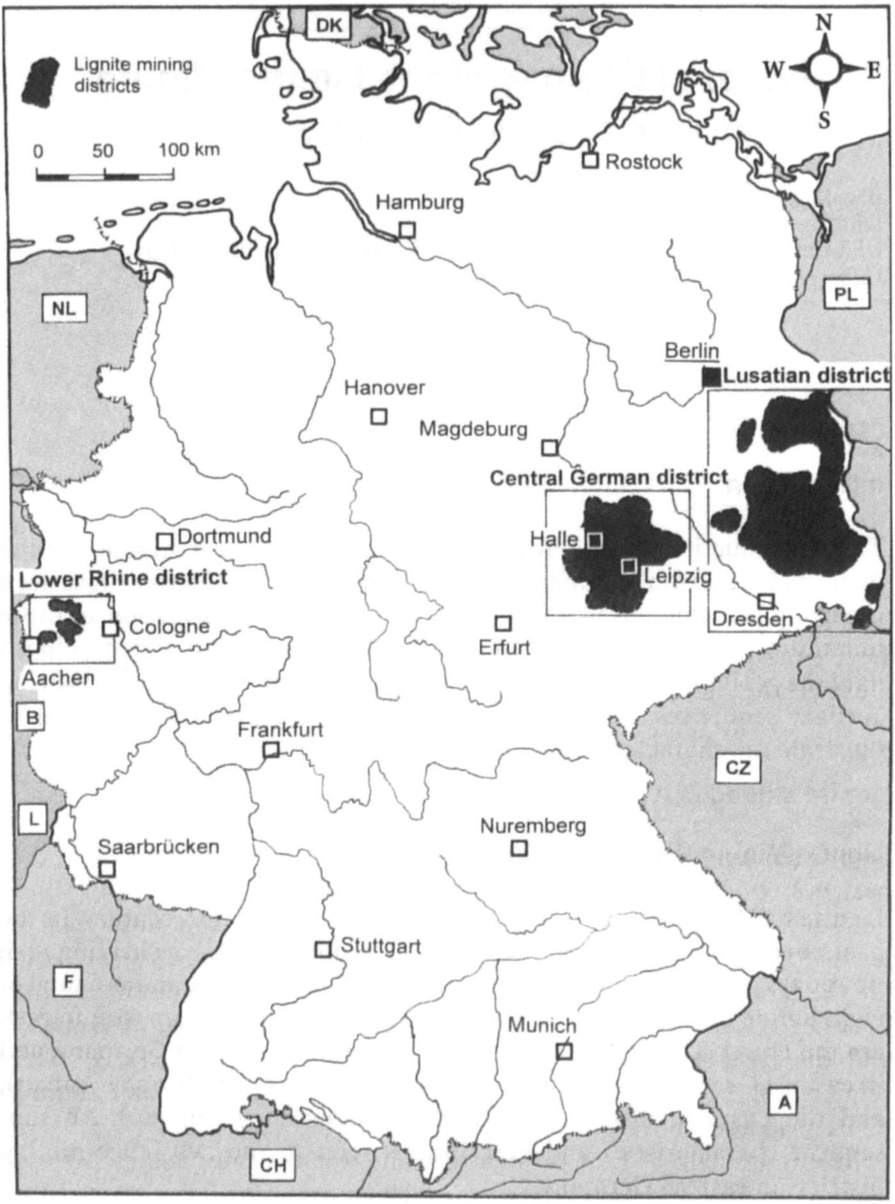

Fig. 2.1. Major lignite mining districts in Germany

2.3 Basic Geology

All lignite deposits in eastern Germany stem from the Tertiary period. They were formed from the Lower Eocene to the Upper Miocene, between 50 and 10 million years ago (Fig. 2.2). The Tertiary was a very active period in the earth's history. The final uplift of the Alps took place and vast areas of Europe were flooded during the transgression of the Rupel Sea. All these activities were accompanied by basaltic volcanism in central Europe. During the Tertiary, most of northern and eastern Germany and Poland was a depression area. Various successive marine transgressions led to the formation of silts and sands. Forests and swamps developed in the marginal parts of the Tertiary basin in areas of freshwater inflow. The remaining peat layers were covered by sands and silts from subsequent transgressions. Thus, the Tertiary strata profile comprises marine, estuarine, and fluviatile sediments such as marine sands, silts, and clays, continental sands and conglomerates (alluvial sediments), estuarine micaceous, carbonous silts and sands, and lignite seams of lenticular or stratiform shape.

In the Quaternary, most of the effects on the lignite-bearing sequences (glaciofluvial erosion, glacigenous foliation) were induced by the masses of ice during the glaciations in the Pleistocene epoch.

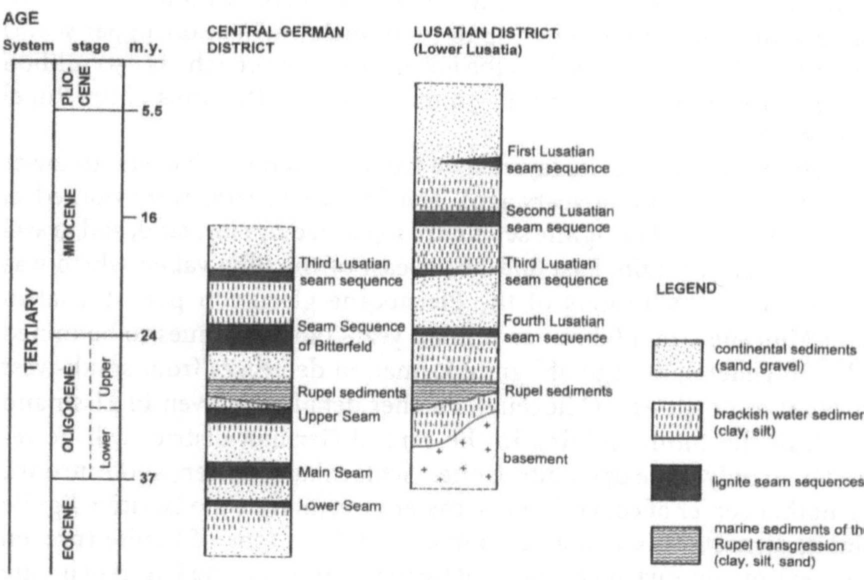

Fig. 2.2. Generalized stratigraphic sequence in the central German and the Lusatian district. *m.y.* Million years. (After Eissmann 1994; Nowel et al. 1995)

2.4 Regional Geology

The oldest lignite deposits in eastern Germany occur in the central German district and date back to the lower Eocene epoch (50 million years ago). Locally restricted lignite seams with intercalated fine-grained sediments can be found in erosional structures such as channels and pockets. These cavities were caused by the leaching of salt, gypsum, and by karst phenomena in the basal rocks, mainly consolidated Triassic and Permian sediments. These lignite stocks are of small scale. The thickness of such lenticular coal accumulations may reach up to 100 m, such as in the Geisel valley. Similar occurrences can be found in the lignite pits of Merseburg (east), Teutschenthal, Amsdorf, and Helmstedt. At these places, salt-bearing coal is present. This involves particular problems in flooding the worked-out open pits.

Most of the lignite is mined from the large, expansive deposits, where several coal seams are present. These deposits are of epicontinental origin, formed at the southern margin of the large Tertiary basin of northern Germany. In the lignite mining district of central Germany, the oldest stratiform coal seams occur in Eocene strata at the southern margin of the White Elster basin. This area contains two lignite layers dating from the middle to upper Eocene epoch (lower seam, 45–40 million years), separated by clay, sand, and gravel. More to the north, around Borna, two younger, minable lignite layers up to 10 m thick (main seam, upper seam) can be found. They date back to the lower Oligocene epoch (35–30 million years). The coal is covered by 20–30 m of marine silty sands of the Rupel transgression.

North of Leipzig, lignite layers date from the upper Oligocene to lower Miocene (30–22 million years ago). The four coal seams were worked in the Bitterfeld area. The lignite seams are separated by clay, sand, and gravel layers of fluvial origin. East and north-east of the Elbe valley, which was formed by the meltwaters of the Pleistocene glaciation period, coal in lower Miocene strata (about 20 million years old) continues to be mined today. In general, the age of lignite formation decreases from south-west (Eocene) to north-east (Miocene). Further details are given in Eissmann (1994). In the future, mining in the central German district will be restricted to only three open pits: United Schleenhain, Profen, and Amsdorf.

Another center of coal mining in eastern Germany is the Lusatian lignite mining district. This vast area contains 13 billion tons of lignite (proven reserves), or one-quarter of the total German reserves. The Lusatian lignite mining district comprises two lithologically different areas: the economically more important area of Lower Lusatia, situated mainly in Brandenburg, and the area of Upper Lusatia in Saxony. Details are given in Nowel et al. (1995).

In Lower Lusatia, lignite is mined in nine open pits. The main source is the second lignite seam dating back to the Miocene epoch (16 million years ago). In central parts of Lower Lusatia, the thickness of the second lignite seam is 10–14 m. The formerly continuous lignite stratum was subject to erosion in the Quaternary, when ice and meltwater disintegrated the coal body, creating isolated 'coal fields.' In the north-western and western part of Lower Lusatia, two more lignite seams occur below the mining target: layers III and IV. Layers I (above the main seam) and IV are present in the center and to the east of the area. All coal seams are of Miocene age. Layers I, III, and IV are only of local economic importance.

In Lower Lusatia, mining is concentrated on three areas: (1) the ice-marginal valley of Baruth with the active open pits at Cottbus and Jänschwalde; (2) deposits close to the 'partition wall of Lower Lusatia,' a terminal moraine of Saale III glaciation [e.g., Seese (east) and Welzow]; (3) the ice-marginal valley of Lusatia with the open-pit mines Senftenberg and Lauchhammer.

Upper Lusatia is of minor importance for lignite mining. Only one deposit is mined at present: the open pit at Berzdorf by Görlitz. The Tertiary sedimentary basins of Upper Lusatia are small. They were formed by tectonic subsidence, resulting from the uplifting of the Lusatian granodiorite massif in the Oligocene/Miocene. The lignite seams are up to 100 m thick. Volcanic rocks of Tertiary age are to be found in the vicinity of the coal.

The most obvious difference in the lignite-bearing sequences of Lusatia and central Germany is the lithology of the cover sediments. In the Lusatian mining district, especially in the ice-marginal valleys, most of the lignite-cover sediments are of glaciofluvial origin. These are pure quartz sands and other sediments of very uniform grain size. Equigranular sediments show unfavorable soil characteristics. One particular danger is liquefaction, which affects the slope stabilities of the restored mine dump areas during flooding. The rapid flooding of the worked-out open pits would help to prevent such effects. Unfortunately, such measures are impeded by the restricted availability of groundwater in this area.

In the lignite mining district of central Germany, in particular in the White Elster basin and the Bitterfeld area, the lithology of the lignite-cover sediments is more favorable for remediation measures. Both the Quaternary and the Oligocene cover sediments exhibit wide variations in their lithologic composition. The cover layers are composed of sediments from preglacial river terraces, basal and lodgment tills, meltwater sands, and clays.

2.5 Environmental Impact of Lignite Mining

Open-pit lignite mining requires the water table to be lowered by up to 100 m and the cover sediments above the lignite seams (including the upper aquifers) to be removed. After closure of the mine, rising water fills the remaining cavities and eventually leads to the formation of the popular mining lakes. These mining lakes act as a link between the various groundwater horizons. Many of these hydraulic short-cuts already exist and their number is set to increase (Glässer 1995).

Depending on the nature of the deeper hard rock aquifers, mineralized groundwater enriched with chloride or sulfate may circulate through the redeposited mine dump sediments and finally contaminate shallow freshwater aquifers. Examples of such a development in the central German district are Merseburg (east) (chloride) and the lignite-mining area south of Leipzig (Triassic and Permian hard rock aquifers; sulfate). Another type of environmental contamination was caused by the injection of phenolic wastewater from low-temperature carbonizing plants into hard rock aquifers. Like the mineralized groundwater, phenolic wastewater spreads out in the redeposited sediments of lignite mining, finally penetrating the freshwater aquifers.

Flooding worked-out open pits triggers and encourages chemical reactions and pollutant migration. During flooding, the flow-rate in the sediments increases considerably, both in the layered and in the dumped, disturbed sediments. Oxygenated water mobilizes and disseminates inorganic and organic components. Owing to the waste deposits contained in these open pits, flooding causes severe problems by dispersing pollutants.

Both lignite mining districts, the Lusatian and the central German district, suffer from the same hydrochemical problem: the oxidation of sulfide minerals in mine dump material and consequently the acidification of mining lakes and the leaching of metals from the sediment. In Lusatia, the cover sediments have a very low chemical buffer capacity and acidic water cannot be neutralized. In central Germany, acidic percolating water is partly buffered by carbonate minerals in the cover sediments. The dissolution of carbonates in the mine dump material affects the stability of slopes and many cases of slope slides are known from mining lakes during flooding.

References

BMU (German Ministry of the Environment, Conservation and Reactor Safety) (1994) Ökologischer Aufbau. Braunkohlesanierung Ost. Bundesumweltministerium, Bonn
Eissmann L (1994) Leitfaden der Geologie des Präquartärs im Saale-Elbe-Gebiet. In: Eissmann L, Litt T (eds) Das Quartär Mitteldeutschlands. Altenburger Naturwiss Forsch 7:11–46
Glässer W (1995) Der Einfluß des Braunkohlenbergbaus auf Grund- und Oberflächenwasser. Geowissenschaften 8/9:291–296
Nowel W, Bönisch R, Schneider W, Schulze H (1995) Geologie des Lausitzer Braunkohlereviers. Lausitzer Braunkohle Aktiengesellschaft, Senftenberg

Part 2
Limnological Case Studies on Acid Lakes

3 Chemical Characteristics of Water and Sediment in Acid Mining Lakes of the Lusatian Lignite District

K. Friese[1], *M. Hupfer*[2] *and M. Schultze*[1]

[1] UFZ Centre for Environmental Research, Leipzig-Halle Ltd., Department of Inland Water Research Magdeburg, Gouvernementsberg 1, 39104 Magdeburg, Germany

[2] IGB-Institute of Freshwater and Fish Ecology, Müggelseedamm 310, 12587 Berlin, Germany

3.1 Introduction

Hard and brown coal mining has a long tradition in central and eastern Europe and covers large areas of mining in Germany, Poland and the Czech Republic (Geller et al., this Vol.). In Germany there are three main districts of lignite surface mining: the Rheinish district near Cologne, the mid-German district around Leipzig and the Lusatian district around Cottbus in the most eastern part of Germany. Surface mining of lignite (brown coal) results in several environmental problems, for example important disturbances of the natural water balance, mass transfer of billions of tons of soil and devastation of nature. Among these, water acidification is an already well-known effect. Sulphide minerals, such as pyrite and marcasite, are commonly associated with coal and most metal ores. Weathering and oxidation of these minerals take place in the host rocks and substrates of the lignite horizons when they are exposed to air. The release of the oxidation products, mainly acidity, iron and sulphate, is known as acid mine drainage (AMD) and has been the subject of intense research for decades (e.g. Singer and Stumm 1970; Lowsen 1982; Nordstrom 1982; van Berk 1987; Morrison et al. 1990; Blowes et al. 1991, Hedin et al. 1994; Wisotzky 1994).

The residual hollow forms of the abandoned surface mines were filled by the rising water table and/or with river water forming the mining lakes. In regions with bedrocks rich in pyrite/marcasite and poor in carbonate, many mining lakes are strongly acidified (Schultze and Geller 1996). In the Lusatian district, more than 100 lakes of various sizes are acidified to pH values below 4 (Pietsch 1979). Detailed information on the chemical variability of these lakes and their sediments is rare. There is no debate that sediment chemistry is important for the chemical and biological development of (mining) lakes (e.g. Salomons and Förstner 1984; Belzile and

Morris 1995; Hamilton-Taylor and Davison 1995). The chemical and biochemical reactions and transformations that take place in the sediments lead to a highly dynamic biogeochemical behaviour of the deposited (trace) elements. The controlling processes include microbially mediated decomposition of organic matter, precipitation and dissolution of minerals, sorption and desorption of elements on biotic and abiotic particles and the diffusion of dissolved components along concentration gradients. As a result of these processes sediments act as sink and source for geogenic and anthropogenically mobilized elements (Belzile and Morris 1995).

Hydro- and geochemical investigations presented here form the basis for further research and the development of remediation and restoration management (Klapper et al. 1996). Any restoration approach for the mining lakes which includes the addition of base (liming, organic additives, etc.) or technological steps adopted from treatment plants (Klapper et al. 1996; Fischer et al., this Vol.), change the physicochemical conditions in the lakes for neutralizing and/or desulphurizing the lake water. It can be expected that changes in redox and pH conditions, which involve transformations of solids and ionic species as well as shifts in adsorption and desorption processes, increase and decrease the mobility of elements, respectively. Therefore, knowledge of the chemical composition of water and sediment in these lakes is required prior to any remediation technology.

This is the first study in this region, which considers not only the water quality but also the sediment composition. In this chapter we will focus on the ionic composition and the significance of the metal concentrations for the lake water. Regarding the sediments, we will give a brief summary of their general characteristics and discuss their metal contents and distribution. Some information about the trophic status and pioneer settlement of biota in these lakes can be found in Klapper and Schultze (1995) and in Nixdorf et al. (1995).

3.2 Sampling Sites and Methods

Sampling for a limnochemical survey of 13 selected lakes in the Lusatian district was performed in July 1993. The area under study covers the local mining districts of “Schlabendorf-Nord”, “Seese-West”, “Weißwasser”, “Senftenberg” and “Lauchhammer” (Fig. 3.1). Most of the lakes are not older than 40 years, showing an isolated situation without any influent or effluent. Due to the secret policy of former East Germany, information about the investigated lakes is rare. The lakes can be divided into a group

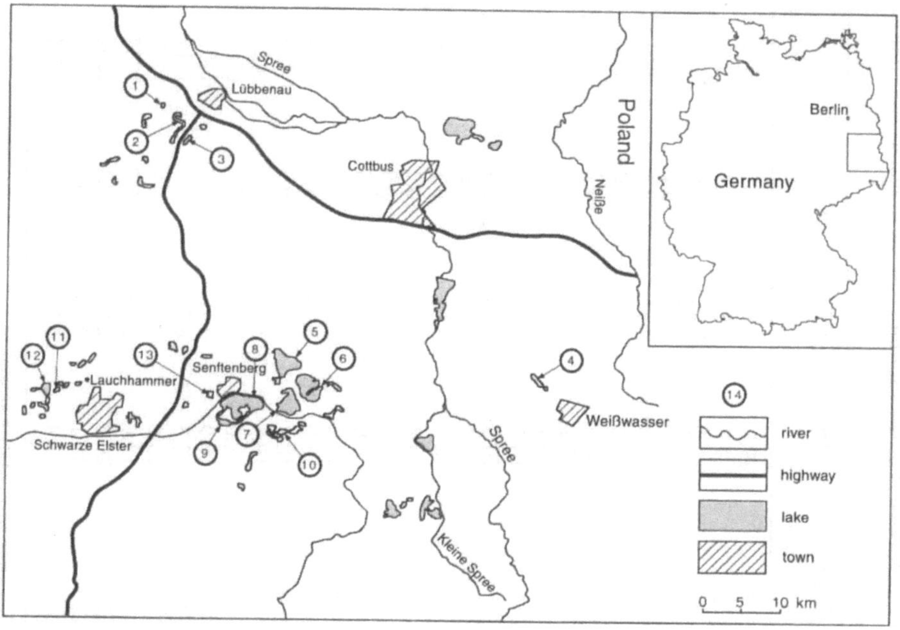

Fig. 3.1. Local map of the Lusatian lignite district showing mining lakes investigated: *1* ML-B, *2* ML-F, *3* ML-4, *4* ML-Halbendorf, *5* ML-Sedlitz, *6* ML-Skado, *7* ML-Koschen, *8* ML-Senftenberger See, *9* ML-Südsee, *10* ML-FKA, *11* ML-78, *12* ML-117, *13* ML-Hörlitz

of mining lakes in a narrower sense without any operation after flooding and a second group of mining lakes which have been used for other purposes during and after mining (Table 3.1). The lake "Senftenberger See" and the lake "Südsee" are parts of a former large mining hole, separated by a spoil dam which today forms an island, according to the mining technology which was used. After flooding this open pit during 1968–1974 a lake separated in two basins was established, strongly acidified by pyrite oxidation in the soil and heaps. Since 1976 the water body of the main basin is exchanged continuously by the throughflow of riverine water (Fig. 3.2a). The input of alkalinity increased the pH of Senftenberger See to neutrality whereas the side basin (Südsee) which does not receive river water remained strongly acidic (Benndorf 1994). Mining lake (ML)-FKA received acid mine drainage in the past and part of it was used as a wastewater disposal site for an aluminium smelter and for urban wastewater discharges as well as for the infill of ash suspension (Fig. 3.2b). In 1990, when the ash-producing power stations beside ML-FKA and the spoiling of the ash were still working, the acidification of the water body stopped to the east of sampling point 3 because of the effluents of the ash spoiling

Table 3.1. Physicochemcial characteristics (temperature pH, O_2 and electric conductivity) of the mining lakes investigated and time of flooding by riverine water or rising water table

No. in Fig. 3.1	Sampling point	Subdistrict	Flooding	Sampling depth (m)	T (°C)	pH	O_2 (mg/l)	EC (µS/cm)
1	ML-B	Schlabendorf-Nord	1963/64	0	19.7	8.12	10.5	1032
				10	9.7	7.50	2.9	1072
2	ML-F	Schlabendorf-Nord	1974–1977	0	19.2	2.55	10.6	3120
				11	9.8	2.55	5.2	3220
3	ML-4	Seese-West	Since 1988[a]	0	19.8	7.32	9.7	1865
				6	11.2	6.40	0.0	4800
4	ML-Halbendorf	Weißwasser	1965–1967	0	18.5	3.00	10.2	1390
				25	6.1	3.05	6.0	1388
5	ML-Sedlitz	Senftenberg	Since 1985[a]	0	20.2	2.85	9.1	2400
				15	16.9	2.89	9.2	2430
6	ML-Skado	Senftenberg	Since 1985[a]	0	20.5	2.65	8.9	2870
				16	8.6	2.76	8.6	2970
7	ML-Koschen	Senftenberg	Since 1974[a]	0	18.8	3.06	9.1	1628
				25	5.9	3.17	10.2	1674
8	ML-Senftenberg	Senftenberg	1968–1974	0	18.5	6.41	8.9	755
				18	8.0	5.92	1.9	830
9	Südsee	Senftenberg	1968–1974	0	18.4	3.48	9.0	833
				15	8.0	3.38	8.7	879
10	FKA-3	Senftenberg	1965–1968	0	18.9	3.06	9.5	1845
	FKA-4	Senftenberg		0	19.5	3.58	10.3	1423
				7.5	11.0	6.12	0	1161
	FKA-1	Senftenberg		0	18.9	3.70	9.9	1405
				11	8.3	6.27	0	1117
11	ML-78	Lauchhammer	1962	0	23.6	2.80	8.9	1729
				6	13.4	2.94	5.7	1752
12	ML-117	Lauchhammer	1966	0	20.8	2.90	9.4	1356
				13	10.1	3.19	0	1177
13	ML-Hörlitz	Lauchhammer	1960	0	23.4	6.82	8.1	1574

T, temperature; EC, electric conductivity. [a] Flooding was interrupted because of unstable slopes.

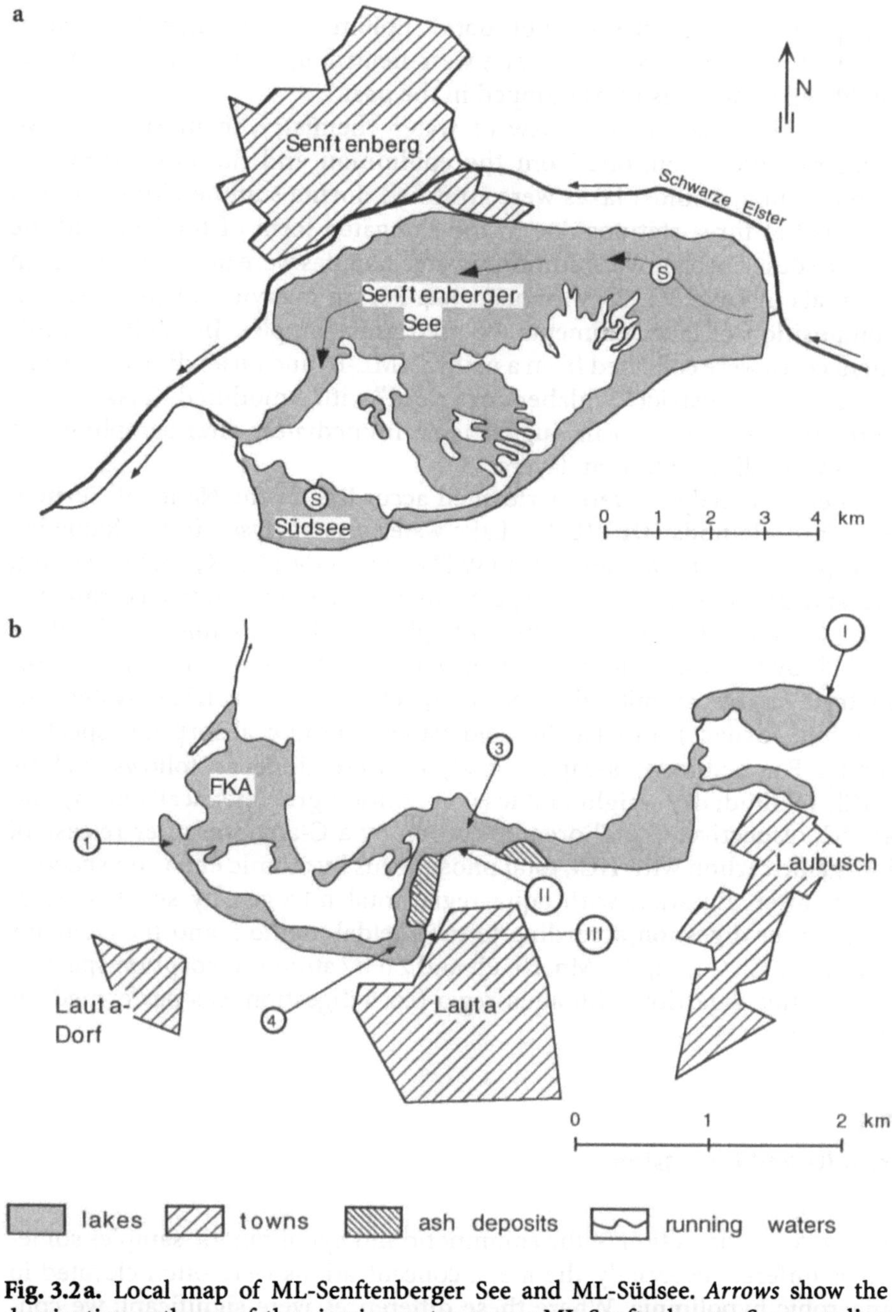

Fig. 3.2a. Local map of ML-Senftenberger See and ML-Südsee. *Arrows* show the dominating pathway of the riverine water across ML-Senftenberger See sampling points (*S*). **b** Local map of ML-FKA. *1*, *3* and *4* sampling points; *I* inflow of acid mine drainage; *II* inflow of wastewater of an aluminium smelter; *III* inflow of urban wastewater. (Redrawn from Benndorf 1994)

(Klapper and Schultze, unpubl. data). Epilimnion and hypolimnion of ML-FKA west of sampling point 3 were neutral up to this time. In ML-4, highly alkaline fly ash was dumped in the past.

In order to get an overview of water chemistry, in most cases two samples were taken, one from the epilimnion and the other from the hypolimnion, as most lakes were stratified during summer. ML-FKA was sampled at three stations due to the elongated form of the lake and the history described above. Additional water samples were taken at 1-m-depth intervals in Lake ML-B in August 1994. For an overview of the chemical composition of lake sediments we used grab samples. In addition, sediment cores were collected from a neutral (ML-B) and an acidic (ML-F) lake in the mining district "Schlabendorf-Nord" with a modified Kajak gravity corer (Uwitec, Austria) in August 1994. Immediately after sampling, the cores were sliced into 1-cm layers.

Chemical analyses were performed according to the National German Standard Methods (DEV/DIN). Lake water was analysed in the following way: pH by in-situ measurement (WTW-electrode); Na, K, Ca, Mg, Fe, Mn, Cu and Zn by flame atomic spectrometry; NH_4^+ by photospectroscopy; SO_4^{2-}, Cl^-, and NO_3^- by ion chromatography; HCO_3^- by titrimetry; Al, Cd, Cr and Pb by flame atomic spectrometry and by atomic absorption spectrometry; As by atomic absorption spectrometry (hydride system or graphite furnace) and Hg by cold vapour atomic absorption spectrometry. For sediment samples, analyses were done as follows: pH by $CaCl_2^-$ method; dry weight and loss on ignition gravimetrically at 105 and 550 °C, respectively; total organic carbon by a C-analyser after release of inorganic carbon with HCl; total phosphorus by atomic emission spectrometry after digestion with aqua regia; total nitrogen by spectrophotometry after digestion according to the Kjeldal method; and the elements Al, As, Cd, Cr, Cu, Fe, Hg, Mn, Ni, Pb and Zn by atomic absorption spectrometry after digestion with aqua regia (cold digestion was performed for mercury).

3.3 Results and Discussion

The water composition of the epilimnetic and hypolimnetic samples sometimes differed; especially, the metal concentrations were often elevated in anaerobic hypolimnia. Where these differences were significant, we considered the different water compositions for the interpretation of the results. In most cases, differences were small and the water chemistry was characterized by mean values.

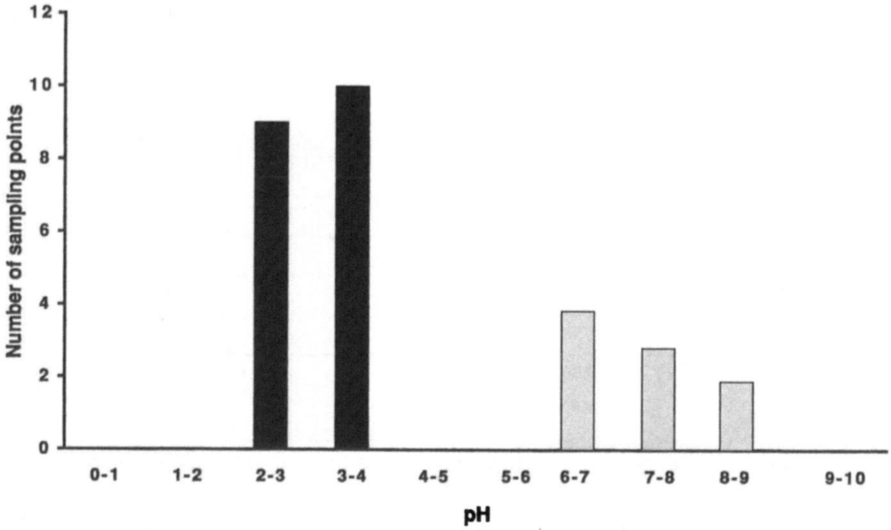

Fig. 3.3. Distribution of pH values in 13 mining lakes (including data from the epi- and hypolimnion). Lake FKA was sampled at three stations from the subsurface. Only from one station (FKA-4) data from the hypolimnion were available

3.3.1
Composition of Lake Water

The lakes under investigation can be divided into two distinct groups according to their pH values (Fig. 3.3): one group with pH values below 4 and the other with pH values between 6 and 8. The observed pH distributions are caused by a $Fe(OH)_x$-buffering system and a CO_2–bicarbonate–carbonate-buffering system, respectively. The $Fe(OH)_x$-buffering system is based on the different species of iron oxyhydroxides and correlates with high concentrations of iron, reaching maximum values of 200 mg/l. The Al-buffering system (pH 4.5 – 5.5), which is well-known from the acidification of soft water lakes by acid precipitation, was not observed in our selection. Nevertheless, this buffering system is also established in some mining lakes of the former East German mining districts (Geller et al., this Vol.).

3.3.1.1
Major Constituents

The water composition of the acid mining lakes is dominated by calcium and sulphate (Fig. 3.4). The ionic strength varies between 12 and 58 mmol/l, showing a negative correlation with increasing pH (Fig. 3.5). Because the

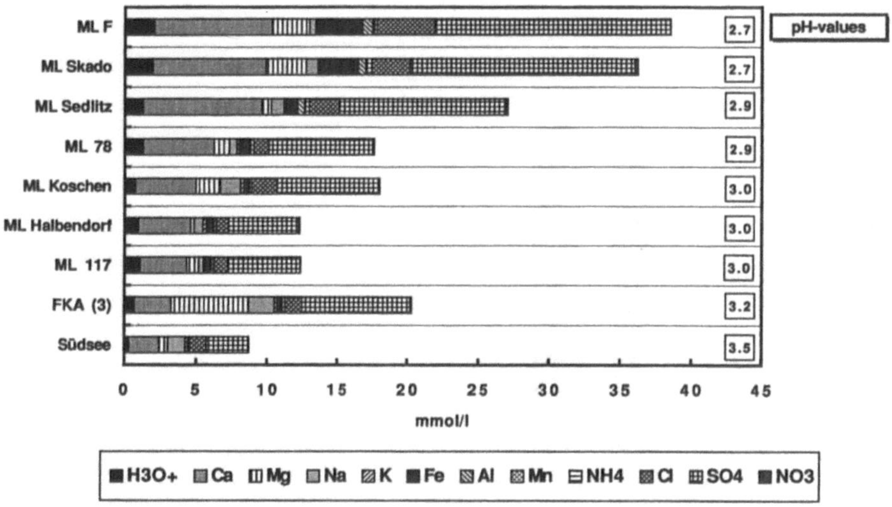

Fig. 3.4. Ionic composition (mmol/l) of acid mining lakes. Concentrations of H_3O^+ were calculated using pH (mean value), neglecting the influence of ionic strength on pH. Measured pH (mean value) of the lake water is given for comparison

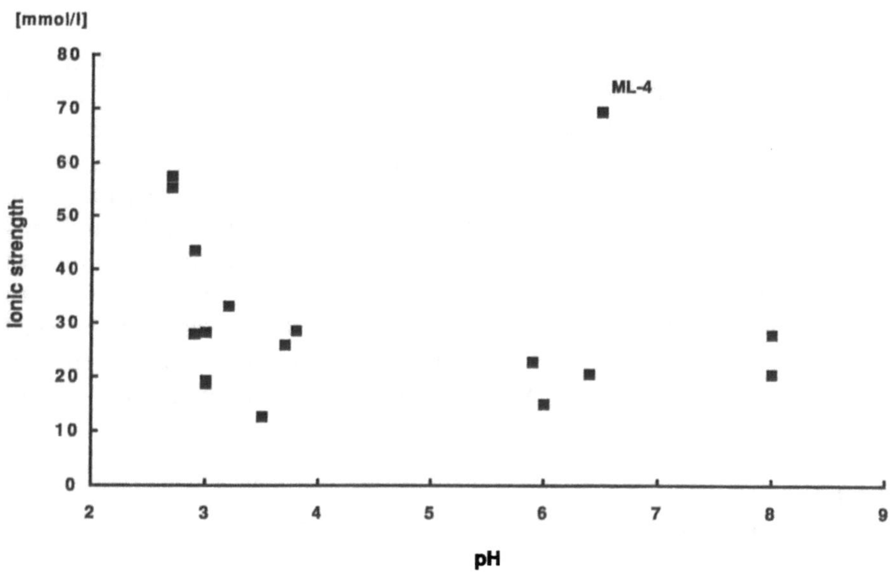

Fig. 3.5. Relation of ionic strength and pH (mean value). Concentrations of dissolved iron and aluminium in the acid mining lakes and mining lake ML-4 (pH 6.5) were not considered for calculating ionic strength (for explanation see text)

precise distribution of dissolved metal species such as Fe^{3+} or $Fe(OH)_2^+$ and others is still unknown, the concentrations of iron and aluminium were not considered for calculating the ionic strength. As a result of the high salt content, the lake water exhibits high values of electric conductivity between 1.5 and 3.7 mS/cm. Negative correlations with pH were observed for iron, manganese, aluminium and zinc (Fig. 3.6) which are caused by their good solubility in a low pH environment (e.g. Stumm and Morgan 1981). The correlation between the pH and the sum of ions may reflect the degree and extent of mass input and pyrite dissolution in the surrounding rocks (soil). The dissolution rate for many minerals is strongly pH-dependent and has a minimum around pH 6–7. Moreover, dissolution rates can be enhanced by accompanying elements such as iron or aluminium (Appelo and Postma 1993). Decreasing pH in these lakes is associated with an increase in dissolved solids as well as iron and aluminium.

Three lakes and the anoxic hypolimnetic water of sampling stations 1 and 4 of lake FKA showed pH values between 6 and 8. These neutral lakes generally possessed lower ion concentrations (Fig. 3.7), resulting in an ionic strength around 20 mmol/l. An exception was lake ML-4, which showed the highest ionic strength among these lakes, at about 70 mmol/l. In these lakes (except ML-4), iron is not evident, and chloride, sodium and hydrogen carbonate are of higher importance. Nevertheless, in most cases calcium and sulphate are still the dominant ions in the neutral lakes. Lake ML-4 was used as a disposal site for fly ash in the past. Probably it was originally acidic, as the water composition indicates, and was neutralized recently by the alkaline fly ashes. The amount of sulphate in the neutral lakes is indicative of pyrite oxidation and weathering in the surrounding environment and we assume that in some cases, for example in the case of lake ML-B, acidity is buffered by more calcareous soil composition. The pH values around 6 in the lakes Senftenberg and FKA (hypolimnion at sampling points 1 and 4) can be attributed to riverine water and wastewater input in the past, respectively. The alkalinity of these waters and presumably microbially mediated anaerobic processes buffer the water of the whole lake or of the hypolimnion, respectively.

3.3.1.2
Trace Metals

In general, in both types of lakes – the highly acidic ones and the more neutral ones – only iron, aluminium, manganese and sometimes zinc concentrations are elevated. Other elements, such as chromium, copper, nickel, lead, arsenic, cadmium or mercury, are of minor importance in the water column (Table 3.2). Elevated metal concentrations of iron and manganese in

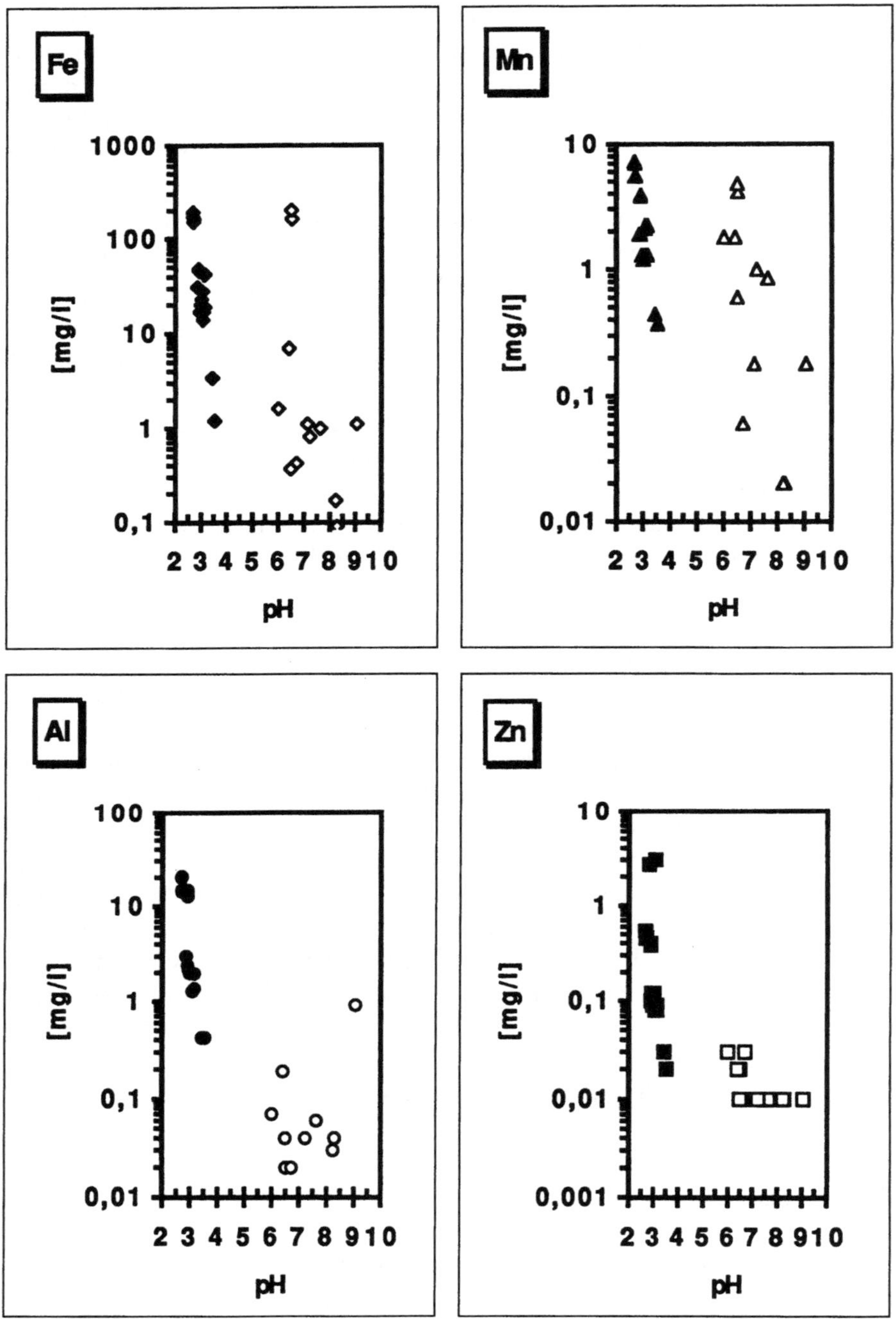

Fig. 3.6. Concentrations of dissolved iron, manganese, aluminium and zinc in mining lakes of the Lusatian lignite district in relation to pH (mean value). Data include epilimnion (*closed symbols*) and hypolimnion (*open symbols*) samples

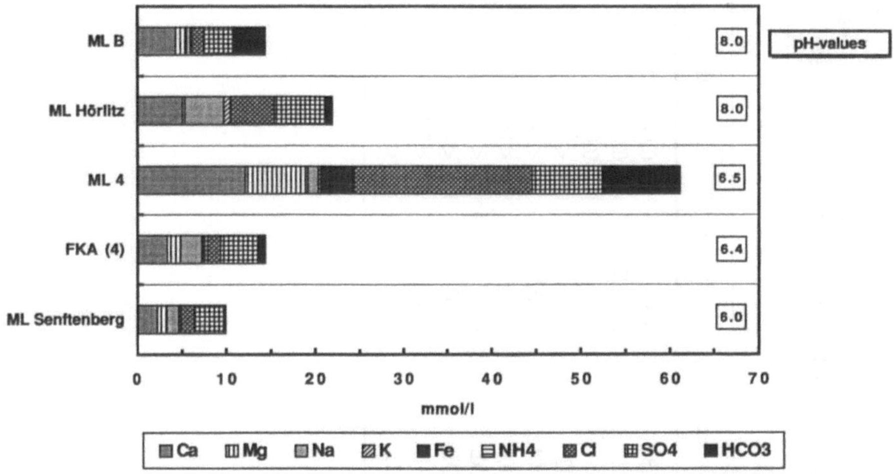

Fig. 3.7. Ionic composition (mmol/l) of neutral mining lakes or neutral sampling points. Measured pH (mean value) of the same stations is given for comparison

the more neutral lakes were only found in anoxic hypolimnia, where the reduced and soluble species lead to higher concentrations. This observation can be attributed to seasonal effects as demonstrated by comparing oxygen and element profiles of the neutral lake ML-B (Fig. 3.8). The oxygen concentration decreases to zero between 8- and 10-m water depth. At the same depth, aluminium, iron and manganese are increased by factors between 3 and 10. Probably this is an effect of iron and manganese oxyhydroxide redissolution from the top sediment layer under anoxic conditions.

3.3.2
Lake Sediments

3.3.2.1
General Parameters

The lakes commonly possess a sediment cover between 15 and 90 cm. The sediments are muddy and fluffy, reaching water contents of up to 90%. They are rich in organic carbon and show loss on ignition (LOI) of up to 29%. Under the extreme living conditions exemplified by low pH and high contents of toxic elements, the microbial degradation of organic matter is inhibited, especially under oxic conditions. Moreover, we have to consider that part of the LOI may be related to loss of volatile sulphur components and of crystalline water of amorphous iron (oxy)hydroxide complexes. According to their sediment pH, the Lusatian lakes can be divided into

Table 3.2. Range of dissolved metal concentrations in the water of the investigated mining lakes, separated for acid and neutral sampling points. All values are given in mg/l. ML-4, a nearly neutral mining lake (pH 6.5), has very high dissolved iron concentrations (about 200 mg/l) and the highest ionic strength (ca. 70 mmol/l) which is more likely for the acid mining lakes. For further explanation see text

Element	Detection limit	Lakes pH <4 Mean ($n = 21$)			Lakes pH 6–8 Mean ($n = 12$)		
			Maximum	Minimum		Maximum	Minimum
Al		7	20	0.4	0.13	0.9	0.006
Fe		59	190	1.2	1	6.9 (200)	0.09
Mn		2.9	7.2	0.4	0.7	1.8	0.06
Zn	0.01	0.5	3	<DL	0.02	0.03	<DL
Cd	0.0005	0.0012	0.0034	<DL	0.0015	0.0026	<DL
Cr	0.005	0.011	0.019	<DL	0.006	0.006	<DL
Pb	0.005	0.007	0.01	<DL	<DL	<DL	<DL
Hg	0.0005	0.0025	0.0028	<DL	<DL	<DL	<DL
As	0.001	0.011	0.013	<DL	0.01	0.01	<DL
Cu	0.05	<DL	<DL	<DL	<DL	<DL	<DL

DL, detection limit.

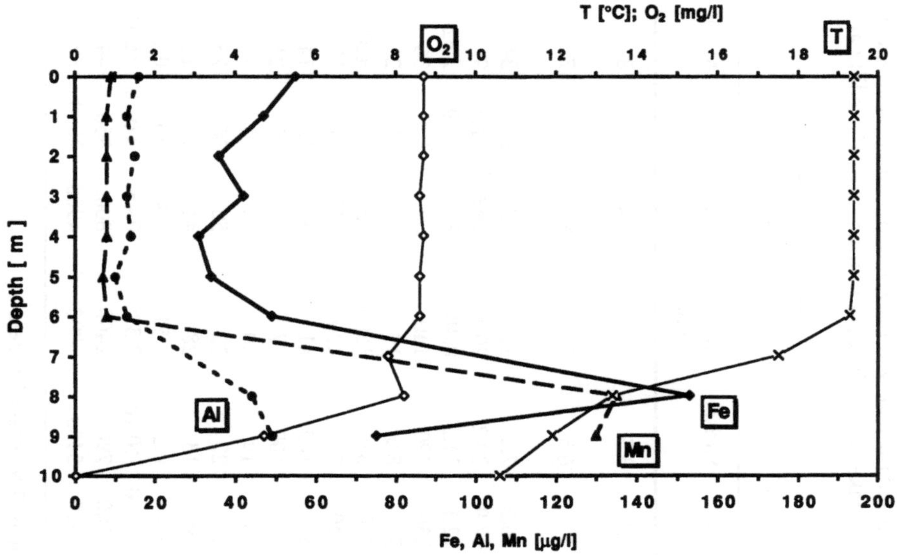

Fig. 3.8. Comparison of depth profiles for temperature (*T*), oxygen (O_2), dissolved aluminium (*Al*), iron (*Fe*) and manganese (*Mn*) in neutral mining lake ML-B. The graph indicates iron and manganese redissolution in the anoxic hypolimnion

three groups: acidic lakes with acidic sediments, acidic lakes with neutral sediments and neutral lakes with neutral sediments (Table 3.3). The occurrence of acid lakes with neutral sediments is really unusual. We assume that in these lakes alkalinity is produced by degradation of organic matter and reduction of sulphate (Bramkamp et al., this Vol.).

3.3.2.2
Metals and Phosphorus

The sediments are rich in iron (up to 290 g/kg) and aluminium (up to 62 g/kg). Manganese, zinc and sometimes arsenic and lead contents are elevated. This is also the case for phosphorus, with maximum concentration of 8 g/kg total P. The highest concentrations of all metals and of total phosphorus are found in lake FKA and lake Senftenberg. In general, phosphorus increases with increasing pH of the sediment, probably as a result of iron-oxyhydroxide precipitation and related phosphorus adsorption (e.g. Uhlmann et al. 1995). This is confirmed by a correlation between iron and phosphorus when excluding the highest values of 2400 and 8000 mg/kg in the lakes Senftenberg and FKA, respectively (Fig. 3.9). Moreover, correlations between total phosphorus and the trace metals zinc, copper and mercury are found if the highest value of 8000 mg/kg in lake FKA is excluded (Fig. 3.10).

Table 3.3. Chemical characteristics of sediments from the investigated mining lakes. pH of the lake water is given for comparison. According to the pH of water and sediment three groups are identified. No sediments were taken from the acid mining lake ML-Sedlitz and the neutral mining lake ML-Hörlitz. For further explanations see text

Lake		Group 1				Group 2				Group 3			
Parameter	Unit	ML-F	ML-Skado	ML-Koschen	ML-117	ML-78	ML-Halbendorf	FKA (3)	Südsee	ML-Senftenberg	FKA (4)	ML-4	ML-B
pH – water		2.7	2.7	3	3	2.8	3	3.1	3.5	6	6.4	6.5	8
pH – sediment		2.7	3	4	4.2	6	6	7	6.5	7	8	6.8	7.1
DW	%	19	27.4	21.3	31.7	49.5	8.2	10.2	7.4	10	6.1	17.4	34
LOI	%	13.7	18.7	18.7	19.8	16.2	23.5	27.9	29.2	20.5	29.6	21.7	9
TOC	%	2.11	6.32	7.05	9.57	8.46	9.03	12.1	11.2	8.48	8.59	7.59	2.52
Fe	g/kg	140	71	140	49	15	220	230	290	140	89	100	36
Al	g/kg	4.9	16	17	19	9.3	17	30	8.9	19	62	21	25
Mn	mg/kg	74	68	74	34	49	69	2300	65	1500	1500	1200	420
Zn	mg/kg	17	23	28	22	13	54	240	74	420	460	71	75
As	mg/kg	11	23	34	18	8.9	50	59	100	41	54	15	14
Pb	mg/kg	22	32	29	18	12	68	36	100	57	10	25	21
Cr	mg/kg	8.8	34	29	38	26	28	35	19	30	25	34	40
Ni	mg/kg	4.4	8.5	8.4	8.5	6.2	11	40	27	72	50	26	26
Cu	mg/kg	1.9	8.1	11	8.1	4.5	9.9	49	13	28	44	17	18
Hg	mg/kg	0.08	0.14	0.14	0.15	0.11	0.22	6.4	0.22	0.51	6.4	0.17	0.12
Cd	mg/kg	<0.10	<0.10	<0.10	<0.10	<0.10	<0.10	<0.10	<0.10	0.32	0.43	<0.10	<0.10
TP	mg/kg	210	300	390	320	190	670	1100	790	2400	8000	860	440
TN	mg/kg	500	1800	2300	2200	400	4800	980	5400	6000	4900	2800	880

DW, dry weight; LOI, loss on ignition; TOC, total organic carbon; TP, total phosphorus; TN, total nitrogen.

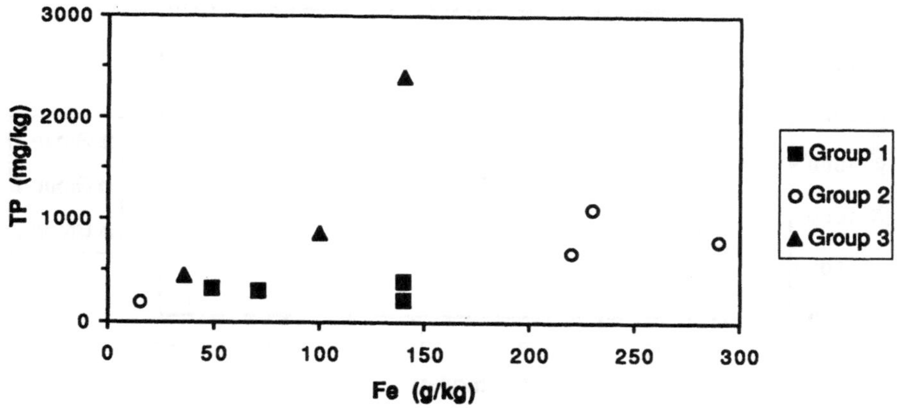

Fig. 3.9. Correlation between iron and total phosphorous of sediments from three groups of lakes [the value of 8000 mg/kg for lake FKA (4) is excluded from the diagram]

In the past, lake FKA was used as a disposal site for urban wastewater discharge and as a wastewater basin for an aluminium smelter. These anthropogenic pollution sources lead to the high phosphorus and aluminium contents in the lake sediments. Lake Senftenberg is neutralized by the throughflow of contaminated river water (River Schwarze Elster), which increases the pH of the lake but also results in enriched metal and phosphorus concentrations in the sediments.

Regarding the low number of lakes under investigation, statistical analysis of the sediment chemistry has to be considered with care. Several contradictory correlations or different slopes of trend lines can be found within the three groups identified. For further discussion we will consider only trends which are relevant for all three groups, neglecting different slopes of correlation lines. Interelement relationships can be found for Zn and Ni, Zn and Cu, Zn and Hg, and for Al and Cu, resulting in correlations between Ni and Cu, Ni and Hg, Cu and Hg, Al and Zn, Al and Hg, respectively. Correlations between pH of the sediment and the trace metals Cu, Ni, Zn, Hg and Al were detected, but not for Fe. This shows a negative correlation with dry weight (Fig. 3.11), which could be a hint for fresh iron precipitates or iron colloids present in the pore waters of the sediments with higher water contents. Arsenic with maximum concentrations of 100 mg/kg shows correlations with the LOI and the amount of total nitrogen (Fig. 3.12). This leads to the assumption that As is bounded to or occurs dominantly in an organic N-form.

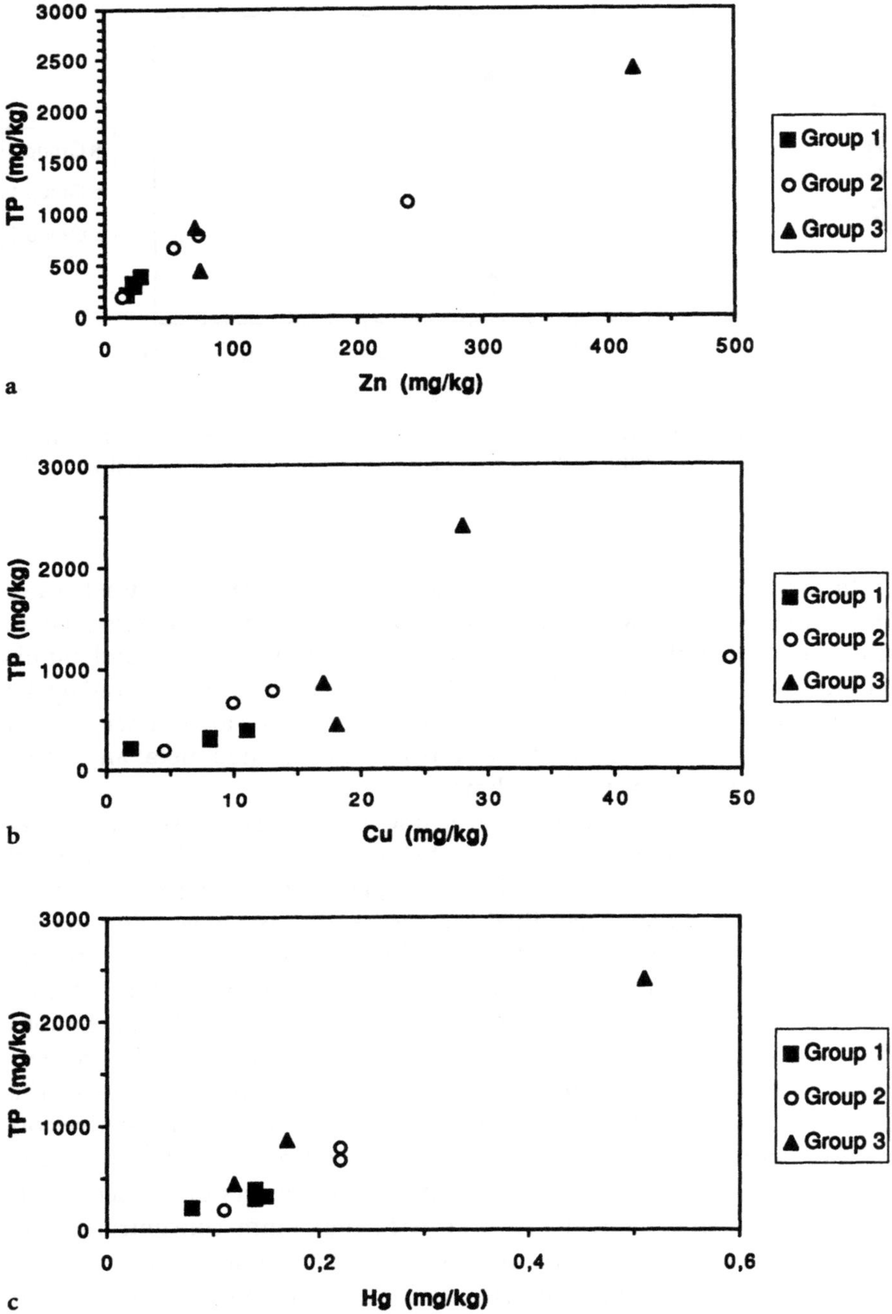

Fig. 3.10. Correlation between **a** total phosphorus and zinc; **b** total phosphorus and copper and **c** total phosphorus and mercury [the value for lake FKA (4) is excluded from all diagrams]

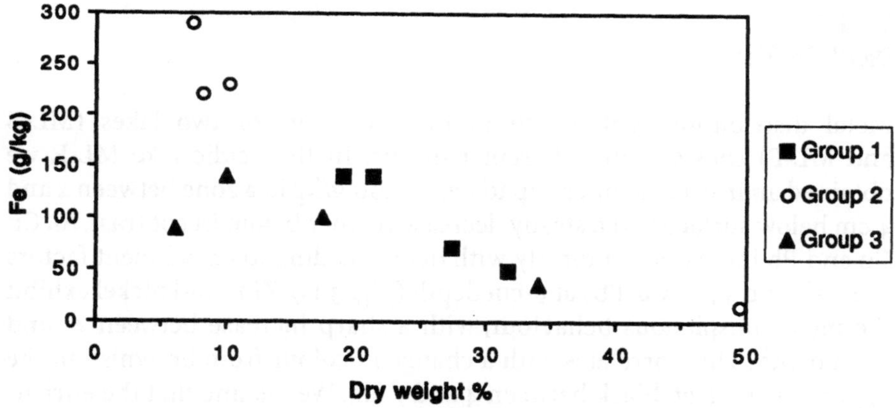

Fig. 3.11. Negative correlation between iron and dry weight

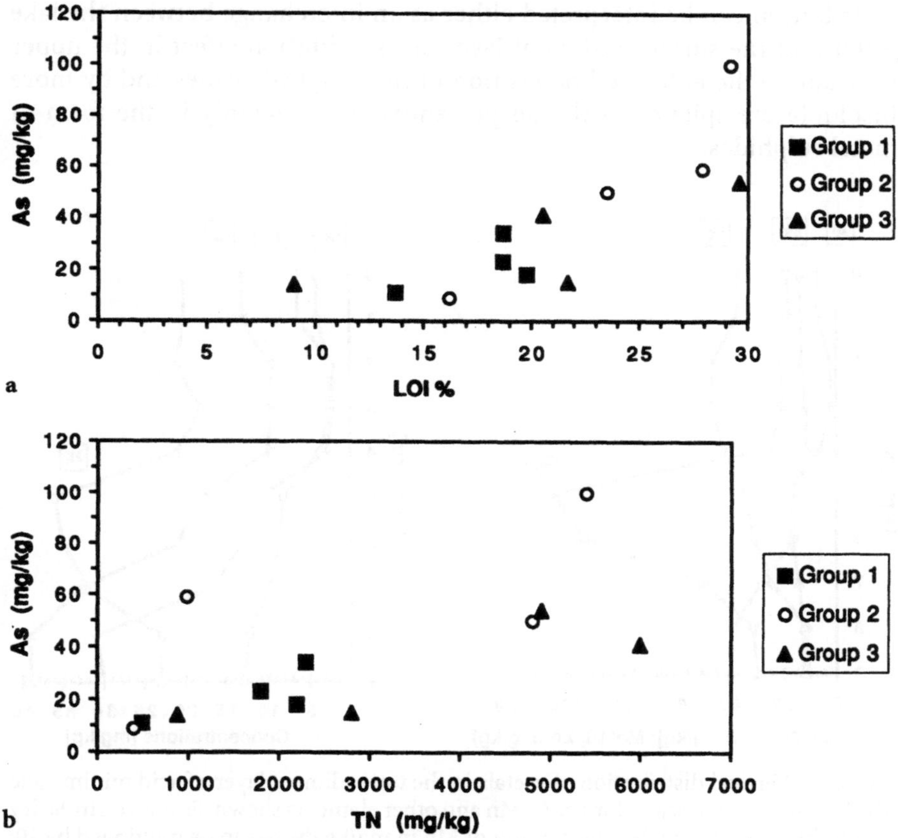

Fig. 3.12. Correlation between a arsenic and loss on ignition (LOI) and between b arsenic and total nitrogen

3.3.2.3
Depth Profiles

Metal distribution with depth in the sediments of two lakes (ML-B and ML-F) shows quite different patterns. In the acidic lake ML-F, we obtained an iron enrichment up to nearly 500 g/kg in a zone between 2 and 4 cm below surface and a steady decrease further below. In contrast, Al, Cr, Cu and Pb increased uniformly with depth, leading to enrichment factors of 4 (Al) and 3 (Cr, Cu, Pb) at 9-cm depth (Fig. 3.13). Zinc and nickel exhibit the most conspicuous behaviour, with a sharp increase between 4- and 7-cm depth. This correlates with a change in colour from brownish in the upper part to deep black between 4 and 7 cm. We assume that the enrichment of these metals corresponds to a change from oxic to anoxic conditions, where anaerobic bacteria are able to reduce sulphur, giving the possibility for metal sulphide precipitation. Therefore the existing element distribution can be interpreted either as an interchange between the lake water and the surface sediment layer or as a dilution effect in the upper part due to the enhanced deposition of iron-oxyhydroxides and by more insoluble precipitates in the deeper anoxic zone, mainly in the form of metal sulphides.

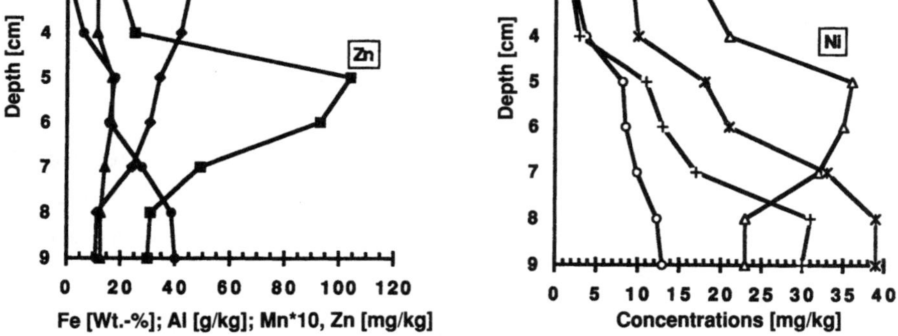

Fig. 3.13. Vertical distribution of metals in the top sediment layer of acid mining lake ML-F. Note different scales for Fe, Al, Mn and other elements shown. Fe is given in % dry weight (dw); Al in g/kg dw; the amount of Mn in mg/kg dw has to be multiplied by 10; Zn, Cu, Pb, Ni and Cr in mg/kg dw. The sharp increase for Zn and Ni between 4- and 7-cm depth is presumably related to formation of sulphides under reducing conditions

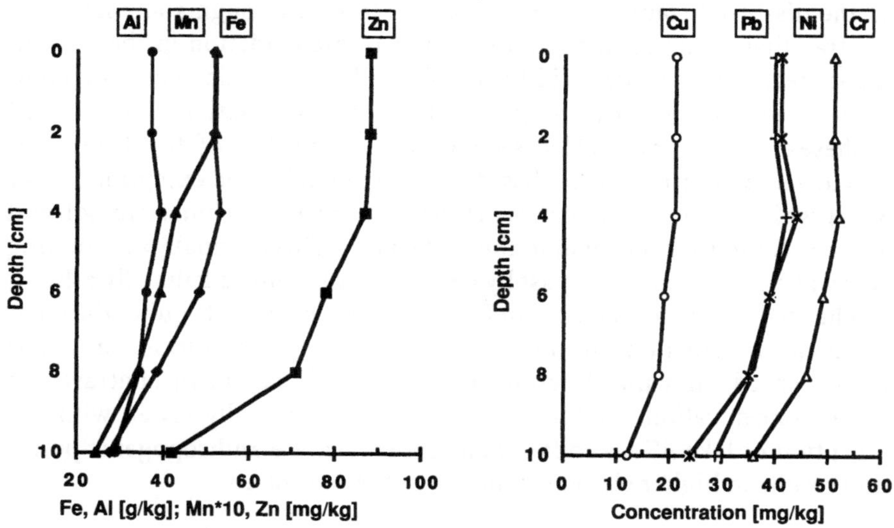

Fig. 3.14. Vertical distribution of metals in the top sediment layer of neutral mining lake ML-B. Note the different scales for Fe, Al, Mn and other elements shown. Fe and Al are given in g/kg dw; the amount of Mn in mg/kg dw has to be multiplied by 10; all other elements are given in mg/kg dw

The metal distribution in the sediment of the neutral lake ML-B shows a different pattern. The iron content is about one order of magnitude lower, in the range of 50 g/kg. This corresponds with the low iron concentration of the lake water. A second observation is a steady decrease with depth of all metals, postulating a concentration gradient (Fig. 3.14). An explanation for this pattern may be the neutral status of the lake which favours iron- and manganese-oxyhydroxide precipitation and related element adsorption in the upper parts of the sediment. This interpretation agrees with the enriched concentrations of iron and manganese observed in the hypolimnion of this lake during summer stagnation (Fig. 3.8) as a result of redissolution of iron and manganese (oxy)hydroxides under reducing conditions.

3.4 Summary

Our survey shows the influence of FeS_2 oxidation and acid mine drainage on the chemical composition of lake water and sediments. Increased input of the mobilized compounds, mainly acidity, sulphate, Al, Fe and Mn, and the good solubility of most of the metals at low pH result in high concentrations of the dissolved compounds in the acidic lakes. Although some lakes are neutral, their ionic composition indicates pyrite oxidation in their catchment area. In these lakes acidity is buffered by more calcareous soil composition, riverine water throughflow or other human activities. As a result of these effects, the neutral lakes show quite lower values of dissolved solids and lower concentrations of dissolved heavy metals. One exception is lake ML-4, which has a circumneutral pH (6.5) but the highest ionic strength and an ionic composition similar to the acidic lakes. This originally acid mining lake was neutralized in the past by deposition of highly alkaline fly ash.

The sediments of acidic and neutral lakes exhibit quite different chemical compositions. In the acidic lakes, high concentrations of metals are observed, showing different behaviour with depth. In contrast, the metal concentrations of the neutral lakes are generally lower, with the exception of lake FKA and lake Senftenberg, where anthropogenic pollution results in higher element contents in the sediments.

Acknowledgements. We would like to express our thanks to A. Fyson, G. Packroff and K. Wendt-Potthoff for stimulating discussions and comments on an earlier draft of the manuscript. Review by an anonymous reviewer greatly improved the original version. The chemical analyses were performed by the IDUS Laboratories in Ottendorf-Okrilla and IFUA Bitterfeld. The investigation was supported financially in part by the German Ministry of Research and Development, Grant No. 0339450B.

References

Appelo CAJ, Postma D (1993) Geochemistry, groundwater and pollution. Balkema, Rotterdam

Belzile N, Morris JR (1995) Lake sediments: sources or sinks of industrially mobilized elements. In: Gunn JM (ed) Restoration and recovery of an industrial region. Springer, New York Berlin Heidelberg, pp 183–193

Benndorf J (1994) Sanierungsmaßnahmen in Binnengewässern: Auswirkungen auf die trophische Struktur. Limnologica 24:121–135

Blowes DW, Reardon EJ, Jambor JL, Cherry JA (1991) The formation and potential importance of cemented layers in inactive sulfide mine tailings. Geochim Cosmochim Acta 55:965–978

DEV-Deutsche Einheitsverfahren zur Wasser-, Abwasser- und Schlamm-Untersuchung (1991) FG Wasserchemie der GDCh, Normenausschuß Wasserwesen, DIN Deutsches Institut für Normung (Hrsg.) 25 Lfg. VCH, Weinheim

Hamilton-Taylor J, Davison W (1995) Redox-driven cycling of trace elements in lakes. In: Lerman A, Imboden D, Gat J (eds) Physics and chemistry of lakes, 2nd edn. Springer, Berlin Heidelberg New York, pp 217–263

Hedin RS, Narin RW, Kleinmann LP (1994) Passive treatment of coal mine drainage. United States Department of the Interior, Bureau of Mines, Information Circular 1994, Pittsburgh, pp 1–35

Klapper H, Schultze M (1995) Geogenically acidified mining lakes – living conditions and possibilities of restoration. Int Rev Ges Hydrobiol 80:639–653

Klapper H, Geller W, Schultze M (1996) Abatement of acidification in mining lakes in Germany. Lakes Reservoirs Res Manage 2:7–16

Lowsen RT (1982) Aqueous oxidation of pyrite by molecular oxygen. Chem Rev 82:461–497

Morrison JL, Scheetz BE, Strickler DW, Williams EG, Rose AW, Davis A, Parizk R (1990) Predicting the occurrence of acid mine drainage in the Alleghenian coal-bearing strata of western Pennsylvania: an assessment by simulated weathering (leaching) experiments and overburden characterisation. Geol Soc Am Spec Pap 248:87–99

Nixdorf B, Rücker J, Köcher B, Deneke R (1995) Erste Ergebnisse zur Limnologie von Tagebaurestseen in Brandenburg unter besonderer Berücksichtigung der Besiedlung im Pelagial. In: Geller W, Packroff G (Hrsg) Abgrabungsseen – Risiken und Chancen. Limnologie aktuell 7. Gustav-Fischer, Stuttgart, S 39–52

Nordstrom DK (1982) Aqueous pyrite oxidation and the consequent formation of secondary iron minerals. In: Kittrick JA (ed) Acid sulphate weathering. Soil Sci Soc Am Spec Publ 10:37–57

Pietsch W (1979) Zur hydrochemischen Situation der Tagebauseen des Lausitzer Braunkohlen-Reviers. Arch Naturschutz Landschaftsforsch Berl 19:97–115

SalomonsW, Förstner U (1984) Metals in the hydrocycle. Springer, Berlin Heidelberg New York

Schultze M, Geller W (1996): The acid lakes of lignite mining district of the former German Democratic Republic. In: Reuther R (ed) Geochemical approaches to environmental engineering of metals. Environmental Science Series. Springer, Berlin Heidelberg New York, pp 89–105

Singer PC, Stumm W (1970) Acid mine drainage: rate-determining step. Science 167:1121–1123

Stumm W, Morgan JJ (1981) Aquatic chemistry, 2nd edn. Wiley, New York

Uhlmann D, Hupfer M, Paul L (1995) Longitudinal gradients in the chemical and microbial composition of the bottom sediment in a channel reservoir (Saidenbach R., Saxony). Int Rev Ges Hydrobiol 80:15–25

Van Berk W (1987) Hydrochemische Stoffumsetzungen in einem Grundwasserleiter beeinflußt durch eine Steinhohlenbergehalde. Besondere Mitteilungen zum Dtsch Gewässerkdl Jahrb 49:1–175

Wisotzky F (1994) Untersuchungen zur Pyritoxidation in Sedimenten des Rheinischen Braunkohlenreviers und deren Auswirkungen auf die Chemie des Grundwassers. Besondere Mitteilungen zum Dtsch Gewässerkdl Jahrb 58:1–153

4 In-Lake Neutralization of Acid Mine Lakes

A. Peine and S. Peiffer

Limnological Research Station, University of Bayreuth,
95440 Bayreuth, Germany

4.1 Introduction

One of the problems caused by strip mining is the generation of acidic lakes. Typically, the huge holes left from strip mining of soft coal are partly filled by mine tailings containing pyrite and other sulphide minerals. The remaining holes are flooded by groundwater forming so-called mine lakes. When in contact with oxygen-rich water the iron-sulphide species of the mine tailings are oxidized to sulphuric acid and ferric iron (Singer and Stumm 1970; Nordstrom et al. 1979). This results in an extreme acidification of the lakes (van Berk 1987; Obermann et al. 1992; Wisotzky 1994).

We propose that in acid mine lake sediments sulphate generated by oxidation of sulphide minerals is reduced to sulphide under anaerobic conditions. Subsequently the inverse process of pyrite oxidation is initiated, i.e. reduced sulphur compounds are again formed in the sediment. Final storage of these products in the sediment retains the alkalinity generated by sulphate reduction in the system, which would lead to partial neutralization of the acidity produced during the oxidation process (Peine and Peiffer 1996). An enhancement of sulphate reduction was found in freshwater sediments receiving high sulphate inputs from acid mine drainage (Herlihy and Mills 1985). Similar effects were demonstrated to occur in soft-water lakes exposed to acid precipitation (Kelly et al. 1987; Schafran and Driscoll 1987; Stumm et al. 1987; Giblin et al. 1990; Sherman et al. 1994).

The development of anoxia is dependent on the input of organic matter as an electron donor (Peiffer 1994; Urban 1994). In acid mine lakes it might be a limiting factor due to their low autochthonous organic carbon production. Accordingly, older mine lakes where organic carbon could accumulate should be more neutralized than younger ones under the same hydrological and hydrochemical conditions.

The objective of this chapter is to present indications for in-lake neutralization processes in five lakes of different ages situated in a former strip-mining area in Bavaria (Germany).

4.2 Site Description

The investigation was performed near the city of Schwandorf (Germany) in a small mining area (Fig. 4.1). The mining area is situated in Miocene sediments. It is surrounded by Cretaceous layers of sand and clay characterized by a high permeability which allows rapid flow of groundwater (Meyer and Mielke 1993). During mining activity, the groundwater level was lowered beneath the mining surface (Oertel 1976). Planned groundwa-

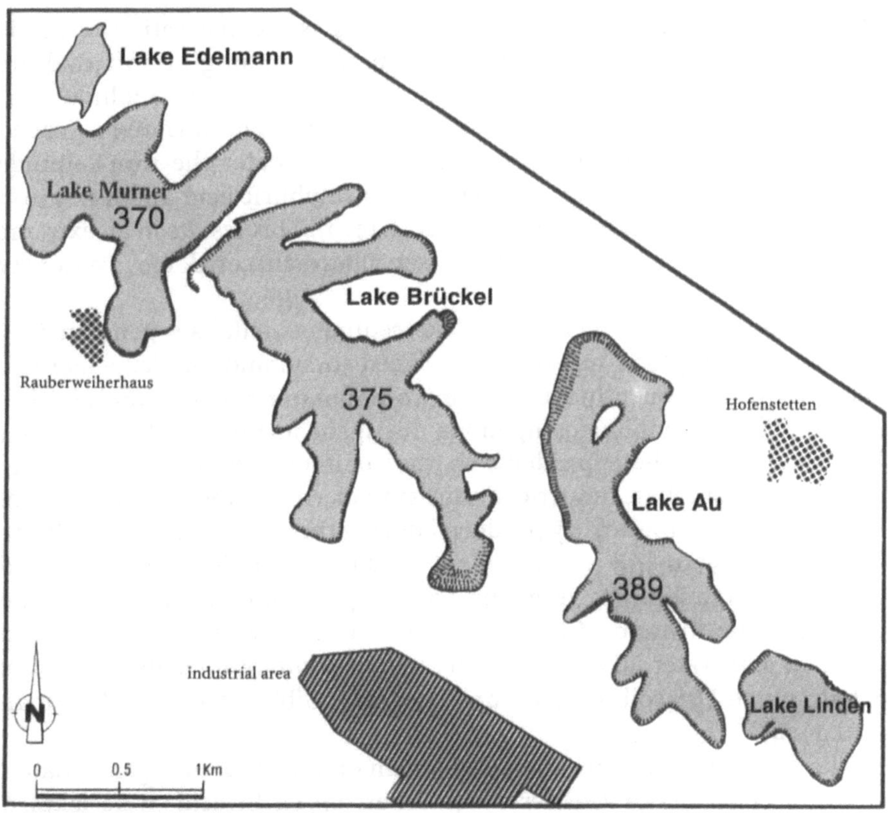

Fig. 4.1. The investigation site. Lakes were flooded consecutively, Lake Edelmann being the oldest, Lake Linden the youngest lake

Table 4.1. Flooding state of the lakes

Lake	Start of flooding	End of flooding	Actual maximum depth (m)	Intended maximum depth (m)	Final lake area (ha)
Edelmann	1964	1979	10	10	90
Murner	1979	1989	45	45	145
Brückel	1980	1994	40	40	121
Au	1982	<1999	26	30	24
Linden	1982	<2002	12	18	11

The symbol $<$ denotes that flooding is planned to be finished before the year specified.

ter flooding of the abandoned mining area started in 1964 and a chain of five lakes had been created consecutively (Table 4.1). All lakes receive groundwater from the same aquifer. To stabilize the lake sides the lake water level is always kept above the groundwater table by removing the groundwater at a certain distance from the lake and then pumping it into the lakes. Lake Edelmann additionally received domestic wastewater of unknown composition for several years. The age of the lakes is determined by the start of flooding and the end of flooding which is dependent on depth and volume of the lakes. Table 4.1 shows the following time sequence of the lakes: Lake Edelmann is the oldest lake, followed by Lake Murner, Lake Brückel, Lake Au, and finally Lake Linden.

4.3 Sampling and Analytical Methods

Water from all five lakes was analysed for pH, conductivity, titration acidity, dissolved sulphate, dissolved aluminium and dissolved iron in June 1995. In Lake Linden, Lake Brückel and Lake Edelmann only surface water was sampled. In Lake Au and Lake Murner depth profiles were measured, in Lake Au at the deepest point of the lake and in Lake Murner in the middle of the lake. In Lake Au sampling reached from the lake surface to just above the sediment at 26-m depth. In Lake Murner the surface sample was lost and therefore the profile begins in 5-m depth and ends in 35-m depth just above the sediment.

pH was measured using an electrode (WTW E50). Dissolved sulphate was measured photometrically after precipitation as $BaSO_4$ (Tabatai 1974). Total iron and aluminium were detected using flame atomic absorption spectrometry (VARIAN/Spectr-AA-20). Ferrous and ferric (after reduction with ascorbic acid) iron was analysed in Lake Au and Lake Murner lake water samples by the phenanthroline method (Frevert 1983). Additionally, titration curves were recorded by adding 0.01 N NaOH solution to the water samples.

Sediment cores were taken from the deepest point in May 1994 in Lake Au, Lake Murner and Lake Edelmann in Plexiglas tubes of 5-cm diameter using a gravity corer. They were sliced under a constant nitrogen flux, freeze-dried, and analysed for total reduced inorganic sulphur (TRIS), acid volatile sulphur (AVS), elemental sulphur (S^0), total iron, reactive iron, total aluminium, carbon content and water content. TRIS refers to pyrite (FeS_2), amorphous FeS and S^0, and AVS only refers to amorphous FeS. TRIS was determined after reduction with $Cr(II)Cl_2$ and heating, and AVS after digestion with HCl and heating (Canfield et al. 1986). Some AVS might have been lost due to oxidation during freeze-drying. We did not analyse AVS in the sediment of Lake Au. In contrast to the other sediments, its colour was brown and indicated absence of significant amounts of black amorphous FeS. The sulphide liberated by extraction of TRIS and AVS was trapped in 0.2 M NaOH solution and then measured photometrically by the methylene blue method (Frevert 1983). S^0 was extracted from fresh sediment by methanol (Ferdelman et al., in prep.). The methanol extract was then filtered and measured by high-pressure liquid chromatography and UV-detector. Reactive iron, which refers to iron monosulphides and iron oxides, was extracted by treatment with acid ammonium oxalate solution (Canfield 1989). For the determination of total iron the sediment was digested with concentrated nitric acid under pressure (Schramel et al. 1980). Iron was measured by flame atomic absorption spectrometry (VARIAN/Spectr-AA-20). Carbon was measured with a carbon analyser (Elementar, ratio EL).

In June 1995 sediment cores were taken by a gravity corer from Lake Au, Lake Murner and Lake Edelmann for pH and E_h measurements. Redox and pH microelectrodes were inserted laterally into holes in the sediment core. The E_h sensor was calibrated against saturated chinhydron solution at pH 4 and 7 (Frevert 1983). The measurements were all related to pH 7 according to Eq. (1), assuming a drop of 59 mV per increase of one pH unit:

$$E_{h,7} = E_h - (7 - pH) \times 59 \text{ mV}. \tag{1}$$

In Lake Au and Lake Murner pore water was sampled in autumn 1994 by means of a dialysis chamber (Höpner 1981) with a cellulose acetate membrane of 0.2-µm pore diameter (SM 11107, Sartorius). The samples were analysed for pH, dissolved sulphate and both ferrous and ferric iron. While the iron species were determined using the same analytical methods as described above, sulphate was measured using ion chromatography (Metrohm 690, Column: PRP X-100).

4.4 Results

4.4.1 Lake Water Composition

In Lake Murner and Lake Au concentrations of all species analysed did not change with depth, indicating that no chemical stratification exists in these lakes. The lowest pH of 2.8 was measured in the youngest Lake Linden. The pH of Lake Au was slightly higher at pH 2.9. Lake Brückel and Lake Murner did not differ and both had a pH of 3.2. The pH of Lake Edelmann was relatively high with pH 5 (Fig. 4.2a). This sequence can be observed with all other parameters: Concentrations of dissolved aluminium (Fig. 4.2b) and dissolved iron (Fig. 4.2c) were not detectable in Lake Edelmann and had highest values in Lake Linden. The lowest concentrations of sulphate (Fig 4.2d) were found in Lake Edelmann, the highest concentrations in Lake Linden. The other lakes ranged in between.

Ferrous iron was detected in Lake Au and Lake Murner in the water column. Although the water was oxic, ferrous iron oxidation seemed to be inhibited by the acid conditions (Fig. 4.2e). We assume that photoreduction of ferric iron is responsible for the increase of ferrous iron in the upper 10 m of the water column in both lakes (McKnight et al. 1988). As sampling of the lake water was done on different days, light intensity was different and absolute values of the upper 10 m are not comparable. It is not possible to distinguish between ferrous iron stemming from groundwater inflow and that produced by photoreduction.

Figure 4.3 presents the titration curves. The dashed line denotes the inflection point which lies at pH $\cong$ 5 for all lakes. It is reasonable to assume that this pH value corresponds to a titration of lake acidity from H^+, Fe^{3+} and Al^{3+}. The highest acidity was found again in Lake Linden with 2.7 mmol/l whereas no acidity from these species could be found in Lake Edelmann (Fig. 4.3).

4.4.2 Pore Water Composition

At the sediment-water interface the redox potential drops to values between −200 and 0 mV in Lake Murner and Lake Edelmann, and to +150 mV in Lake Au (Fig. 4.4), indicating reducing conditions. The difference of the redox potential between the sediments of the two older lakes and the younger lake must be attributed to the existence of two completely different anaerobic microbial metabolisms. The pH values of the pore

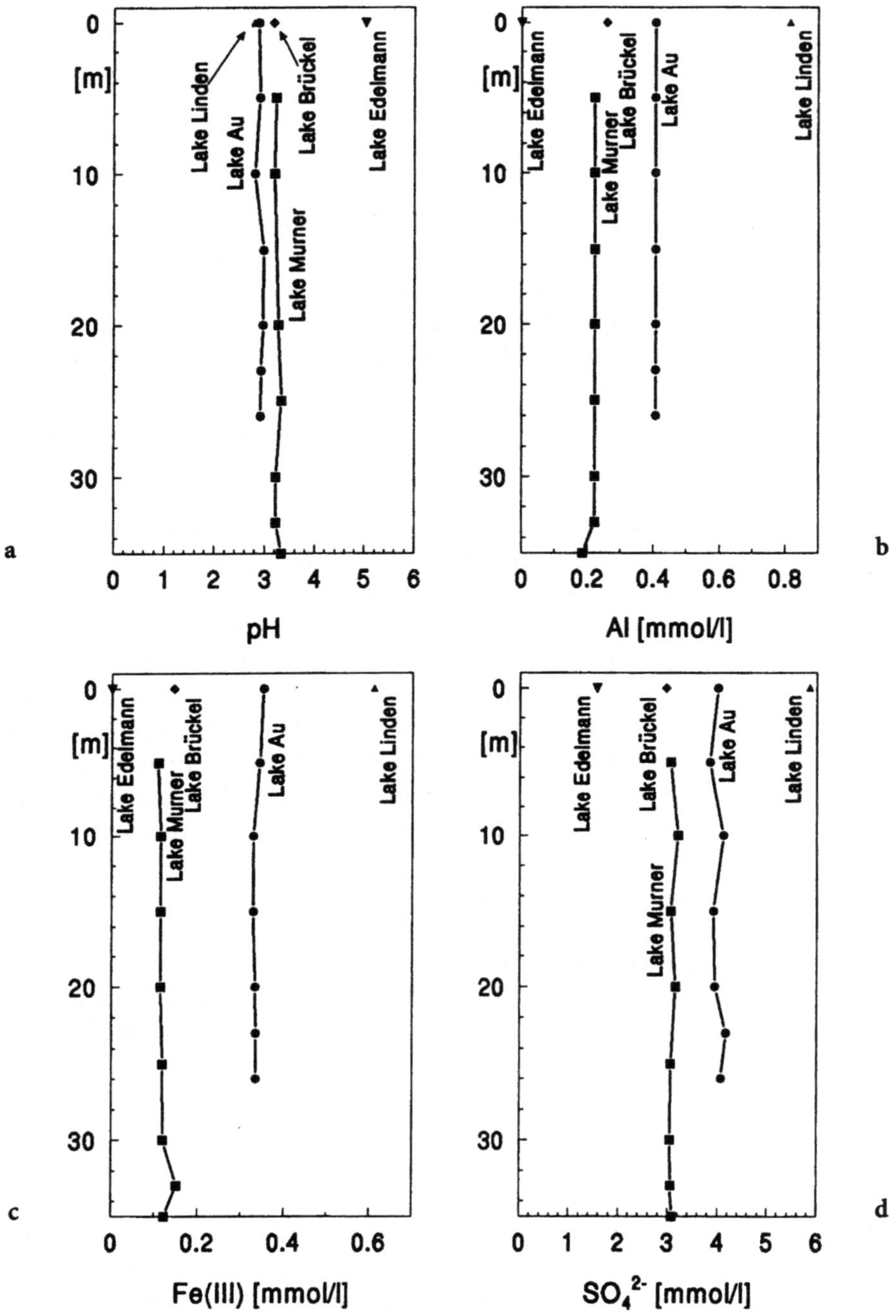

Fig. 4.2 a–d. Profiles of lake water concentrations of Lake Au and Lake Murner. Surface water concentrations of Lakes Linden, Brückel and Edelmann. **a** pH; **b** aluminium; **c** ferric iron; **d** sulphate

Fig. 4.2 e ferrous iron

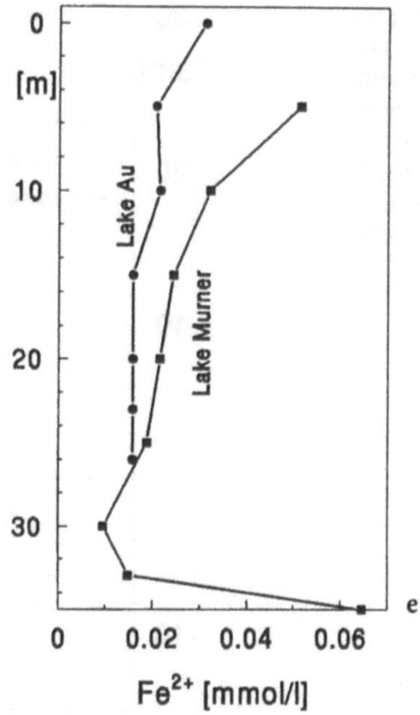

Fig. 4.3. Titration curves recorded by adding 0.01 N NaOH solution to the surface water samples. The *dashed line* denotes the inflection point which lies at pH ≅ 5

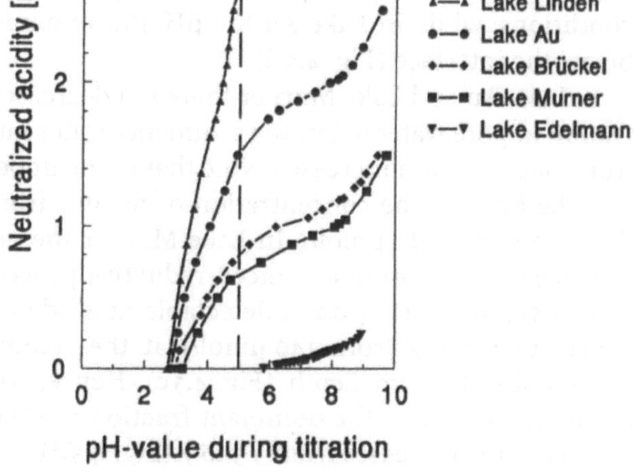

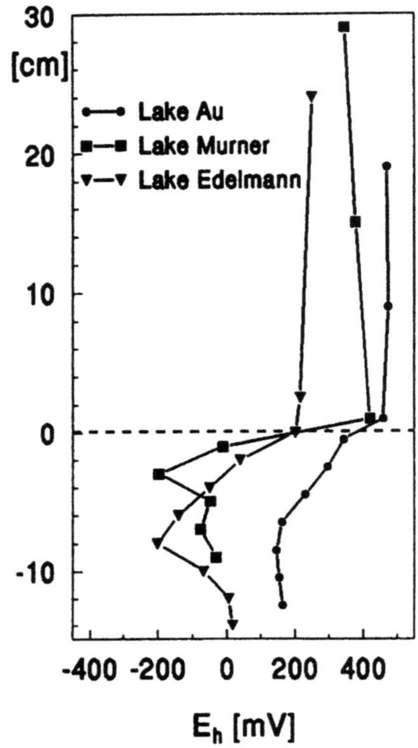

Fig. 4.4. E_h profiles at the sediment–water interface of Lakes Au, Murner and Edelmann. Measurements were performed by a microelectrode inserted laterally into the sediment cores taken in June 1995

water increased in all investigated lakes at the sediment–water interface compared with the lake water. In Lake Edelmann (with maximum pH 7.4, data not shown) and Lake Murner (maximum pH 6.9) the pH values indicate that the alkalinity gain in the pore water leads to near-neutral conditions, while in Lake Au the pH still remains low a few centimetres below the interface (Fig. 4.5 d).

In Lake Au and Lake Murner there is a decrease in the sulphate concentration in pore water below the sediment–water interface, while the ferrous iron concentration increases. Note that in the upper layers of the sediment of Lake Murner the concentration of ferrous iron is distinctly above the detection limit of 1 µmol/l. In Lake Murner the gradients of both species are steeper, indicating enhanced reductive processes relative to Lake Au (Fig. 4.5 a, b). Ferric iron is detectable at all depths, with concentrations slowly increasing from 140 µmol/l at the sediment–water interface to 500 µmol/l at 13-cm depth (Fig. 4.5 c). Hence, particularly in the upper 10 cm, ferric iron is the dominant fraction of total dissolved iron. In this depth the pH increases from 3.5 to 7 (Fig. 4.5 d).

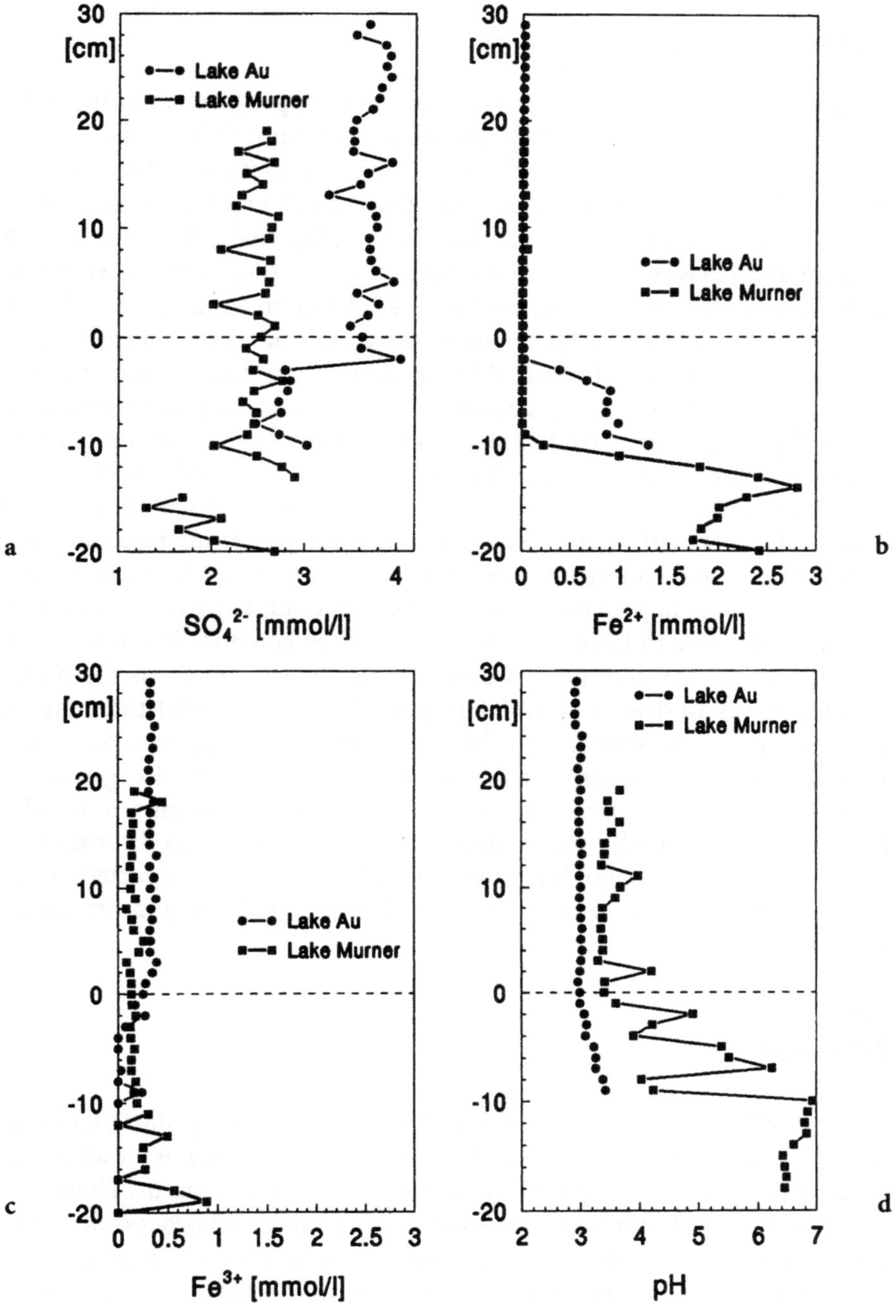

Fig. 4.5 a–d. Concentrations of sulphate, ferrous and ferric iron, and pH at the sediment-water interface of Lake Au and Lake Murner. Samples were taken by a dialysis chamber in autumn 1994

4.4.3
Sediment Composition

A large amount of reactive iron is found in the upper 5–6 cm of the sediments of Lake Au, Lake Murner and Lake Edelmann (Fig. 4.6a). In Lake Edelmann maximum values of 136 g/kg reactive iron (all values refer to mass of dry substance) were found, in Lake Murner 47 g/kg and in Lake Au 67 g/kg. TRIS was found in all three lakes (Fig. 4.6b). In Lake Au the content of TRIS was very low with a maximum of 1 g/kg. The highest value TRIS was 72 g/kg in Lake Murner. In Lake Edelmann a maximum value of 21 g/kg TRIS was found. Pyritic sulphur was estimated by subtraction of AVS and S^{o} from TRIS. In Lake Edelmann and Lake Murner a significant fraction of the TRIS is AVS (79 and 61% respectively; the percentages refer to the depth-integrated contents). In Lake Murner there is also elemental sulphur (32% of TRIS). In Lake Au most of the TRIS should be FeS_2.

In Lake Edelmann and Lake Murner the organic carbon content was high in the first centimetres below the sediment–water interface. In Lake Murner increased concentrations were found only in the first 3 cm, with 23% (w/w) in the upper centimetre. In Lake Edelmann up to 6% was found in the upper 5 cm. In Lake Au the organic carbon content did not increase significantly at the sediment–water interface with respect to the background values in deeper layers (Fig. 4.6c). We suppose that these layers enriched in organic matter stem from autochthonous sedimentation. This is confirmed by measurements of the water content, which changes at the same depth as the organic carbon content in Lake Edelmann and Lake Murner (Fig. 4.6d). The water content is a good measure to discriminate between fresh sediment of high porosity and older, more compacted material. In Lake Au water content does not change significantly in the upper sediment layer.

4.5
Discussion

There is little data available from the time when flooding started, but as geological background and mining procedure were the same for all lakes, we assume that all lakes started from the same chemical status. Therefore the results suggest that the progress of the neutralization process depends on the age of a lake. pH values increase and acidity and sulphate, which is a major species responsible for acidification, decrease from youngest Lake Linden to oldest Lake Edelmann. However, the sulphate concentrations in all lakes exceed values usually measured in soft water lakes (Cook et al. 1986).

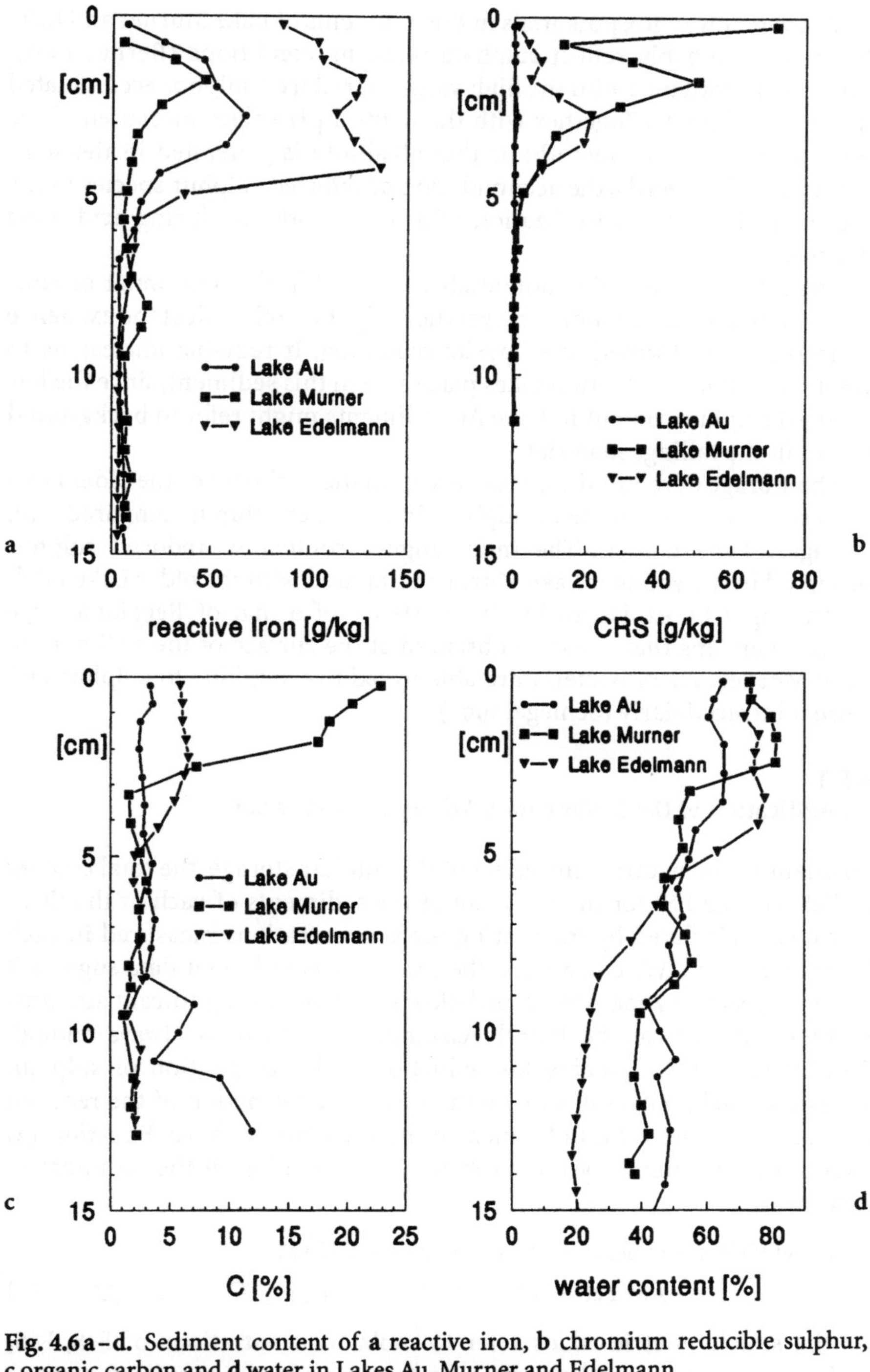

Fig. 4.6 a–d. Sediment content of **a** reactive iron, **b** chromium reducible sulphur, **c** organic carbon and **d** water in Lakes Au, Murner and Edelmann

The very low redox potentials in the sediments of Lake Murner and Lake Edelmann probably reflect sulphate reducing conditions (Berner 1963). This is in accordance with the high values of reduced sulphur accumulated in these sediments. Together with the neutral pH values measured, these findings support the hypothesis that alkalinity is generated in the sediments. In other words, the accumulation of reduced sulphur species in the sediments leads to neutralization of acidity produced during acid mine drainage.

The intermediate redox potentials measured in the pore water of Lake Au point to reducing conditions, yet they might merely reflect the existence of ferrous iron formed after pyrite oxidation. It remains unclear as to whether sulphate reduction takes place also in this sediment, since the low reduced sulphur content in Lake Au sediments might refer to background values in the tailings material.

The storage of reduced sulphur species in the sediment of the older lakes corresponds to a decrease in sulphate in the water column compared with youngest Lake Linden. The much higher content of reduced sulphur observed in the younger Lake Murner compared with the older Lake Edelmann might be explained by the existence of a mat of *Beggiatoa*-type microorganisms that could be observed at the surface of the sediment of Lake Murner. These bacteria are able to oxidize sulphide to sulphur and store it intracellularly (Schlegel 1992).

4.5.1 Quantification of the Sedimentary Acidity Neutralization

To quantify the neutralizing effect of the sulphur storage, the total content of FeS, FeS_2 and S^o for the top 6 cm of the sediments of each of the three lakes was calculated by integrating the concentrations measured in each layer (Table 4.2). We considered the top 6 cm as sediment data suggest it to be the reactive zone. Moreover, below this depth no significant amounts of these species can be found. Assuming that there is always enough FeOOH and SO_4^{2-} available for reduction and precipitation of sulphur, we established an overall stoichiometry for the formation of the reduced sulphur using the values obtained in the calculation above. Equation (2) shows the stoichiometry for sulphate transformation in the sediment of Lake Murner:

$$68\ FeOOH + 107\ SO_4^{2-} + 214\ H^+ + 211.5 <CH_2O> \rightarrow 61\ FeS + 7\ FeS_2 + 32\ S^o + 211.5\ H_2CO_3 + 141\ H_2O . \quad (2)$$

As the stoichiometric ratios refer to molar concentrations of FeS, FeS_2 and S^o as measured in the sediments, the amount of neutralized protons

Table 4.2. Content of reduced sulphur species in the reactive sediment layer of Lakes Edelmann, Murner and Au. The stoichiometric numbers refer to measured ratios of $FeS:FeS_2:S^0$ in sediment and are derived according to stoichiometries such as given in Eq. (2). These stoichiometries are the base to calculate proton consumption per centimetre squared

Lake	FeS (mol/cm²)	FeS_2 (mol/cm²)	S^0 (mol/cm²)	Stoichio-metries	H^+ consumed (mol/cm²)
Edelmann	27.9×10^{-6}	5.7×10^{-6}	1.6×10^{-6}	79:17:4	81.5×10^{-6}
Murner	45.6×10^{-6}	5.0×10^{-6}	24.4×10^{-6}	61:7:32	160.1×10^{-6}
Au	–	1.1×10^{-6}	–	0:100:0	4.2×10^{-6}

per centimetre squared can be calculated from the stoichiometries after relating them to the surface of the sediment core (19.6 cm²). Table 4.2 lists the total FeS, FeS_2 and S^0 contents together with the stoichiometries obtained and the protons consumed per centimetre squared.

The values in Table 4.2 confirm that in Lake Au storage of sulphur is low compared with the other lakes. This might be due to carbon limitation. Furthermore, the low pH in the sediment could inhibit sulphate reduction. Sulphur storage in younger Lake Murner is higher than in older Lake Edelmann. It is unclear as to whether the high acidity consumption in Lake Murner is due to the observed microbial mat on the sediment. Until now, at least optically, no microbial mat was found on other cores taken from Lake Murner.

In order to compare the neutralization capacity of the investigated mine lakes with that of lakes acidified by acid precipitation, we related the consumed protons to the years passed since flooding of the lakes and thus obtained the proton neutralization rate per year. Such an extrapolation from one core is highly problematic due to sediment heterogeneity. Moreover, sampling at the deepest point of the lake may overestimate the carbon turnover. Due to focusing effects, carbon accumulation may be higher compared with the rest of the lake (Davis 1973). Nevertheless, annual rates may provide at least a rough estimate of the alkalinity gain and can be compared with other lake systems.

Table 4.3 shows that the neutralization rates in acid mine lakes have the same order of magnitude as the rates of other lakes. Only the value of Lake Au is very low, probably due to the aforementioned reasons. Apart from sulphate reduction, the fate of aluminium and iron plays a major role in the decrease in acidity. While aluminium forms due to proton-promoted weathering of the tailings, dissolved iron stems mainly from the pyrite itself. Originally in the ferrous form (Stumm and Morgan 1981), its oxida-

Table 4.3. Comparison of in-lake neutralization rates by sulphate reduction in sediment of acidified lakes. The upper three lakes were exposed to acid deposition. Rates for the mining lakes of this study were calculated by relating proton consumption presented in Table 4.2 to age of the lakes

Neutralization ($mmol\ m^{-2}\ year^{-1}$)	Reference
124	Cook et al. (1986)
74–104	Psenner (1988)
10–159	Giblin et al. (1990)
	This work:
27	Lake Edelmann
107	Lake Murner
4	Lake Au

tion is accelerated upon increase in pH (Stumm and Lee 1961) and therefore great amounts of ferric iron form. Figures 4.2e and 4.5b indicate that, in both the acidic pore and lake water, there is still a certain portion of the dissolved iron in the ferrous form. As indicated by the titration curves of the lake water (Fig. 4.2), a pH increase to pH > 5 leads to neutralization of the acidity from these species and subsequent precipitation as the corresponding (hydro)oxides. The alkalinity necessary for the pH increase may stem from reduction of sulphate so that we obtain in total:

$$Fe^{2+} + Al^{3+} + 0.25\ O_2 + 3\ SO_4^{2-} + 6 <CH_2O> + 5.5\ H_2O \rightarrow Fe(OH)_3 + Al(OH)_3 + 3\ H_2S + 5\ H_2CO_3 + HCO^{3-}. \quad (3)$$

This is in accordance with the iron oxide precipitates found in the sediments of Lake Au, Lake Murner and Lake Edelmann. To quantify the amount of precipitated iron we integrated the reactive iron content in each sediment layer over the top 6 cm of the sediments and related it to the sediment surface. In Lake Au $8 \times 8 \times 10^{-5}$ mol/cm^2 was obtained, in Lake Murner $5 \times 9 \times 10^{-5}$ mol/cm^2 and in Lake Edelmann $25 \times 9 \times 10^{-5}$ mol/cm^2. As can be seen from Eq. (3), the precipitation reduces the net alkalinity gain to one-sixth of what one would expect from sulphate reduction alone. The main effect of this reaction is the consumption of acidity provided by the two Lewis acids Fe^{2+} and Al^{3+}. Unfortunately, we did not measure aluminium oxalate so we cannot distinguish recently precipitated aluminium from total aluminium. However, in all lakes the ratio of total aluminium to total iron content in the sediment is about 10:1. Therefore we assume that consumption of alkalinity by precipitation of aluminium hydroxide is higher than by precipitation of iron hydroxide.

Indeed, the main difference of the water chemistry of the lake water between the five lakes is the amount of metal acidity (Fig. 4.2). It is highest in the lake water of the youngest Lake Linden, decreasing in the older Lakes Au, Brückel and Murner and is not detectable in the oldest Lake Edelmann. In this lake the pH of the surface water is beyond the inflection point of the titration curves. The conditions for precipitation of hydroxides should have been favourable in Lake Edelmann, as it received water rich in organic carbon for several years. Thus, the alkalinity production by sulphate reduction was not carbon-limited in this lake.

4.6 Conclusions

There are two main processes responsible for the recovery of acid mine lakes: Sulphate reduction with burial of reduced sulphur species in the sediment and precipitation of aluminium and iron oxides. As long as there is still dissolved iron and aluminium in the water column, the base buffer capacity is high and the pH increase is small. However, alkalinity production in the sediments leads to a buildup of a ferric oxide (and probably also aluminium oxide) buffer in the sediments. This is the case in all lakes of which the sediment composition was studied and probably also in Lake Linden and Lake Brückel. In the oldest Lake Edelmann no dissolved iron and aluminium are detectable in the water column, so that recovery of the lake has progressed significantly.

It is unclear as to whether the precipitation takes place already in the lake water or whether it is localized at the sediment-water interface. From the pore water data it may be concluded that the deeper layers of the sediments are anoxic and oxygen penetrates into the sulphidic sediments. Eventually the groundwater gradient induced by the pumping activities influences this pattern, since lake water probably flows into the aquifer. The sediment might be envisioned as an interface between the groundwater in contact with the tailings and the lake water, which, in this special case, prevents penetration of oxygen into the sulphidic aquifer. More generally, one may hypothesize that sedimentation of organic matter initiates anaerobic alkalinity generating processes that cause two effects:

- oxidation of ferrous to ferric iron due to the pH increase.
- precipitation of ferric and aluminium oxides.

These effects should be self-promoting as long as there is enough supply of organic carbon to maintain sulphate reduction. Thus, a barrier might be built up at the sediment-water interface which prevents further penetration of oxygen-rich water in the tailings and subsequent production of acidity in the pore waters of the aquifer due to neutralization processes in this barrier. The recovery of the lake water should therefore be only a function of the mean retention time of the lake, which, however, can be very long in these groundwater lakes.

Acknowledgements. We would like to thank Mr. Ehrenstrasser and Dr. Wischert of the Bayernwerk AG for their support of our study. In particular, we are grateful to 14 students of geoecology who helped collect lake water data during a field course.

References

Berner RA (1963) Electrode studies of hydrogen sulfide in marine sediments. Geochim Cosmochim Acta 27:563–575
Canfield DE (1989) Reactive iron in marine sediments. Geochim Cosmochim Acta 53:619–632.
Canfield DE, Raiswell R, Westrich JT, Reaves CM, Berner RA (1986) The use of chromium reduction in the analysis of reduced inorganic sulfur in sediments and shales. Chem Geol 54: 149–155
Cook RB, Kelly CA, Schindler DW, Turner MA (1986) Mechanisms of hydrogen ion neutralization in an experimentally acidified lake. Limnol Oceanogr 31(1): 134–148
Davis MB (1973) Redeposition of pollen grains in lake sediment. Limnol Oceanogr 18:45–52
Frevert T (1983) Hydrochemisches Grundpraktikum. UTB Birkhäuser, Basel
Giblin AE, Likens GE, White D, Howarth RW (1990) Sulfur storage and alkalinity generation in New England lake sediments. Limnol Oceanogr 35(4):852–869
Herlihy AT, Mills AL (1985) Sulfate reduction in freshwater sediments receiving acid mine drainage. Appl Environ Microbiol 49(1):179–186
Höpner T (1981) Design and use of a diffusion sampler for interstitial water from fine grained sample. Environ Technol Lett 2:187–196
Kelly CA, Rudd JWM, Hesslein RH, Schindler DW, Dillon PJ, Driscoll ST, Gherini SA (1987) Prediction of biological acid neutralization in acid sensitive lakes. Biogeochemistry 3:129–140
McKnight DM, Kimball BA, Bencala KE (1988) Iron photoreduction and oxidation in an acidic mountain stream. Science 240:637–640
Meyer RKF, Mielke H (1993) Geologische Karte von Bayern 1:25000, Erläuterungen zum Blatt Nr. 6639 Wackersdorf. Bayerisches Geologisches Landesamt, München
Nordstrom DK, Jenne EA, Ball JW (1979) Redox equilibria of iron in acid mine waters. In: Jenne EA (ed) Chemical modeling in aqueous systems. American Chemical Society, Washington. ACS Symp Ser 93:51–79
Obermann P, van Berk W, Wisotzky F, Krämer S (1992) Zwischenbericht über die Untersuchungen zu den Auswirkungen der Abraumkippen im Rheinischen Braunkohlenrevier auf die Grundwasserbeschaffenheit – Pyritoxidation im Tagebau Inden I der Rheinbraun AG und Chemische Beeinflussung des Grundwassers durch Braunkohlenabraumkippen. Landesamt für Wasser und Abfall NW, Düsseldorf

Oertel F (1976) Der Braunkohlenbergbau in der Oberpfalz/Bayern. IX. Weltbergbaukongreß, Düsseldorf. Braunkohle 5:151–155
Peiffer S (1994) Reaction of H_2S with ferric oxides. In: Baker LA (ed) Environmental chemistry of lakes and reservoirs. ACS Advances in Chemistry Series No 237, Washington, DC
Peine A, Peiffer S (1996) Neutralisation processes in acid mine lake sediments. Arch Hydrobiol Spec Issues Adv Limnol 48:261–267
Psenner R (1988) Alkalinity generation in a soft-water lake: watershed and inlake processes. Limnol Oceanogr 33(6):1463–1475
Schafran GC, Driscoll CT (1987) Comparison of terrestrial and hypolimnetic sediment generation of acid neutralizing capacity for an acidic Adirondack lake. Environ Sci Technol 21(10):988–993
Schlegel HG (1992) Allgemeine Mikrobiologie, 7th edn. Thieme Verlag, Stuttgart, 559 pp
Schramel P, Wolf A, Seif R, Klose BJ (1980) Eine neue Apparatur zur Druckveraschung von biologischem Material. Fresenius Z Anal Chem 302:62–64
Sherman LA, Baker LA, Wier EP, Brezonik PL (1994) Sediment pore-water dynamics of Little Rock Lake, Wisconsin: geochemical processes and seasonal and spatial variability. Limnol Oceanogr 39(5):1155–1171
Singer PC, Stumm W (1970) Acidic mine drainage: the rate-determining step. Science 167:1121–1123
Stumm W, Lee GF (1961) Oxygenation of ferrous iron. Ind Eng Chem 53(2):143–146
Stumm W, Morgan JJ (1981) Aquatic chemistry, 2nd edn. Wiley, New York, 780 pp
Stumm W, Sigg L, Schnoor JL (1987) Aquatic chemistry of acid deposition. Environ Sci Technol 21(1):8–13
Tabatai MA (1974) Determination of sulfate in water samples. Sulphur Inst J 10:11–13
Urban N (1994) Retention of sulfur in lake sediments. In: Baker LA (ed) Environmental chemistry of lakes and reservoirs. ACS Advances in Chemistry Series 237, Washington, DC
van Berk W (1987) Hydrochemische Stoffumsetzungen in einem Grundwasserleiter – beeinflußt durch eine Steinkohlenbergehalde. Besondere Mitteilungen zum Deutschen Gewässerkundlichen Jahrbuch Nr. 49, Landesamt für Wasser und Abfall des Landes Nordrhein-Westfalen, Düsseldorf
Wisotzky F (1994) Untersuchungen zur Pyritoxidation in Sedimenten des Rheinischen Braunkohlenreviers und deren Auswirkungen auf die Chemie des Grundwassers. Besondere Mitteilungen zum Deutschen Gewässerkundlichen Jahrbuch Nr. 58, Landesumweltamt Nordrhein-Westfalen, Essen

5 Acid Generation and Metal Immobilization in the Vicinity of a Naturally Acidic Lake in Central Yukon Territory, Canada

Y. T. J. Kwong[1] *and J. R. Lawrence*[2]

[1] Klohn-Crippen Consultants Ltd., Suite 114, 6815-8th Street N. E., Calgary, Alberta T2E 7H7, Canada

[2] National Hydrology Research Institute, 11 Innovation Boulevard, Saskatoon, Saskatchewan S7N 3H5, Canada

5.1 Introduction

Sulphide oxidation in mine waste is one of the primary causes of anthropogenic production of acid lakes and drainage (Geller et al., this Vol.). In addition to the toxic effects of acid itself, the environmental hazards of acid drainage rest with the commonly elevated content of dissolved heavy metals (Allard et al. 1987; Blowes and Jambor 1990; Axtmann and Luoma 1991; Alpers et al. 1994; Deniseger and Kwong 1996). These metals may be released directly from the sulphides upon oxidation or derived indirectly by acid leaching of the associated geologic materials. Sulphide oxidation is, however, also a natural weathering process which can lead to the occurrence of natural acid drainage in the absence of mining activities (van Everdingen et al. 1985; Kwong and Whitley 1993). By studying a natural acid drainage system, one can better understand the various factors controlling acid generation in sulphide-rich material and the metal attenuation mechanisms occurring in nature. In turn, one can apply the relevant knowledge in devising appropriate strategies for the abatement of acidification and metal contamination in mining lakes. It is the purpose of this chapter to document the observations made in the vicinity of a naturally acidic lake in central Yukon Territory, Canada. Field observations coupled with analyses on water chemistry, sediment geochemistry and microbiology of a suite of water and sediment samples are used to explain the acid generation and metal immobilization processes occurring at the site. The implications of the findings for rehabilitating acidic mining lakes in general are briefly discussed.

5.1.1
Location and Geological Setting of Clear Lake

Clear Lake is a small acidic lake about 700 m long, 200 m wide and up to 9 m deep, located at the north end of the massive sulphide-rich Selwyn Basin in central Yukon Territory (Fig. 5.1, inset). High metal contents in the lake sediments (19,000 ppm Zn, 1.2 ppm Ag and 20–40 ppm Cu) have led to the discovery of the Clear Lake deposit buried adjacent to the lake (Indian and Northern Affairs Canada 1993). The Clear Lake deposit is a 27-million t exhalative massive sulphide lens occurring in Devono-Mississippian carbonaceous argillite, siltstone, chert and tuff of the Earn Group. The geological reserve includes 6.1 million t grading 11.34% Zn, 2.15% Pb and 40.8 g/t Ag at a cutoff of 7% combined Zn–Pb. The strata-bound deposit, approximately 1000 m long and up to 100 m thick, dips steeply to the east and is covered by 5–26 m of glacial overburden. Away from the sulphide lens, Clear Lake and its drainage system are surrounded by calcareous rocks such that acidic conditions prevail only in the close vicinity of Clear Lake and the massive sulphide subcrop (Fig. 5.1). Clear Lake lies in the Pelly River ecoregion of the Yukon Territory which

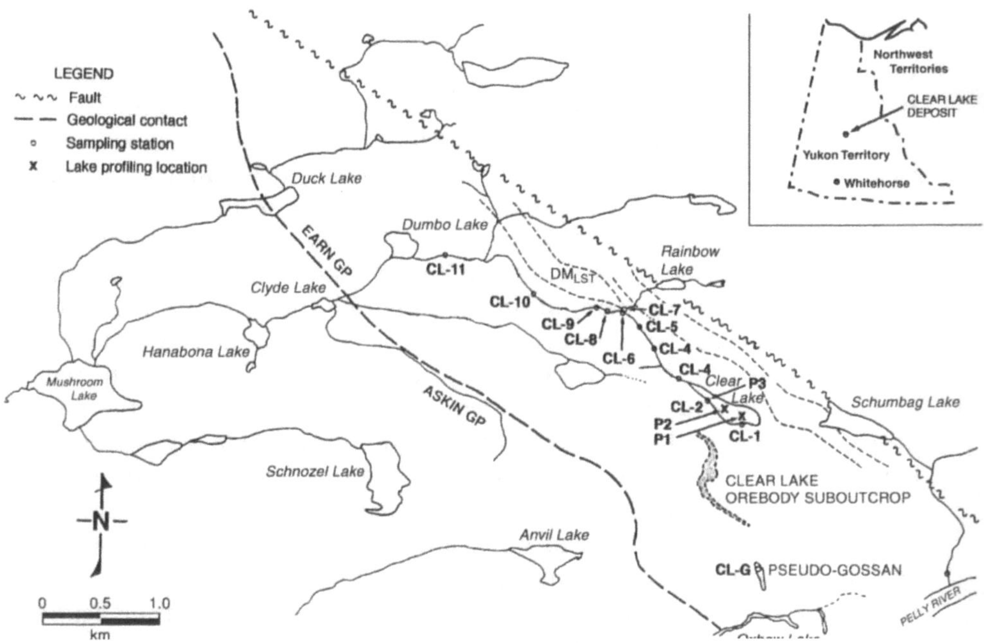

Fig. 5.1. Geological map of the Clear Lake area showing sampling locations

is characterized by widespread occurrences of discontinuous permafrost; the mean annual temperature and precipitation are −5 °C and 300 mm, respectively (Oswald and Senyk 1977). Rolling hills and plateaus exceeding 1000 m in elevation, separated by deeply cut broad valleys (down to 600 m), dominate the regional physiography. Except during the snow freshet in late May or early June, water flow through the Clear Lake drainage system is generally very slow.

5.2 Materials and Methods

5.2.1 Field Sampling

Located in a remote area, the study site could only be accessed by an aircraft. The field investigation and sampling described here were conducted as part of a pre-mine environmental assessment exercise carried out by Environment Canada in collaboration with the property owner. Three days were scheduled for visual examination of diamond drill cores as well as a reconnaissance survey of the water quality of the Clear Lake drainage system and one day was devoted to the limnology of Clear Lake. Thus, comprehensive sampling was limited to only a few selected sites.

In the reconnaissance water quality survey of the Clear Lake drainage system, sampling stations were set up along the receiving stream of Clear Lake at 100–200-m intervals up to 2 km downstream from the lake. At each station, the stream water pH, Eh, temperature and conductivity were measured using a YSI Water Quality monitor. Water samples were collected at 200-m intervals or where prominent changes in water pH or conductivity were noted. Two 100-ml samples were collected in polyethylene bottles at each sampling site. The one for dissolved metal analysis was collected by passing water through a portable, polyethylene Millipore filtering unit with a disposable 0.45-mm cellulose acetate filter. The one for anion analysis was not filtered. All samples for dissolved metals analysis were acidified with 2 ml ultra-pure nitric acid in the camp at the end of the day. Where fine-grained sediments were observed at a water sampling site with notable change in stream water pH from the previous station, a 500-g grab-sediment sample was collected using a plastic scoop and stored in plastic bags. Additional water and sediment/soil samples were collected at selected sites away from the Clear Lake drainage system and accessible by helicopter. All samples were kept cool and shipped for various analyses within 2 days after the completion of field sampling.

Sampling in Clear Lake was conducted aboard a 4-m-long rubber boat. The morphology of the lake was defined by making six transacts across the lake and the depth sounder readings recorded using a Lorance X-16 Depth Chart Recorder. At three selected locations, a water quality profile was obtained using a submersible water-quality monitor, Aquamate-1000 (Applied Microsystems, Sidney, British Columbia), equipped with a depth sensor and probes for measuring temperature, pH, dissolved oxygen and salinity, as well as specific gravity. Water samples were collected at depths where obvious changes in physical and chemical properties were noted. Attempts were also made to sample lake sediments using the Phleger Corer, a gravity-driven coring device. However, the lining of the lake bottom by up to 1.5 m of a moss resembling *Fontinalis antipyretica* (Mackenzie-Grieve, pers. comm. 1995) prevented successful retrieval of sediments for analysis.

To assess the microbiology of the Clear Lake drainage system, water and sediment samples from Clear Lake and the adjacent drainage area were collected in sterile vials, retained cool and shipped to the laboratory for subsequent analyses.

5.2.2 Laboratory Analysis

Polished thin sections were made from selected diamond drill cores and examined under an Olympus BH-2 petrographic trinocular microscope. Photomicrographs of selected areas in these sections were taken with the attached Olympus SC35 Type 12 camera. Selected sulphides were also analysed for their major and trace element content using a JEOL Model 660 SuperProbe.

Dissolved metals in collected water samples were determined by inductively coupled plasma atomic emission spectroscopy (ICP-AES) and the anions by ion chromatography according to published procedures (Environment Canada 1978). Selected samples were also checked for their trace element content to sub-ppb level by inductively coupled plasma mass spectrometry (ICPMS). The major and trace element compositions of the sediments and soils were measured by ICP-AES after digestion overnight in an $HF–HCl–HNO_3$ mixture. The fractionation of selected trace elements in the fine particles in the sediments (<80 mesh) was determined according to the procedure described by Tessier et al. (1979). QA/QC protocols were carried out as described by Environment Canada (1978) and random duplicate samples were analysed at a frequency of one for every ten samples.

The microbial populations in the water and sediment samples were analysed using a combination of most probable number (MPN) and plate count techniques. All analyses were performed in duplicate on replicate subsamples. The MPN was determined by reference to the table of Cochran (1950) for use with ten-fold dilutions and five tubes per dilution. Total aerobic heterotrophs were enumerated using 10% tryptic soy agar and the same medium with the pH adjusted to 3.5; numbers of actinomycetes were assessed using actinomycete isolation agar at pH 7 and 3.5. For MPN determinations the media were added to 24 well microtitre plates (2 ml per well) and inoculated with 0.1 ml of the appropriate dilution of sediment or water. Plates were incubated at 23 °C 2 in polyethylene bags and checked for positive results over a 12-week period. Autotrophic populations enumerated were autotrophic sulphur oxidizers with pH optimum of less than 5.4 (*Thiobacillus thiooxidans*), thiosulphate oxidizers with pH of 5.4, and iron oxidizers (*Thiobacillus ferrooxidans*). The sulphur oxidizers were enumerated using ATCC medium 125 supplemented with 1% flowable sulphur or thiosulphate as described by Germida (1985). Iron oxidizers were detected using ATCC medium 64. Sulphate reducing populations (SRB) were enumerated using peptone iron agar (Difco, Detroit, Michigan). In all cases, control wells were included to detect false positives.

5.3 Results

5.3.1 Mineralogy of the Clear Lake Deposit

As revealed by core logging and subsequent petrographic analysis, sulphide minerals occurring in the Clear Lake ore lens are laminated and consist largely of framboidal pyrite and local concentrations of sphalerite and galena. In places, the pyrite, which was recrystallized to varied extents, is slumped and fragmented (Fig. 5.2a). Rare coarse granular pyrite is often intensively fractured (Fig. 5.2b). The reactive nature of the pyrite in the deposit was demonstrated by the formation of melanterite ($FeSO_4 \cdot 7\,H_2O$) coatings on exposed grains in drill cores stored for less than a year on the surface. Tuffaceous rocks intercalated with the sulphides were largely altered to a soft grey clay. Argillites in the hanging wall and footwall of the massive sulphide lens were intensively silicified up to a thickness of 90 m. Irregular pyrite stringers and masses also commonly occur in these argillites. The silicification effectively depletes the argil-

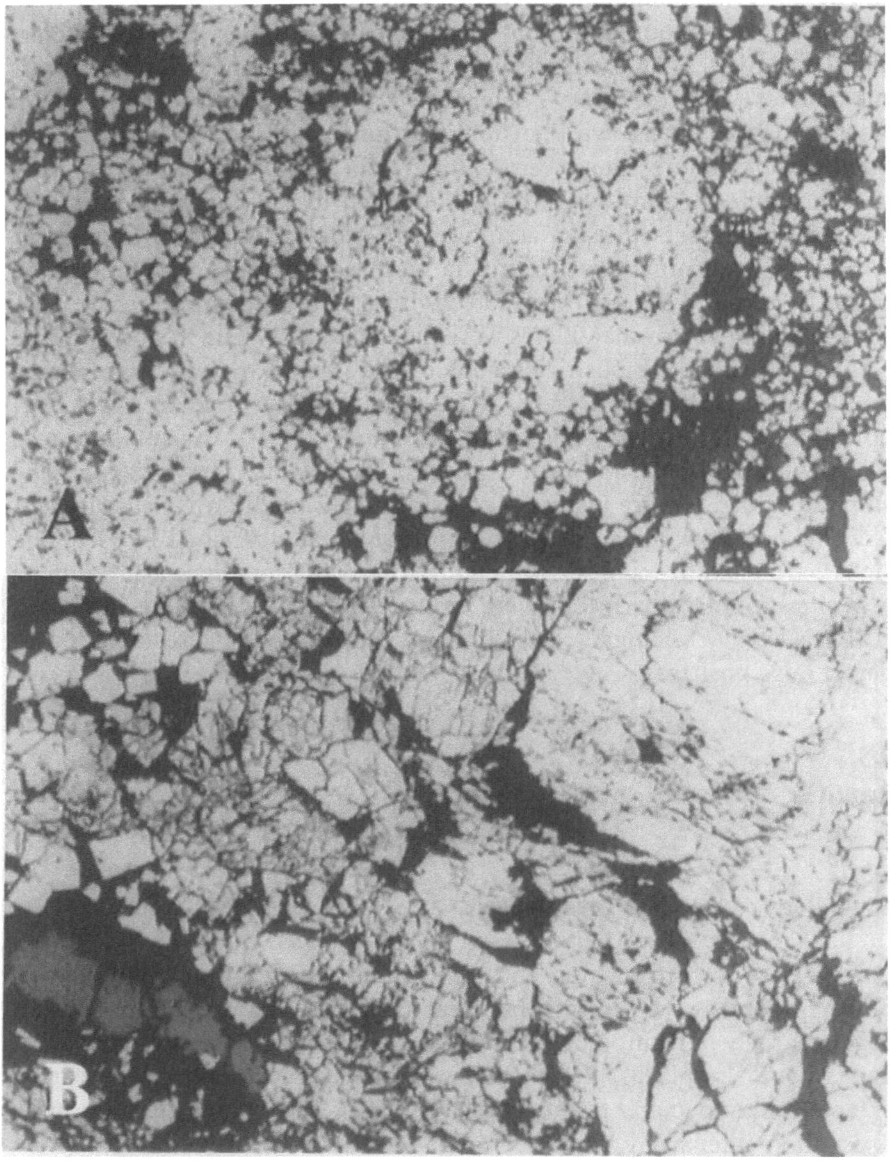

Fig. 5.2. Common textures of pyrite in the Clear Lake massive sulphide deposit. **A** As individual or clusters of framboids recrystallized to varied degrees and locally showing slumped features. **B** Coarser grains are highly fractured. H & E. × 50

lites of any acid-neutralizing capacity (Kwong 1993). This, coupled with the reactivity of the pyrite, renders the sulphide deposit particularly susceptible to acid generation under an oxygenated environment. External to the massive sulphide lens, it is not uncommon to find 40% disseminated pyrite in the overlying black shale. The black shale, with weak bedding planes acting as conduits for the transport of water and oxygen, also provides an ideal environment for acid generation through microbial sulphide oxidation (Kwong 1993).

5.3.2
Water Chemistry

The variations in water temperature, pH and dissolved oxygen concentration with depth at three locations in Clear Lake are shown in Fig. 5.3. The detailed chemistry of a suite of lake water samples is given in Table 5.1. Water at the shallow end of the lake (<3 m, site P3) was well oxygenated, probably as a result of photosynthesis occurring in the aquatic weeds as well as thorough mixing by wind action. At the deeper portions of the lake (sites P1 and P2), the dissolved oxygen concentration dropped sharply with depth, starting at about 4 m, coincident with a drop in water temperature. The lake water pH, however, remained constant at 2.8 to a depth of about 6 m, where it started to increase rapidly. At a depth of 8 m, amid the moss lining near the lake bottom, the pH had risen to 4.6 and the dissolved oxygen concentration dropped to 1.52 mg/l. Compared with the concentrations measured near the lake surface, the dissolved iron concentration (presumably as ferrous iron) had increased seven-fold in response to the developing reducing conditions, and the dissolved zinc concentration decreased five-fold as a result of increasing pH. Dissolved silica, chromium, sulphate and, to a lesser extent, arsenic also increased notably, while dissolved aluminium and cadmium decreased significantly towards the deeper part of the lake. Anoxic conditions were suspected to occur close to the sediment–water interface such that pyrite and sphalerite would be stable in the lake sediments. However, the moss lining in the lake bottom prevented successful retrieval of sediment cores at any of the sampling locations. Pore water chemistry in the bottom sediments is thus not available.

The detailed water chemistry of samples collected along the Clear Lake drainage system is given in Table 5.2. Included in this table is also the pore water chemistry of a saturated soil forming part of a pseudo-gossan outcrop (station CL-G). Surface water in Clear Lake had a uniform pH of 2.8, and dissolved zinc, aluminium and sulphate concentrations of 2.8, 11 and 1210 mg/l, respectively. In comparison, the corresponding values in

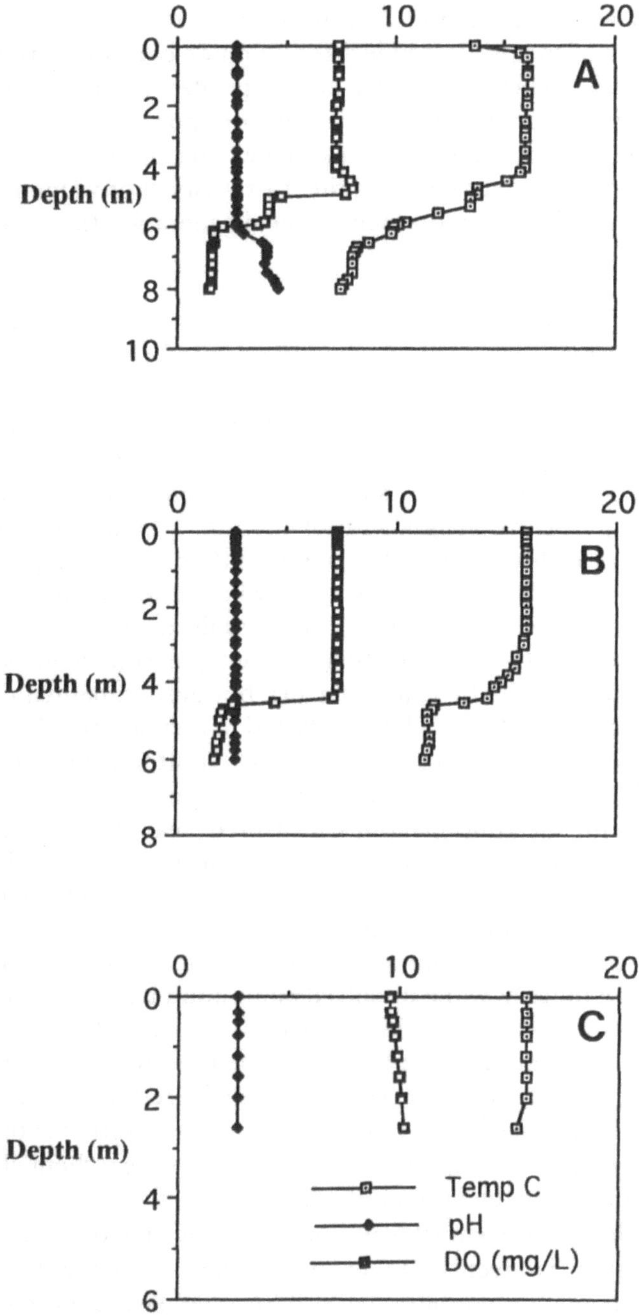

Fig. 5.3. Variations in temperature, pH and dissolved oxygen (*DO*) with depth at three locations in Clear Lake corresponding to **A** P1, **B** P2 and **C** P3 in Fig. 5.1

Table 5.1. Variation of water chemistry with depth in Clear Lake (analysis by inductively coupled plasma mass spectrometry)

Sample location	P1	P1	P2	P2	P3
Water depth	0.2 m	8.5 m	0.2 m	5.5 m	0.2 m
pH	2.8	4.6	2.8	2.8	2.8
Ag (µg/l)	<0.02	0.02	<0.02	0.04	<0.02
Al (mg/l)	12.46	6.44	10.32	12.04	12.68
As (µg/l)	0.01	0.27	0.07	0.24	0.05
Ba (µg/l)	15.11	9.68	15.19	12.64	15.9
Be (µg/l)	1.37	1.21	0.55	0.62	0.76
Ca (mg/l)	149.8	144.8	128.6	129.8	149.5
Cd (µg/l)	8.01	0.42	7.58	9.33	7.63
Co (µg/l)	6.03	5.18	5.42	6.39	6.05
Cr (µg/l)	31.1	125.5	26	42	31.8
Cu (µg/l)	18.4	5.15	16.1	21.1	18.1
Fe (mg/l)	49.72	367.2	45.19	91.57	49.57
K (mg/l)	9.14	8.03	9.07	9.29	9.14
La (µg/l)	5.16	4.53	5.1	6	5.14
Li (µg/l)	8.9	9.29	6.23	6.51	9.17
Mg (mg/l)	49.31	36.36	36.08	35.92	47.07
Mn (mg/l)	1.18	1.43	1.07	1.15	1.19
Mo (µg/l)	0.12	0.54	0.1	0.07	4.34
Na (mg/l)	5.2	3.9	4	3.8	5.5
Ni (µg/l)	31.7	22.5	27.9	33.1	32.5
P (µg/l)	263	355	195	213	308
Pb (µg/l)	0.66	1.56	0.48	0.54	0.73
Rb (µg/l)	12.99	12.93	12.7	13.15	13.11
Sb (µg/l)	<0.01	0.01	<0.01	<0.01	<0.01
Si (µg/l)	43	6200	21	340	51
Sr (µg/l)	830.3	808	823.8	848.2	833.9
Ti (µg/l)	27.38	32.81	23.24	27.14	32.78
V (µg/l)	<0.05	0.06	<0.05	0.07	<0.05
Zn (µg/l)	2586	543.3	2484	2973	2531
SO_4 (mg/l)	1503	2443	1526	1766	1495

the pore water of saturated soils in the upper portion of the pseudo-gossan outcrop (station CL-G) south of the massive sulphide subcrop were 1.9, 72, 324 and 8222 mg/l, respectively. The highly acidic pore water from station CL-G was the only sample that contained detectable dissolved lead (0.54 mg/l). Acidification of the Clear Lake drainage system was limited to the vicinity of the massive sulphide deposit such that the watercourse farther than 1 km downstream from the lake was characterized by neutral pH values, low sulphate and negligible zinc and alumini-

Table 5.2. Variation of water chemistry along the Clear Lake drainage system

Station	CL-1	CL-2	CL-3	CL-4	CL-5	CL-6	CL-7	CL-8	CL-9	CL-10	CL-11	CL-G
Ag (mg/l)	< 0.02	< 0.02	< 0.02	< 0.02	< 0.02	< 0.02	< 0.02	< 0.02	< 0.02	< 0.02	< 0.02	< 0.02
Al (mg/l)	16.9	17.1	7.78	8.44	0.12	< 0.1	< 0.1	< 0.1	< 0.12	< 0.1	< 0.1	324
As (mg/l)	< 0.1	0.1	< 0.1	< 0.1	< 0.1	< 0.1	< 0.1	< 0.1	< 0.1	< 0.1	< 0.1	1.04
B (mg/l)	0.10	0.10	0.10	0.80	0.10	0.06	< 0.02	0.06	0.04	0.04	0.02	0.54
Ba (mg/l)	0.018	0.034	0.058	0.022	0.066	0.074	0.048	0.038	0.070	0.036	0.046	< 0.002
Be (mg/l)	< 0.02	< 0.02	< 0.02	< 0.02	< 0.02	< 0.02	< 0.02	< 0.02	< 0.02	< 0.02	< 0.02	0.014
Ca (mg/l)	188.6	190.0	172.8	198.8	179.4	154.2	96.2	114.2	118.4	110.2	131.0	406.0
Cd (mg/l)	0.016	0.014	< 0.01	< 0.01	< 0.01	< 0.01	< 0.01	< 0.01	< 0.01	< 0.01	< 0.01	0.65
Co (mg/l)	< 0.01	< 0.01	< 0.01	< 0.01	< 0.01	< 0.01	< 0.01	< 0.01	< 0.01	< 0.01	< 0.01	2.66
Cr (mg/l)	0.016	0.048	0.02	< 0.01	0.78	0.056	< 0.01	< 0.01	0.078	< 0.01	0.016	1.54
Cu (mg/l)	< 0.01	< 0.01	< 0.01	< 0.01	< 0.01	< 0.01	< 0.01	< 0.01	< 0.01	< 0.01	< 0.01	0.288
Fe (mg/l)	41.6	42.0	2.74	20.0	0.324	0.236	0.242	0.526	0.592	0.114	0.096	1018
K (mg/l)	10	10	6	4	6	4	< 4	< 4	< 4	< 4	< 4	10
Mg (mg/l)	96.8	97.8	95.6	124.6	116.6	104.6	55.6	79.6	81.2	79.8	96.0	126.8
Mn (mg/l)	1.23	1.25	1.02	1.41	1.44	3.20	< 0.002	0.01	0.052	0.026	0.03	2.02
Mo (mg/l)	< 0.02	< 0.02	< 0.02	< 0.02	< 0.02	< 0.02	< 0.02	< 0.02	< 0.02	< 0.02	< 0.02	< 0.02
Na (mg/l)	10.2	10.4	11.0	13.8	18.8	18.4	10.8	14.8	15.0	15.2	24.8	14.4
Ni (mg/l)	< 0.04	< 0.04	< 0.04	< 0.04	< 0.04	< 0.04	< 0.04	< 0.04	< 0.04	< 0.04	< 0.04	0.16

Table 5.2 (continued)

Station	CL-1	CL-2	CL-3	CL-4	CL-5	CL-6	CL-7	CL-8	CL-9	CL-10	CL-11	CL-G
P (mg/l)	<0.2	<0.2	<0.2	<0.2	<0.2	<0.2	<0.2	<0.2	<0.2	<0.2	<0.2	3.8
Pb (mg/l)	<0.1	<0.1	<0.1	<0.1	<0.1	<0.1	<0.1	<0.1	<0.1	<0.1	<0.1	0.54
Sb (mg/l)	<0.1	<0.1	<0.1	<0.1	<0.1	<0.1	<0.1	<0.1	<0.1	<0.1	<0.1	<0.1
Se (mg/l)	<0.1	<0.1	<0.1	<0.1	<0.1	<0.1	<0.1	<0.1	<0.1	<0.1	<0.1	0.12
Si (mg/l)	0.62	0.64	4.22	1.36	3.22	3.24	4.36	3.08	3.26	3.62	3.94	35.6
Sn (mg/l)	<0.1	<0.1	<0.1	<0.1	<0.1	<0.1	<0.1	<0.1	<0.1	<0.1	<0.1	<0.1
Sr (mg/l)	0.742	0.750	0.756	1.004	0.906	0.790	0.444	0.600	0.620	0.588	0.766	0.396
Ti (mg/l)	<0.004	<0.004	<0.004	<0.004	<0.004	<0.004	<0.004	<0.004	<0.004	<0.004	<0.004	1.04
V (mg/l)	<0.02	<0.02	<0.02	<0.02	<0.02	<0.02	<0.02	<0.02	<0.02	<0.02	<0.02	0.80
Zn (mg/l)	2.68	2.72	0.856	0.946	0.258	0.118	<0.004	0.022	0.038	0.030	0.022	71.8
SO_4 (mg/l)	1208	1206	954	1230	940	758	220	546	528	508	576	8222
pH field	2.8	2.8	3.4	3.2	5.9	6.9	7.4	4.9	7.1	7.5	7.7	1.9
pH lab.	2.77	2.76	3.35	3.06	6.99	6.89	7.71	7.84	7.93	8.06	8.06	2.04
Eh (mV)	566	569	-15	544	197	95	14	151	42	84	67	677
Cond. (µS/cm)	1807	1822	1451	1605	1220	1070	665	955	946	964	1037	>2000
Temp. (°C)	19.8	20.1	17.8	14.1	12.1	11.9	13.6	14.3	14.7	14.7	12.4	23.9

Cond., conductivity.

Table 5.3. Major and trace element composition of selected sediments from the Clear Lake drainage system

Sample location	CL-8 stream bed	CL-8 stream bank	CL-10 stream bed	CL-G pseudo-gossan
% Al_2O_3	2.14	14.34	10.81	6.06
% Fe_2O_3	51.31	7.94	9.11	17.44
% CaO	0.712	3.606	3.058	0.395
% MgO	0.090	1.813	1.211	0.558
% K_2O	0.05	1.68	1.15	1.45
% Na_2O	0.076	2.733	1.306	0.989
% TiO_2	0.01	0.42	0.22	0.58
% MnO	0.050	0.044	0.146	0.013
% P_2O_5	0.05	0.18	0.25	0.26
% C	14.2	5.06	14.64	0.81
% S	0.54	0.19	0.29	1.11
% LOI	41.3	11.3	34.9	8.35
Ba (μg/g)	31	626	510	998
Be (μg/g)	0.1	1.2	2.0	0.7
Cd (μg/g)	12	2	4	4
Co (μg/g)	2	10	14	6
Cr (μg/g)	10	30	39	86
Cu (μg/g)	10	22	23	27
La (μg/g)	9	19	20	27
Ni (μg/g)	22	16	39	16
Pb (μg/g)	2	6	8	12
Sr (μg/g)	37	563	280	100
Th (μg/g)	3	4	3	3
V (μg/g)	16	74	41	158
Y (μg/g)	14	20	31	9
Zn (μg/g)	357	256	1281	106
Zr (μg/g)	4	72	47	55

um levels. A particularly interesting observation in the data presented in Table 5.2 is the local increase in pH value and decrease in dissolved zinc and sulphate concentrations at Station CL-3 (Fig. 5.1). This station is located close to the western end of Clear Lake where a local, densely vegetated wetland has developed. Reducing conditions prevailed here as evidenced by the negative field Eh measurement. It should also be noted that the field pH measured at station 8 differed significantly from the measurement of a water sample in the laboratory. Instrumental malfunctioning might be the cause, but the influence of a localized acidic seepage was more likely, considering that the stream bed (grey) showed a distinctly different coloration and geochemistry (Table 5.3) from those of the stream bank (reddish brown) at this location.

5.3.3 Sediment Geochemistry

Geochemical analyses of a small number of stream sediments and soil samples at the gossan outcrop are shown in Table 5.3, and the zinc partition pattern in three samples is depicted in Fig. 5.4. There is a general increase in sediment zinc content in the downstream direction in the vicinity of Clear Lake. Two sediment samples from station 8 (1 km from the lake) contained an average zinc concentration of 306 µg/g, while a sample from station 10 (1.7 km from the lake) contained 1281 µg/g. The results of sequential extraction analysis conducted on these samples indicate that the bulk of the sediment zinc is associated with the iron and manganese oxides fraction of the sediment (58 and 75% for the samples from stations 8 and 10, respectively). This suggests that the sediment zinc is largely derived by deposition of dissolved zinc through sorption onto the precipitating iron and/or manganese oxides. In contrast, the soil sample from the upper portion of the pseudo-gossan with a pore-water pH of 1.9 contains only 146 µg/g Zn and the bulk of it (77%) occurs as residual sulphides that are incompletely leached by the acidic water.

5.3.4 Microbiology

The results of the microbial population analysis are shown in Table 5.4. Water samples from Clear Lake contained very low numbers of aerobic heterotrophs, both bacteria and actinomycetes, regardless of the pH of the medium. These water samples also had detectable levels of sulphur-oxidizing autotrophs and iron oxidizers (10^2–10^3/ml); however, no thiosulphate oxidizers or sulphate-reducing bacteria (SRB) were detected. Samples from below Clear Lake with pH 3.1, 5.9 and 6.9 showed increased aerobic heterotrophic populations up to 10^4/ml and populations of all sulphur and iron oxidizers and SRBs. Sediment samples from these latter locations showed increased heterotrophic populations and the highest populations of all groups measured, with up to 10^5 sulphur and thiosulphate oxidizers, 10^3 iron oxidizers and 10^4 SRB/g.

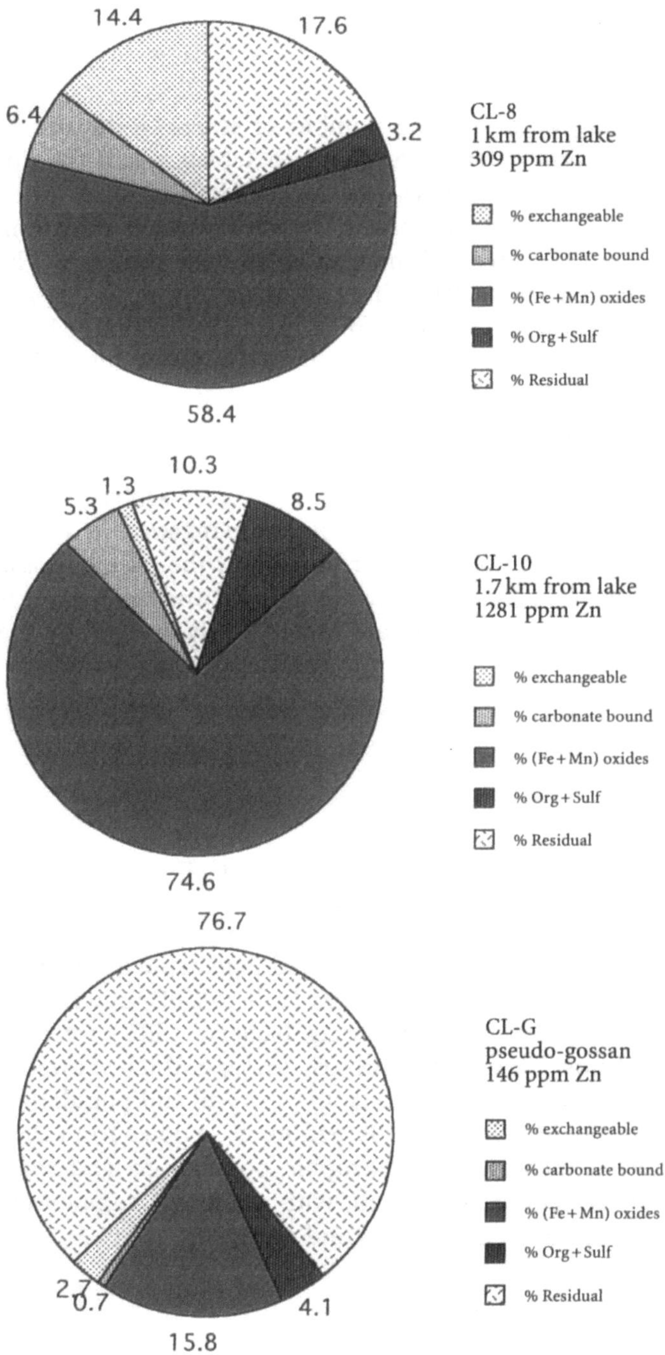

Fig. 5.4. The results of zinc fractionation studies at three locations in the Clear Lake drainage system

Table 5.4. Microbial population analysis

Sample location and description	10% TSA cfu	10% TSA @pH 3.5 cfu	AIA cfu	AIA @pH 3.5 cfu	Sulphur oxidizers MPN	Thiosulphate oxidizers MPN	Iron oxidizers MPN	SRB MPN
1. Water at P1, 0 m	0.3	ND	ND	0.7	ND	ND	ND	ND
2. Water at P2, −0.2 m	0.7	2.3	3	ND	150	ND	ND	ND
3. Water at P2, −5.5 m	0.7	0.7	7600	ND	230	ND	2800	ND
4. Water at P3, −0.2 m	1.3	0.3	ND	ND	230	ND	230	ND
5. Water at CL-3	190	160	40	ND	640	150	230	22
6. Water at CL-4	130	8300	71,000	ND	79,000	45	2300	33
7. Water at Cl-5	36,000	80	10,000	ND	8200	170	ND	85
8. Water + sediment at CL-5	8,800,000	10,000	640,000	20,000	140,000	870,000	2800	11,000
9. Sediment at CL-5	4,500,000	3200	420,000	ND	8200	350,000	2300	60,000

TSA, tryptic soy agar; AIA, actinomycete isolation agar; SRB, sulphate-reducing bacteria; cfu, colony forming units per millilitre or per gram; MPN, most probable number per millilitre or per gram; ND, no data.

5.4 Discussion

5.4.1 Acidification of Clear Lake and Metal Immobilization in Its Drainage System

The acidification of Clear Lake is not caused by mining activities; it is the result of natural near-surface sulphide oxidation and entrainment of the acidity in subsequent groundwater discharge. Given the access of the weathering elements (i. e. water and oxygen), the highly reactive pyrite in the massive sulphide lens and the surrounding rocks can readily oxidize to generate acid. Dictated by the local topography and orientation of the stratigraphy, oxygenated groundwater percolating through the sulphide body or part of it must flow in a northeasterly direction and discharge into Clear Lake. In spite of the occurrence of abundant calcareous rocks in the vicinity, rocks enclosing the sulphide deposit are devoid of any acid neutralization capacity because of intense silicification. Although pyrite is occasionally associated with sulphides having low electrode potentials (i. e. galena and sphalerite), its rate of oxidation is apparently not retarded. This is in great contrast to observations made in the Keno Hill mining district farther north where galvanic interactions in the pyrite–sphalerite–galena assemblages have produced zinc contamination of the mine drainages but no acidification (Kwong 1995; Kwong et al. 1997). The explanation probably lies in the differences in sulphide morphology and relative abundance of sulphides occurring at the two localities. In the Keno Hill mining district, the silver mineralization is hosted by relatively narrow quartz veins instead of a massive sulphide lens. Here, although pyrite is still the most abundant sulphide mineral occurring in the graphitic country rocks and some vein material, it is typically well crystalline and devoid of structural defects. In addition, it seldom occurs in excess of about 10 % in any rock. By comparison, as shown in Fig. 5.2, framboidal pyrite and highly fractured pyrite characterize the Clear Lake deposit. In spite of the fact that the Clear Lake deposit is located in a discontinuous permafrost zone with long and cold winters, the rapid acquirement of a pH value of less than 2 in the pore water of saturated soils above the pseudo-gossan attests to the involvement of acidophilic bacteria in the sulphide oxidation process. The confirmed presence of an active microbial community in the sediments from the area attests to the plausibility of the contention.

Based on the water chemistry data presented in Section 5.3.2, it is evident that zinc is the only metal contaminant of concern in the Clear Lake drainage system. However, it appears that only a limited amount of zinc has been leached from the sulphide deposit. Given a pH value of 2.8 in Clear Lake, a dissolved zinc concentration of 2700 µg/l is unexpectedly low. This implies that the bulk of the zinc and lead mineralization still lies well below the regional water table where sulphide oxidation is precluded. Alternatively, the physico-chemical settings near the sediment–water interface at the bottom of the lake may differ greatly from the near-surface environment, such that the sulphides are stabilized in the lake sediments. This is supported by the decrease in dissolved oxygen and increase in pH with depth in the lake water column. The generally gradual increase in pH and decrease in dissolved zinc and sulphate concentrations along the drainage downstream from Clear Lake are brought about by dilution and neutralization reactions with carbonate minerals. The small but abrupt changes in water chemistry at station 3 indicate the occurrence of sulphate reduction in the small wetland that further attenuates aqueous metal transport and generates alkalinity in the drainage system. The operation of sulphate reduction is confirmed by the detection of authigenic elemental sulphur and framboidal pyrite (Fig. 5.5) as well as sphalerite in the sediment sampled from this station. Sulphate-reducing bacteria, identified in the microbial analysis as one of the more abundant members in the local microbial community, probably mediate the process because the rate of abiotic sulphate reduction is well known to be extremely slow under atmospheric conditions (Berner 1971). Sulphate reduction could also be occurring locally in the bottom of Clear Lake, as indicated by the smell of hydrogen sulphide when the near-shore bottom sediment and vegetation were stirred by a probing stick. However, the lush moss growth (up to 2 m thick) on the lake bottom prevented successful retrieval of the underlying sediment for detailed analysis.

5.4.2 The Role of Bacteria in Acid Generation and Metal Immobilization Processes

The distribution of heterotrophic microbial populations in the Clear Lake drainage system followed the change in pH within the system, with very low detectable populations in the region with pH <3 and increasing populations as pH increased along the drainage downstream. Previous studies have shown a significant adverse effect of Zn on the biomass, number and activity of bacteria at concentrations of 0.5 mg/l (Dean-Ross 1990). Zn levels within this drainage system exceeded this level and likely had a negative impact on the activity of heterotrophic populations. Similarly, Wassel and Mills (1983) found that acid drainage produced a reduction in the numbers of heterotrophic bacteria and community

Fig. 5.5. Evidence of sulphate reduction in a local wetland in the Clear Lake drainage system. **A** Authigenic framboidal pyrite (*white globules*) precipitated among plant residues. **B** Elemental sulphur (*white*) coating a matrix of diatoms. Both **A** and **B** are backscatter electron micrographs of stream sediment sampled at station CL-3 (Fig. 5.1)

diversity in both sediment and water. Our observations also indicated the presence of active sulphur- and iron-oxidizing populations within the Clear Lake drainage system. These populations of sulphur oxidizers were found in conjunction with SRB populations, particularly in the sediments. This indicated that microorganisms were intimately involved in the oxidation of pyrite within the Clear Lake deposit and might have played a role in the attenuation of metals and acid through sulphate reduction in the sediments within the lake and in the sediments downstream of the lake. Indeed, at the sediment–water interface conditions likely exist in which reduced sulphur compounds are adsorbed into the sediments through reaction between reduced sulphur compounds and metals and organics. The challenge to remediation is to enhance the role of the sediments as a sink for metal contamination in the overlying waters. This may be accomplished through enhancing the activity of SRB populations within the drainage system or at specific locations.

5.4.3 Implications for Rehabilitation of Acidic Mining Lakes

Although the Clear Lake deposit has not been mined, the acidification of Clear Lake has resulted from a similar process to that of other lakes acidified by mining development. Based on the observations made in the vicinity of Clear Lake, it is apparent that a few important parameters must be taken into consideration when designing rehabilitation schemes for mining lakes in general.

Firstly, the local hydrology was a very important factor leading to the acidification of Clear Lake and determining the extent of acid neutralization downstream from the lake. Clear Lake is located in a rather restricted sub-basin with a very slow surface flow system. Without the discharge of groundwater from the adjacent massive sulphide lens, the lake would not have been acidified. If the surface drainage system had been more dynamic, the effects of acidification would have extended much farther downstream than the currently observed 1-km watercourse. This is because a more rapid flow would reduce the contact time between the acidic water and materials lining the stream bed and banks; the efficiency of any potential acid neutralization reactions is thus greatly reduced. Similarly, metal attenuation through the sorption process is also hampered. Thus, it is essential that prior to any lake remediation work, the local and regional hydrology is fully characterized.

Secondly, given the size of Clear Lake relative to the size of the adjacent massive sulphide deposit, it is debatable whether or not the lake can be rehabilitated without the sulphide deposit first being removed. Even if this could be done, the cost would be disproportionately high. Given the poor bio-productivity of the lake, it may be more economical to reserve the lake as a potential locale for disposing of highly reactive mine waste in future mining development. For this purpose, the lake may have to be enlarged artificially to increase its holding capacity and steps taken to assure the prevalence of anoxic conditions at the lake bottom. In a similar fashion, for the rehabilitation of mining lakes in general, the size factor, which affects the balance between abatement cost and environmental cost, needs to be addressed. If the extent of acid generation in the source area dictates that perpetual treatment is required in the receiving lake, depending on the cost involved, it may not be worthwhile to reclaim the affected lake. Alternative water- or land- use should always be considered in addition to technological viability.

Lastly, the cyclic nature of sulphide oxidation and sulphate reduction observed in the Clear Lake area suggests that engineering ingenuity may be required to ensure sustainable good water quality in a rehabilitated mining lake. At Clear Lake, framboidal pyrite is being oxidized in the massive sulphide body, but is regenerated in small wetland areas in the Clear Lake drainage system. Sulphate reduction is also suspected to occur in the lake bottom. In large lakes, cycling of metals at the sediment–water interface and resuspension of undesirable particulate matter in the water column are not uncommon phenomena. To ensure consistent water quality, the plausibility of creating a buffer zone of appropriate thickness by installing, for example, a sand or gravel cover at the lake bottom should perhaps be tested. In shallow lakes such as Clear Lake, aquatic weeds may also serve to stabilize the aquatic chemistry by modulating the local environment and suppressing metal influx from the bottom sediments. The challenge lies in identifying and maintaining the lush growth of the appropriate species at a specific site.

5.5 Conclusions

The acidification of Clear Lake is the result of natural sub-surface oxidation of reactive pyrite in an adjacent sulphide lens and the entrainment of the acidity thus produced in subsequent groundwater discharge. Dissolved metal, particularly zinc, is immobilized in the lake bottom and along the downstream drainage mainly through adsorption onto iron and manganese oxides and/or through sulphide precipitation. Both the acid-generation and sulphate-reduction processes are apparently enhanced by a local, active microbial community. The Clear Lake case study demonstrates that site-specific parameters control the extent of acidification effects. A closer examination of these parameters suggests that, in developing strategies for the rehabilitation of mining lakes in general, in addition to technological viability, environmental and economic costs as well as alternative water/land-use should be carefully considered.

Acknowledgements. The senior author (Y. T. J. Kwong) would like to thank the organizers of the International Workshop on Abatement of Geogenic Acidification in Mining Lakes, 4–6 September 1995, Magdeburg, Germany, for the invitation to present this paper. The Clear Lake project was partially funded by the Natural Resources and Environment Branch, Indian and Northern Affairs Canada, Ottawa. The cooperation of and assistance by staff from Total Energold Resources Ltd. and the Environmental Protection Service of Environment Canada at Whitehorse, Yukon, in field sampling and investigation are gratefully acknowledged. We also thank George Swerhone and Brij Verma of the National Hydrology Research Institute for assistance in microbial analysis.

References

Allard B, Bergstrom S, Brandt M, Karlsson S, Lohm U, Sanden P (1987) Environmental impacts of an old mine tailings deposit – hydrochemical and hydrological background. Nord Hydrol 18:279–290

Alpers CN, Nordstrom DK, Thompson JM (1994) Seasonal variation of Zn/Cu ratios in acid mine water from Iron Mountain, California. In: Alpers CN, Blowes DW (eds) Environmental geochemistry of sulfide oxidation. ACS Symposium Series 550. American Chemical Society, Washington, DC, pp 324–344

Axtmann EV, Luoma SN (1991) Large-scale distribution of metal contamination in the fine-grained sediments of the Clark Fork River, Montana, USA. Appl Geochem 6:75–88

Berner RA (1971) Principles of chemical sedimentology. McGraw-Hill, New York

Blowes DW, Jambor JL (1990) The pore-water geochemistry and mineralogy of the vadose zone of sulfide tailings, Waite Amulet, Quebec, Canada. Appl Geochem 5:327–346

Cochran WG (1950) Estimation of bacterial densities by means of the "most probable number". Biometrics 5:105–116

Dean-Ross D (1990) The response of attached bacteria to zinc in artificial streams. Can J Microbiol 36:561–566

Deniseger J, Kwong YTJ (1996) Risk assessment of copper-contaminated sediment in the Tsolum River near Courtenay, British Columbia. Water Qual Res J Can 31(4):725–740

Environment Canada (1978) A handbook of analytical methods. Inland Waters Directorate, Ottawa

Germida JJ (1985) Modified sulfur containing media for studying sulfur oxidizing micro-organisms. In: Caldwell DE, Brierley JA, Brierley CL (eds) Planetary ecology. Van Nostrand Reinhold, New York, pp 333–344

Indian and Northern Affairs Canada (1993) Yukon Minefile – Northern Cordilleran Mineral Inventory. File 105L 045 on Disk 3 of 6. Exploration and Geological Services Division, Whitehorse, Yukon

Kwong YTJ (1993) Prediction and prevention of acid rock drainage from a geological and mineralogical perspective. MEND Report 1.32.1, Canada Centre for Mineral and Energy Technology (CANMET), Energy, Mines and Resources Canada, Ottawa, 47 pp

Kwong YTJ (1995) Influence of galvanic sulfide oxidation on mine water chemistry. Proc Sudbury 1995, Mining and the Environment, May 28 – June 1, 1995, Sudbury, Ontario, Canada. CANMET, Ottawa, vol II, pp 477 – 483

Kwong YTJ, Whitley WG (1993) Heavy metal transport in northern drainage systems. In: Prowse TD, Ommanney CSL, Ulmer KE (eds) Proc 9th Int Northern Research Basin Symp/Worksh, 14 – 22 Aug 1992, Whitehorse, Yukon Territory, Canada, National Hydrology Research Institute Symposium No 10, pp 305 – 322

Kwong YTJ, Roots C, Roach P, Kettley W (1997) Post-mine metal transport and attenuation in the Keno Hill mining district, central Yukon, Canada. Environ Geol 30 (1/2): 98 – 107

Oswald ET, Senyk JP (1977) Ecoregions of Yukon Territory. Environment Canada, Canadian Forestry Service, Ottawa

Tessier A, Campbell PGC, Bisson M (1979) Sequential extraction procedure for the speciation of particular trace metals. Anal Chem 51: 844 – 851

Van Everdingen RO, Shakur MA, Michel FA (1985) Oxygen- and sulfur-isotope geochemistry of acidic groundwater discharge in British Columbia, Yukon, and District of Mackenzie, Canada. Can J Earth Sci 22: 1689 – 1695

Wassel RA, Mills AL (1983) Changes in water and sediment bacterial community structure in a lake receiving acid mine drainage. Microb Ecol 9: 155 – 169

6 Geochemical Behaviour of Submerged Pyrite-Rich Tailings in Canadian Lakes

T.F. Pedersen[1], *J.J. McNee*[2,*], *D.H. Flather*[2,*], *B. Mueller*[1] *and C.A. Pelletier*[2]

[1] Department of Earth and Ocean Sciences, University of British Columbia, Vancouver, British Columbia V6T 1Z4, Canada
[2] Rescan Environmental Services, Vancouver, British Columbia, Canada

6.1 Introduction

Sulphide-rich mine tailings have been discharged to marine and lacustrine waters in many parts of the world for centuries, often with little regard for potential long-term chemical consequences. Somewhat fortuitously, this strategy may have limited overall environmental impact, because recent evidence suggests that chemical reactivity of tailings is in fact inhibited by storage underwater. Pedersen et al. (1993) showed, for example, that fresh pyrite-rich tailings discharged to Anderson Lake, Manitoba, did not release metals to the overlying water column, at least in the summer season. Similarly, pore water studies conducted more that a decade ago in pyrite and chalcopyrite-bearing tailings on the floor of a British Columbian fjord (Pedersen 1985) revealed that excess copper was not released from the submerged deposits, at least during the period shortly after deposition. Such studies imply that the permanent storage of sulphide-rich tailings underwater offers considerable promise as a technology to inhibit to a very large degree the release of acid and metals from such materials to the environment.

This implication requires more rigorous testing before it can be accepted, however. Outstanding concerns include: (1) oxidation of tailings particles during the time window between the cessation of discharge and covering of the deposits by natural sedimentation; (2) oxidation of barely submerged tailings, where the oxygen fugacity may be relatively high and wind-driven turbulence may cause episodic resuspension; (3) seasonal effects, where, for example, overturn of a lacustrine water column might promote oxidation of sulphides present in the surface sediments.

* *Present address*: Lorax Environmental Services, Vancouver, British Columbia, Canada.

In this chapter, new data collected from detailed recent studies in Anderson Lake, Manitoba, and Buttle Lake, British Columbia are used to shed light on these three concerns. The data show that metals are not released to the water column in Anderson Lake in either winter or summer, even from very shallow tailings deposits, and that the decade-old abandoned tailings in Buttle Lake exist in a chemically benign state. These results add to the growing body of information that suggests that tailings storage underwater, if properly conducted, may be an environmentally sound disposal option.

6.2 Study Areas and Background

Anderson Lake is located approximately 2 km south of the town of Snow Lake, central Manitoba, and is a horseshoe-shaped, small (6-km-long), shallow, mesotrophic to eutrophic typical Precambrian Shield lake. The basin was floored originally by organic-rich sediments (Pedersen et al. 1993). The maximum depth of the lake is about 8 m, and there is little outflow. About 9×10^6 t of mill tailings from the processing of various copper–lead–zinc massive sulphide ores were deposited in Anderson Lake between 1979 and February 1994, when mining ceased. The pyrite-rich tailings were discharged to the lake at the surface via a floating pipeline (30 cm in diameter) which was moved seasonally toward shallower depths (3 m) in the summer and greater depths in the winter (up to 6 m). The moving discharge prevented buildup of tailings in any one area and helped to maintain a minimum water depth of 0.6 m over the deposits. On occasion the tailings built up above the lake surface and had to be moved in order to keep them submerged. The tailings consist mainly of silt-sized pyrite, quartz and feldspar, with accessory amounts of pyrrhotite, galena, chalcopyrite and sphalerite. The mill circuits were run at high pH (~ 10–11) maintained by adding lime, and the tailings stream was therefore alkaline as it left the mill. The mill reopened and tailings deposition in the lake resumed in late summer 1995.

Water quality in the lake during recent years was very poor, largely as a result of the influx of acid rock drainage that flowed into the basin from an old roadway built of waste rock and tailings along the north shore, as well as input of acid drainage that was pumped from a nearby abandoned mine and conducted to the lake via the tailings pipeline. Over the past few years, outflow from the lake contained ~ 800 mg l^{-1} of SO_4^{2-} and ~ 600 μg l^{-1} of dissolved Zn, and the pH of the lake water measured at various sites ranged from 6.8 to 7.4 [Hudson Bay Mining and Smelting Co. (HBMS), unpubl. data]. Water quality data collected by HBMS prior to the start of tailings discharge in 1979 indicated an ambient pH of 8.1–8.3 and dissolved metals levels similar to those in other Precambrian Shield lakes.

Buttle Lake is a large (35 km long by 1 km wide with a maximum depth of 87 m) water body which occupies a U-shaped valley in an area of high relief on central Vancouver Island, British Columbia. Like many lakes in the coastal region of British Columbia, Buttle Lake is oligotrophic – nutrient levels are relatively low (Deniseger et al. 1990). Between 1967 and mid-1984, some 5.5×10^6 t pyrite-rich, Zn-, Cu- and Pb-bearing tailings were discharged to the south basin of the lake via a raft-supported outfall that was submerged below the thermocline. Polyacrylamide flocculants were used to assist settling of the solids. The tailings consisted of sand- and silt-sized silicate gangue minerals and residual copper, iron, lead and zinc sulphides. Base metal concentrations in the tailings solids ranged widely, but averaged 7000, 1300 and 900 mg kg^{-1} for Zn, Cu and Pb, respectively. Lime (CaO) was used in significant quantity in the mill; approximately 1 kg t^{-1} ore was added to the tailings to raise the pH in the milling circuits and to enhance coagulation in the thickening tanks.

Dissolved zinc, copper, lead and cadmium concentrations in Buttle Lake water began to increase shortly after the mine commenced operation, and peaked in 1981, reaching levels in south basin surface waters as high as 370 μg l^{-1} Zn, 40 μg l^{-1} Cu, 25 μg l^{-1} Pb and 3.6 μg l^{-1} Cd (Deniseger et al. 1990). The high metal loadings were derived from acid mine drainage which percolated from a waste rock area into a stream beside the mine site and thence to the lake. Pedersen (1983) showed that the tailings on the lake floor were not releasing dissolved metals into the overlying water column during the period when the lake water was badly contaminated. In 1983, a surface and groundwater collection and treatment system was installed to capture the bulk of the acid drainage and metal levels subsequently declined significantly.

Buttle Lake waters are near neutral, with pH values ranging from extremes of ~7.1 to ~8.6. The highest pH values are observed in surface or near-surface samples, which may reflect carbon dioxide consumption by phytoplankton; the lowest pH levels are generally seen in the deep waters and probably result from addition of carbon dioxide from oxidative degradation of sinking organic detritus. Concentrations of dissolved metals measured in October 1993 were the lowest seen in the south basin of Buttle Lake in the last 20 years. Zinc levels ranged from minima of ~5 μg l^{-1} in surface or near-surface waters to maxima of ~24 μg l^{-1} in the deepest samples at 35 m depth. Copper concentrations were 2 μg l^{-1} or less, and varied little with depth. Dissolved lead was not detected anywhere in the water column (< 0.2 μg l^{-1}).

6.3 Sampling and Analytical Methods

Interstitial water and sediment samples were collected in 1993 in Anderson Lake using the coring, extrusion and centrifugation approach described by Pedersen et al. (1993). Two sampling campaigns took place: the first through the ice in April (the "Winter Survey") and the second in August (the "Summer Survey"). Duplicate cores were collected and processed at two sites in the Winter Survey: the first near the point of tailings discharge (site B, the "Pure Tailings site", cores 3 m apart) and the second at a distal location where natural sediments predominate (site A, the "Natural Sediments site", cores 12 m apart). These two sites along with a third (the "Shallow Tailings site") were occupied in the summer. The third site, some 300 m north of site A, was only 1 m deep, and was located near the centre of the lake in an area particularly susceptible to wave-driven turbulence.

Sediment cores were raised from Buttle Lake in October 1993 from three sites: the first near the old tailings discharge point in the south basin (the "Outfall site"), the second about 1.3 km to the north also in the south basin (the "Distal site") and the third about 6 km to the north in the tailings-free central basin (the Natural Sediments site). Duplicate cores were collected at the Distal site only. In addition to the coring, acrylic dialysis-membrane array samplers ("peepers") were emplaced by divers in Anderson Lake in the Summer Survey and in Buttle Lake. The arrays, which were pushed manually into the bottom, consisted of rectangular acrylic plates nearly 1 m long and 25 cm wide into which two banks of cells (70 × 8 × 14 mm) spaced 1 cm apart vertically had been milled. The spatial resolution in the central section of each peeper is 5 mm which is obtained by offsetting a series of smaller wells (32 × 8 × 14 mm). Prior to deployment, the cells were filled with ultra-pure water and covered with a 0.45-μm pore-size polysulphone filter membrane (in sheet form) which was in turn overlain by an acrylic face-plate. A 1.6-mm-thick expanded porous polyethylene frit was placed on top of the acrylic plate. The membrane/plate/frit assembly was held in place with nylon screws. Each peeper was stored in a sealed acrylic box containing ultra pure water. N_2 gas was bubbled through each box for several days, right up to the time when the peepers were removed to be inserted in the sediments. This technique is expected to have greatly diminished the concentration of O_2 dissolved in the plastic, which could theoretically affect the chemistry of anoxic pore waters (Carignan et al. 1994). The peepers were left inserted in the sediments for 2 weeks, which is more than adequate to ensure full diffusive equilibrium between the pore water and the (initially) ultra pure water in the cells of the array (Carignan 1984). Following recovery by divers (in Buttle Lake) or by pulling on an attached cord (Anderson Lake), the peepers were sampled in a nitrogen-filled glove bag by removing the frit and carefully piercing the membrane with Eppendorf pipettes. Extracted samples were immediately acidified to pH 2 with ultra pure HNO_3.

Dissolved metals were analysed in all pore water samples by graphite furnace atomic absorption spectrophotometry (Fe and Mn, following dilution with acidified ultra pure water) or inductively coupled plasma mass spectrometry (Zn, Cu and Pb, on neat pore solution spiked with an indium internal standard). Sulphate and nitrate concentrations were assayed by ion chromatography, and sulphide colourimetrically (Cline 1969). The chemical composition of solid-phase (core) samples was determined by Analytical Services Laboratory (Vancouver, British Columbia) using acid digestion – atomic absorption spectrophotometry – while total carbon, nitrogen and sulphur contents were determined at the University of British Columbia by elemental analysis using a Carlo-Erba CNS analyser. Organic carbon is reported as the difference between total C and carbonate C determined colourmetrically (Pedersen et al. 1993).

6.4 Results

6.4.1 Anderson Lake

6.4.1.1 *Water Column*

The water column in Anderson Lake during the Summer Survey was well oxygenated (6–7 mg l^{-1} O_2), warm (19 °C) and well mixed, as indicated by the lack of a distinct thermocline. Previous fieldwork indicates that this is the common state of the lake in the summer season. However, during the preceding winter the water column was very well stratified (Fig. 6.1). At the Natural Sediments site, relatively high oxygen contents (up to 4.6 mg l^{-1}) were limited to a cold water layer 20 cm thick immediately under the 0.5-m-thick ice. The oxygen concentration decreased sharply between 60- and 80-cm depth, where the temperature increased from <1 to >6 °C. A mid-depth warm layer (6.4 °C) between 90 and 250 cm below the ice surface had a near-zero oxygen content. Up to ~1.5 mg l^{-1} oxygen occurred in a deeper 1-m-thick aerobic zone, which was slightly

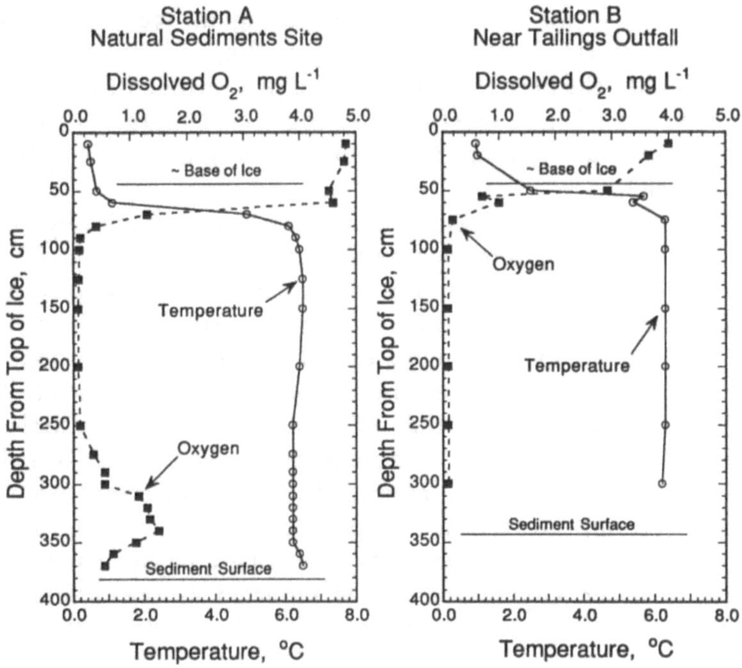

Fig. 6.1. Oxygen and temperature profiles in the water column at stations A and B in Anderson Lake in April 1993. The depth scale is relative to the ice surface, not to the bottom

cooler than water in the mid-depth layer. Oxygen was again depleted in a slightly warmer (by ~0.2 °C) 20-cm-thick layer near the bottom.

Less complex profiles characterized the water column near the tailings outfall at station B (Fig. 6.1). Oxygenated water was confined to a thin horizon immediately below the base of the ice and to the water in the 50-cm-deep hole through the ice, which must have been aerated (at both sites) when the holes were cut. As at station A, the oxygen-depleted layer at depth was much warmer than the relatively oxygenated waters above.

6.4.1.2
Sediments

The duplicate cores collected ~3 m apart at the Natural Sediments site (site A) in the Winter Survey in Anderson Lake were visually and compositionally very similar. Both cores were veneered by a 2- to 3-mm-thick cap of rusty-brown non-cohesive fine-grained sediment, which contained abundant organic fragments. Small copepods (*Daphnia*?) were abundant in the water immediately above the sediment–water interface and were

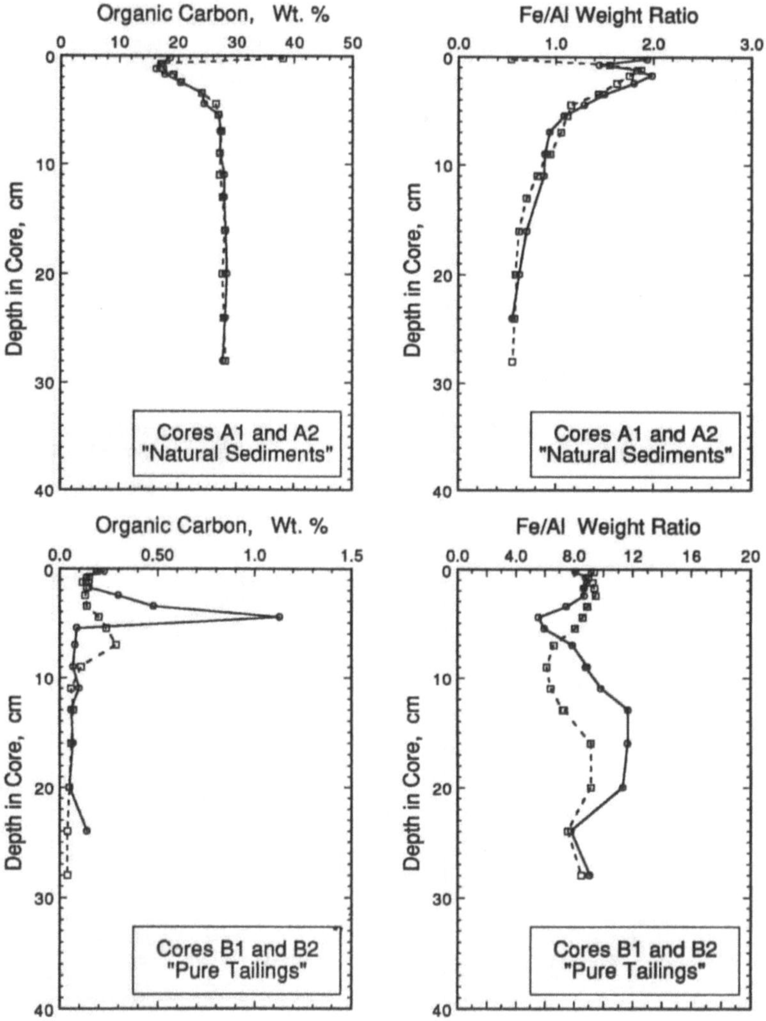

Fig. 6.2. Sedimentary organic carbon concentrations and Fe/Al weight ratio profiles in duplicate cores from stations A and B in Anderson Lake in April 1993. *Open circles* represent cores A1 and B1; *open squares* represent cores A2 and B2

observed burrowing into the rust-coloured surface layer. The surface veneer was underlain by about 10 cm of charcoal-grey fine-grained gelatinous ooze, which graded into the brown, gelatinous, homogeneous, organic-rich ooze that characterized the lower 40 cm of both cores. Small (presumably) methane bubbles developed below 30-cm depth in both cores shortly after collection. Both cores raised about 12 m apart at the Pure Tailings site (site B) consisted of undifferentiated fine-grained grey

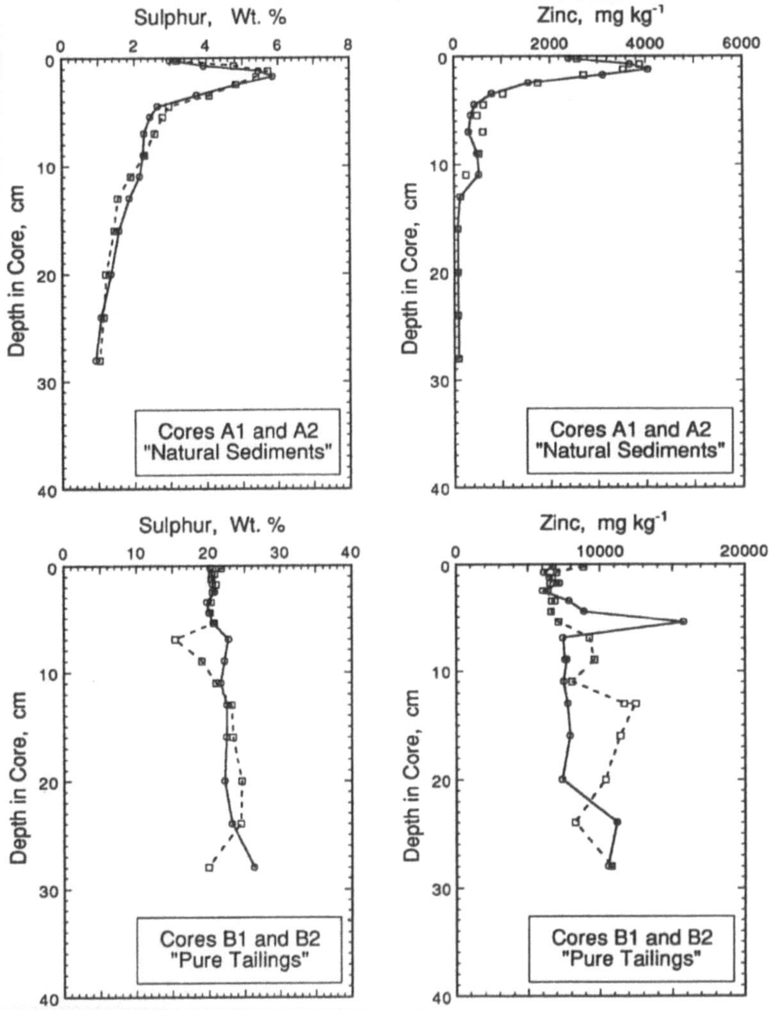

Fig. 6.3. Sedimentary sulphur and zinc concentration profiles in duplicate cores from stations A and B in Anderson Lake in April 1993. *Open circles* represent cores A1 and B1; *open squares* represent cores A2 and B2

tailings which contained a very high proportion of glistening, very fresh-looking pyrite grains. The lake floor at this location was irregular, with differences of more than 1-m depth between sites only 12–15 m apart. Observations made by divers during the August 1993 survey of the lake confirmed that the sediment surface near the tailings outfall is hummocky. Lobes of tailings appear to account for the observed differences in local relief.

Below a depth of 12 cm, cores A1 and A2 consist of tailings-free, organic-rich sediments, as shown by the organic carbon (C_{org}), Fe/Al, S, Zn, Cu and Pb distributions (Figs. 6.2–6.4). The upper 12 cm are composed of a mixture of tailings and natural sediments, with metals concentrations reaching maxima between 1- and 2-cm depth. The marked compositional similarity of cores A1 and A2 is not matched by the pair of cores from station B. Organic carbon, Fe/Al, S, Zn, Cu and Pb distributions in cores B1 and B2 show differences with depth (Figs. 6.2–6.4), which apparently reflect significant local-scale lateral inhomogeneities in the tailings deposit in the lake. This inhomogeneity is not inconsistent with the variable composition of the ore feedstock, the hummocky terrain in the vicinity of the outfall and the frequent changing of the location of the discharge point.

Petrologic examination of a set of polished sections prepared from sediment subsamples taken from core A1 indicates that the lithogenic fraction of the natural sediments is composed typically of silt-sized quartz grains with accessory feldspar, biotite, hornblende and rare carbonate. Framboidal pyrite occurs in all samples at this site, including the upper 0.5 cm of the deposits. The framboids are typically fine-silt-sized and occurred either free or incorporated within organic flocs. Crystalline sulphides, consisting mainly of pyrite, with some pyrrhotite and some chalcopyrite, are also observed in the top 12 cm of the cores, but not at greater depths. These silt-sized grains were all fresh, regardless of composition.

Examination of polished sections from subsamples of the tailings at station B (core B1) revealed a very high sulphide content with approximately equal proportions of pyrite and pyrrhotite. The silicate fraction consisted predominantly of quartz and feldspars with accessory hornblende, chlorite and biotite. Considerable variations in texture of the tailings was observed with depth in the core, which presumably represent changes in feedstock, milling procedures and hydraulic influences during deposition. A very high concentration of sphalerite (~3.5%) was observed in the sample AL-9 (5–6 cm deep, core B1); this is consistent with the very high zinc concentration of $\sim 16 \times 10^3$ mg kg^{-1} measured in a split of the same sample (Fig. 6.3). No authigenic framboids were observed in these deposits, and without exception the sulphide grains were consistently fresh. No physical evidence of leaching or oxidation was noted.

During the Summer Survey, one core was collected by divers at each of the Natural Sediments and Tailings sites by hand inserting butyrate barrels. A third core was collected by a diver from the turbid Shallow Tailings site (~1.5 m deep). The sediments at this location consisted of a highly porous lag deposit of pyrite-rich indurated clasts composed entirely of tailings grains, and ranging from <1 to >10 mm in size. All

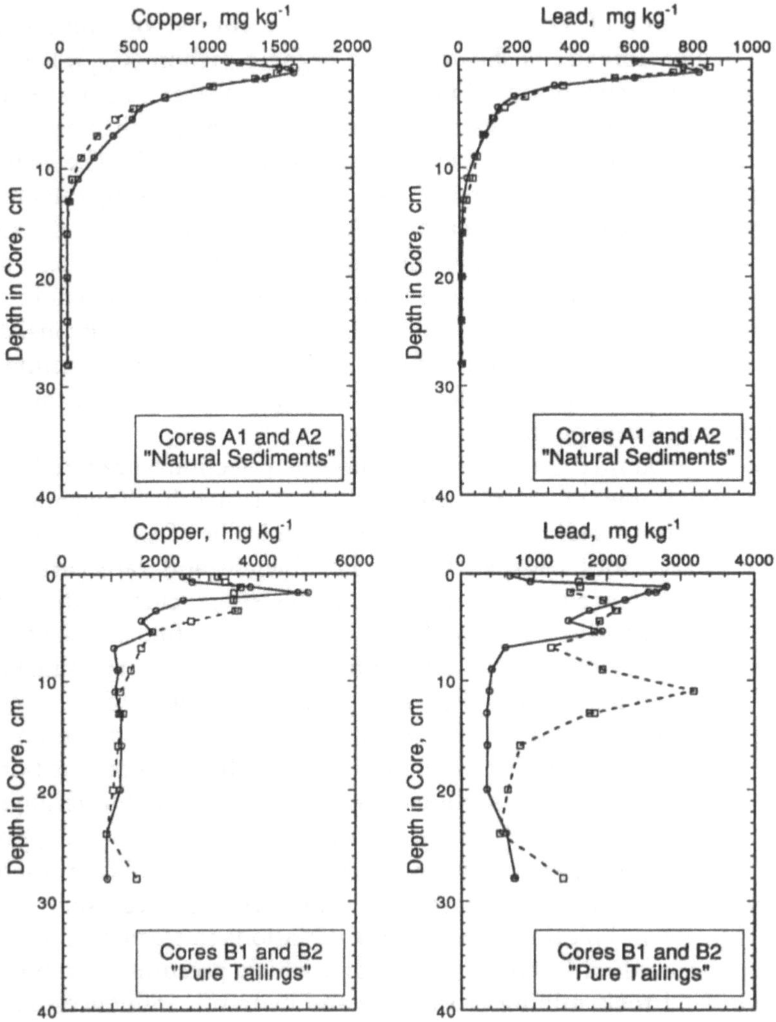

Fig. 6.4. Sedimentary copper and lead concentration profiles in duplicate cores from stations A and B in Anderson Lake in April 1993. *Open circles* represent cores A1 and B1; *open squares* represent cores A2 and B2

pyrite particles, either discrete or within the clasts, appeared fresh and unaltered.

Each of the cores was capped with a rust- to orange-coloured ferruginous veneer. At the Natural Sediments site this consisted of a 2-mm thick, gelatinous, rust-coloured layer, with abundant faecal casts and several protruding worm tubes on the upper surface. Burrows with oxidized haloes were observed down to 10-cm depth, suggesting active bioturbation in this interval. At the Pure Tailings and Shallow Tailings sites, the ferruginous layer was 2–3 mm thick and consisted of filamentous bacteria in a matrix of rust- to brown-coloured flocculant material, presumably iron oxyhydroxides. Based on a brief microscopic examination, the bacterium was probably *Sphaerotilus* or *Leptothrix* spp. (W. Mohn, University of British Columbia, pers. comm.). *Sphaerotilus* forms filamentous sheaths inside which the cells divide and are extruded out the ends of the sheath. Both genera require oxygen, and both are associated with iron oxyhydroxides. In the case of *Sphaerotilus*, the iron appears to precipitate abiologically on the sheath as the bacterium degrades the organic moiety of iron–organic complexes. No bacterial mat was evident on the tops of the cores collected from this area during the Winter Survey which may reflect the dearth of oxygen in the bottom waters at that time.

With the exception of the bacterial mat, the natural sediment core collected in the summer was almost identical to the duplicate cores raised from the same station in winter. Such similarity was not matched at the Tailings site, where the Summer-Survey core consisted of a vertically varying admixture of tailings and natural sediments. The pair of Winter Survey cores collected from the same location consisted of pure tailings of somewhat variable composition, but natural sediments were not a significant component. These contrasts indicate that near the discharge point in Anderson Lake, there can be marked variations in the extent of deposition of the tailings over short distances.

6.4.1.3
Interstitial Waters

6.4.1.3.1
Winter Survey

All filtered pore water samples from all four cores collected in the winter were analysed for a suite of dissolved metals, including Fe, Zn, Cu and Pb. Sulphate, nitrate and total dissolved sulphide were measured in selected aliquots where volume permitted. Dissolved iron profiles are shown in Fig. 6.5 along with concentrations measured in a sample collected 5 cm above the bottom by peristaltic pumping from the ice surface. The "duplicate" profiles are similar in form at each site: high concentrations characterize the pore waters in the upper several centimetres, particularly in the station B cores; below the upper 10 cm, concentrations fall to very low values. Dissolved zinc concentrations at the Natural Sediments site sharply decrease with depth in the upper 2 cm of the deposits (Fig. 6.5),

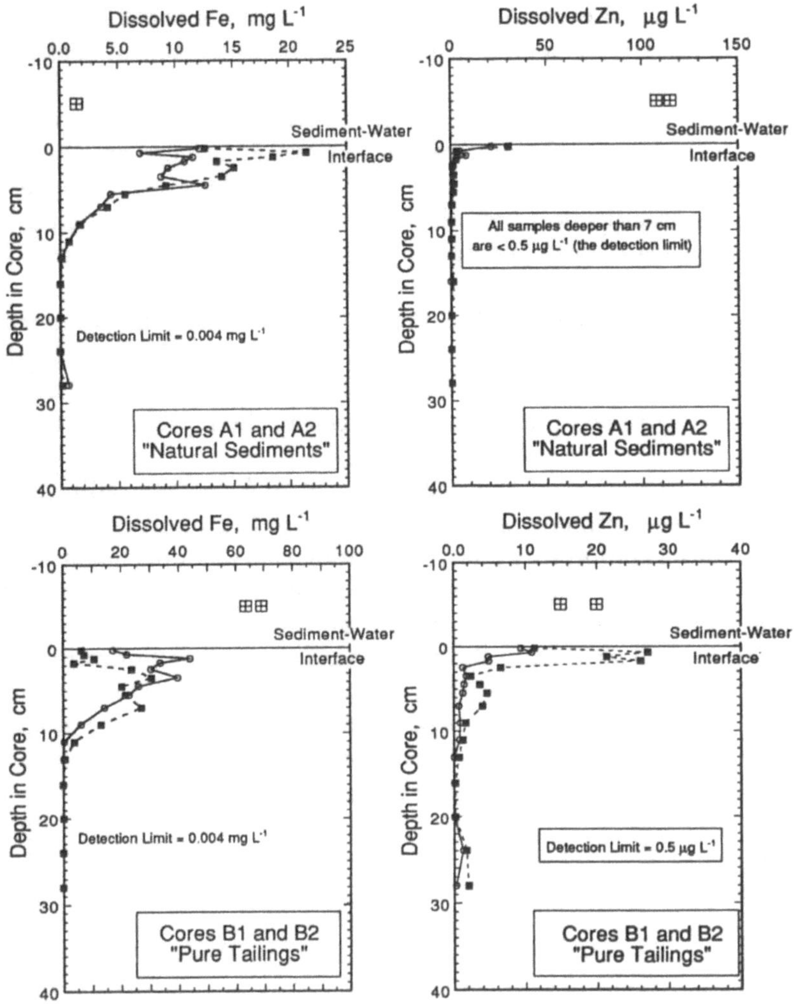

Fig. 6.5. Dissolved (< 0.2-mm filtered) iron and zinc concentration profiles in duplicate cores from stations A and B in Anderson Lake in April 1993. *Open circles* represent cores A1 and B1; *solid squares* represent cores A2 and B2; *crossed squares* above the interface represent the concentration in the lowermost sample (~ 5 cm above the bottom) collected by peristaltic pumping

similar to the tailings in core B1, whereas in B2, a peak in concentration (~ 25 µg l^{-1}) occurs between 0.5 and 2-cm depth. Dissolved copper (Fig. 6.6) is essentially undetectable (< 0.12 µg l^{-1}) in all pore waters extracted from the Pure Tailings cores (as well as in the overlying bottom water). In cores A1 and A2, maximum concentrations of 2.8 µg l^{-1} are observed in the upper 2 cm, with very low or undetectable values at

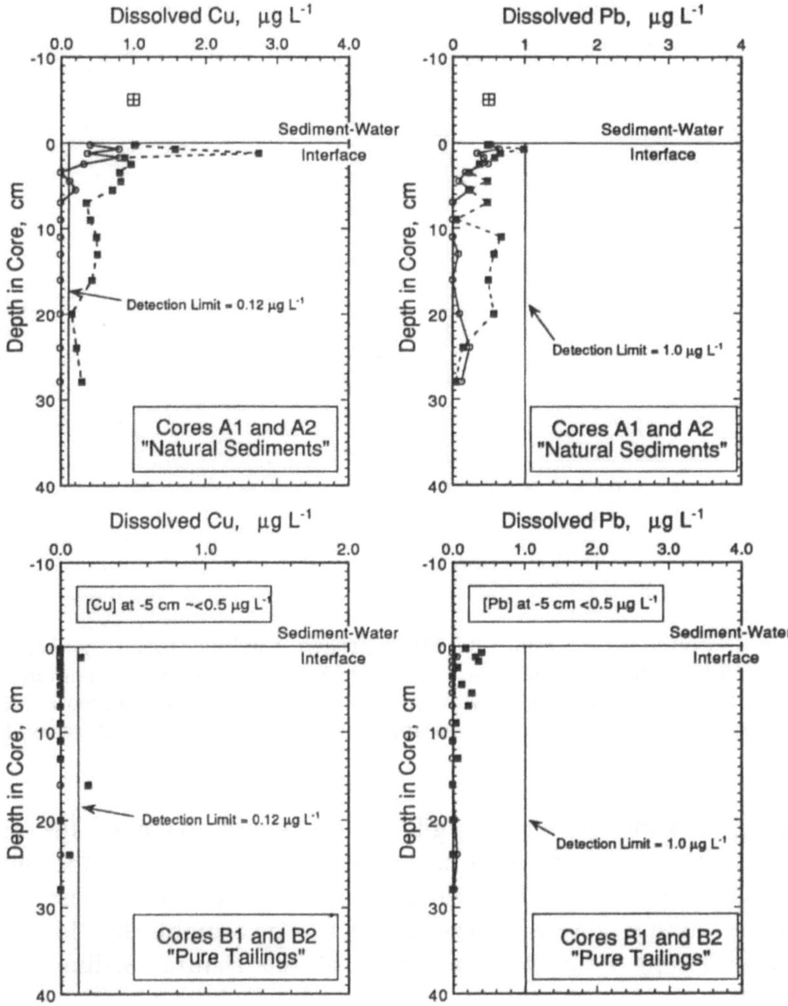

Fig. 6.6. Dissolved (< 0.2-mm filtered) copper and lead concentration profiles in duplicate cores from stations A and B in Anderson Lake in April 1993. *Symbols* as in Fig. 6.5. Copper was not detectable in the bottom water at site B

greater depths. Lead was not detectable in any of the pore water samples at either station (Fig. 6.6).

Sulphate distributions in pore waters from the two locations showed considerable contrast. In the natural sediments, SO_4^{2-} decreases with depth from values of ~900 mg l^{-1} at the surface to ~400–600 mg l^{-1} at the base of the cores (Fig. 6.7). Concentrations in the tailings increase with depth,

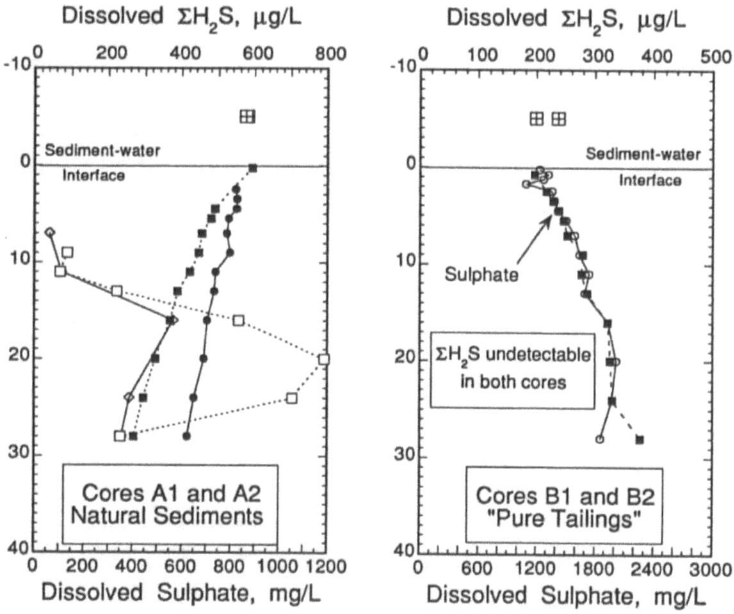

Fig. 6.7. Dissolved (< 0.2-mm filtered) hydrogen sulphide (*open symbols*) and sulphate (*filled symbols*) concentration profiles in duplicate cores from station A and sulphate profiles from station B in Anderson Lake in April 1993. *Circles* represent core A1; *squares* represent core A2. H_2S was not detectable in pore waters at station B

reaching maxima of ~2000 mg l^{-1} in the lower 10 cm of the cores, which is about twice the level in the local bottom waters at the time of core collection. Dissolved sulphide was not detectable in pore waters from the tailings cores, in contrast to the measurable but low levels (up to ~800 µg l^{-1} as H_2S) recorded at the Natural Sediments site (Fig. 6.7).

6.4.1.3.2
Summer Survey

Dialysis-array ("peeper") samples from sets of peepers emplaced at all three sampling locations were analysed for ΣH_2S and a suite of dissolved metals, including Fe, Zn, Cu and Pb. Pore waters extracted from the single cores raised at the Natural Sediments and Tailings sites were also analysed in the same way.

Shallow Tailings Site. Dissolved iron profiles from two peepers emplaced 4 m apart are shown in Fig. 6.8. The "duplicate" profiles show significant

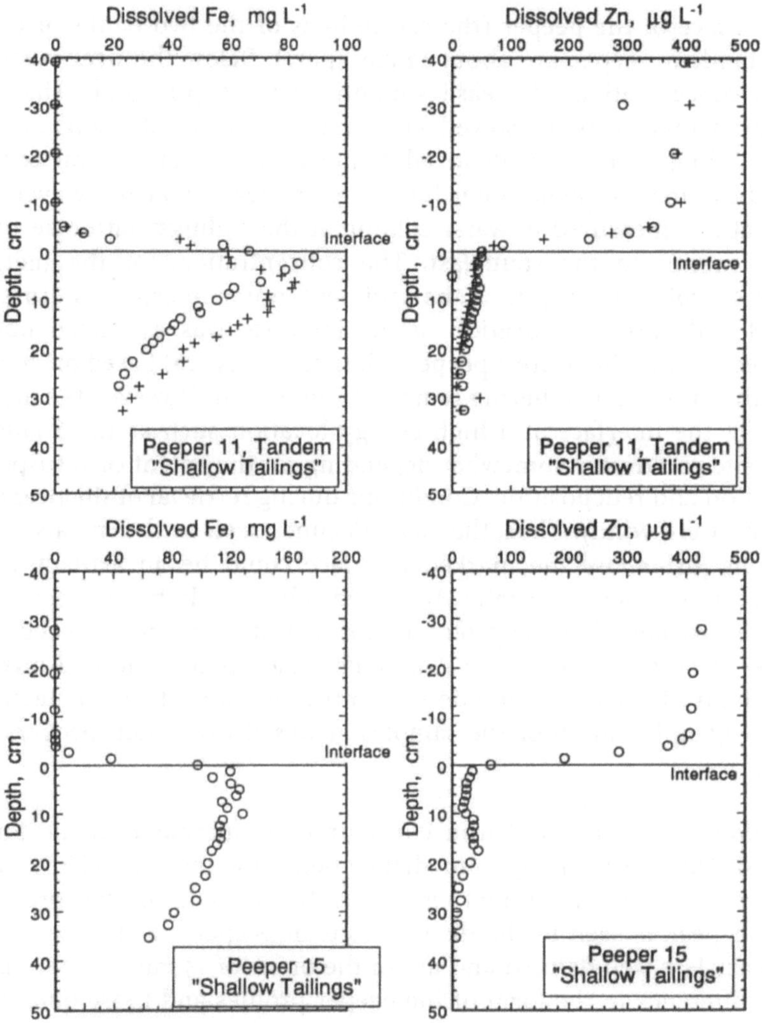

Fig. 6.8. Dissolved iron and zinc distributions in peeper pore waters at the Shallow Tailings site in August 1993. *Open circles* represent samples from the left cell bank

differences both in form and absolute concentrations which may reflect short-scale lateral inhomogeneity at this site. Extremely high dissolved iron concentrations occur at shallow subsurface depths at both locations, reaching ~90 and ~120 mg l^{-1} within 2 cm of the interface in two of the peeper cell banks; a maximum of ~80 mg l^{-1} occurs in the right cell bank of peeper 11. The offset between the two profiles in peeper 11 may reflect uneven topography at the site, which could produce uneven immersion across the face of the peeper (the centre lines of the two banks of cells in each tandem peeper are about 12 cm apart). Below the near-surface maxima, concentrations decrease smoothly with depth, more sharply in peeper 11 than in 15. Dissolved Zn, Cu and Pb concentrations at the Shallow Tailings site are high in all samples more than several centimetres above the sediment–water interface. The levels match the average concentrations measured in water column at the Tailings station several hundred metres to the southeast. The concentrations of the metals decrease sharply with depth as the sediment–water interface is crossed (Figs. 6.8 and 6.9). The location of the interface was estimated upon peeper recovery to the nearest peeper cell (nearest 1.3 cm) based on staining and the location of adhering sediments on the frit. However, the mean location of the interface at a high-energy location such as the Shallow Tailings site could vary somewhat depending on the extent of resuspension, erosion and redeposition. Conditions during retrieval of the peepers at this site were windy. Thus, the steep declines seen in the metals concentrations just *above* the marked interface could be an artifact, and reflect recent exposure of shallowly buried cells shortly before recovery. It is more probable that the profound concentration decreases occur in the upper few centimetres of the sediments rather than in the well-mixed water column. Dissolved sulphide concentrations were below detection limit ($< \sim 17\ \mu g\ l^{-1}$) in all of the samples analysed from both peepers at this site.

Pure Tailings Site. Dissolved iron concentrations measured at the Pure Tailings site increase sharply immediately below the interface (Fig. 6.10). A shallow maximum of $\sim 4\ mg\ l^{-1}$ is seen in the top 2 cm in the core, but no similar peak is seen in the distributions in peepers 12 and 13, where maxima are less well defined and are in the order of $15\ mg\ l^{-1}$. There is a marked difference in the form of the peeper profiles and those obtained from extraction of pore waters from the cores collected in summer 1990 (Pedersen et al. 1993) and in winter and summer 1993. The core profiles have invariably shown a near-surface maximum and a sharp depletion in dissolved iron at depth. The difference between the core and peeper results for iron is thought to reflect a design limitation of the dialysis arrays. As noted above (Sect. 6.4.1.2), the sediments at the Tailings site in August 1993 were capped with a very thin ferruginous bacterial mat which was soft but cohesive. It is possible that during emplacement of the peepers, some of this material adhered to or was trapped by the faceplates or frits, and was carried downward into the (anoxic) deposits at depth. During the 14-day equilibration period, the advected iron oxides would have progressively dissolved, and "excess" iron would have diffused

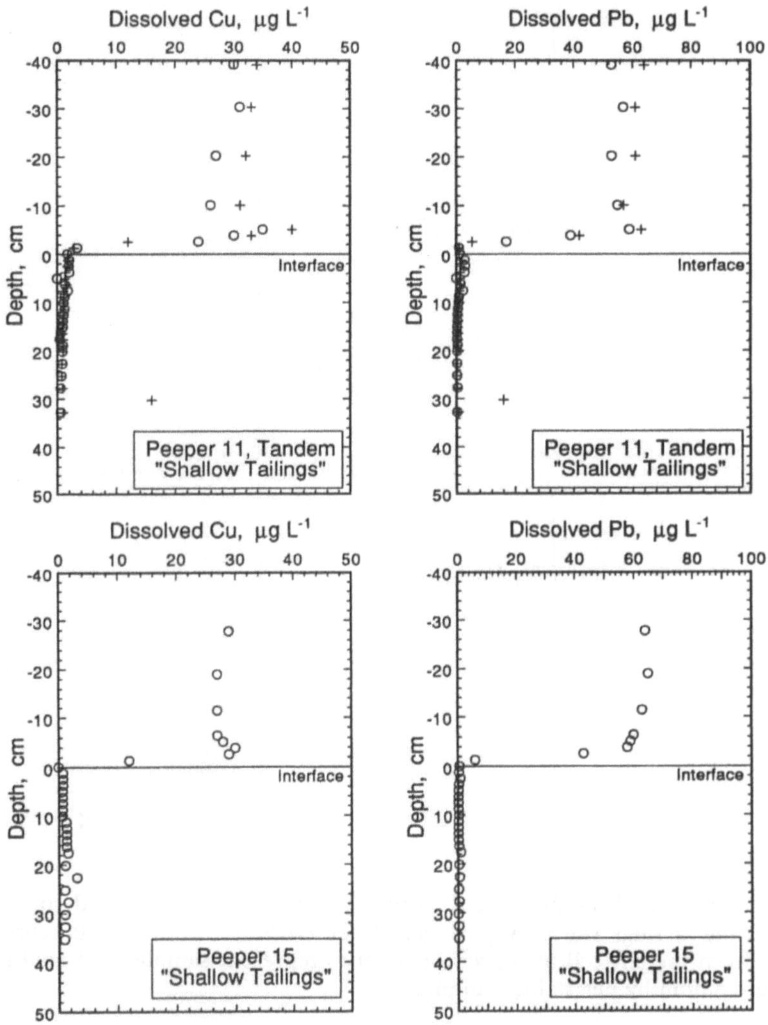

Fig. 6.9. Dissolved copper and lead distributions in peeper pore waters at the Shallow Tailings site in August 1993. *Open circles* represent samples from the left cell bank

into the cells at depth, resulting in false profiles. This hypothesis can explain the contrast between the pairs of profiles from each peeper as well as the rather wide overall differences and the large range in absolute concentrations. If this suggestion is correct, the "snap-shot" profiles yielded by the core samples should be more representative of the true distribution of dissolved iron in the tailings pore waters. The much lower maximum absolute concentration observed in the core samples is con-

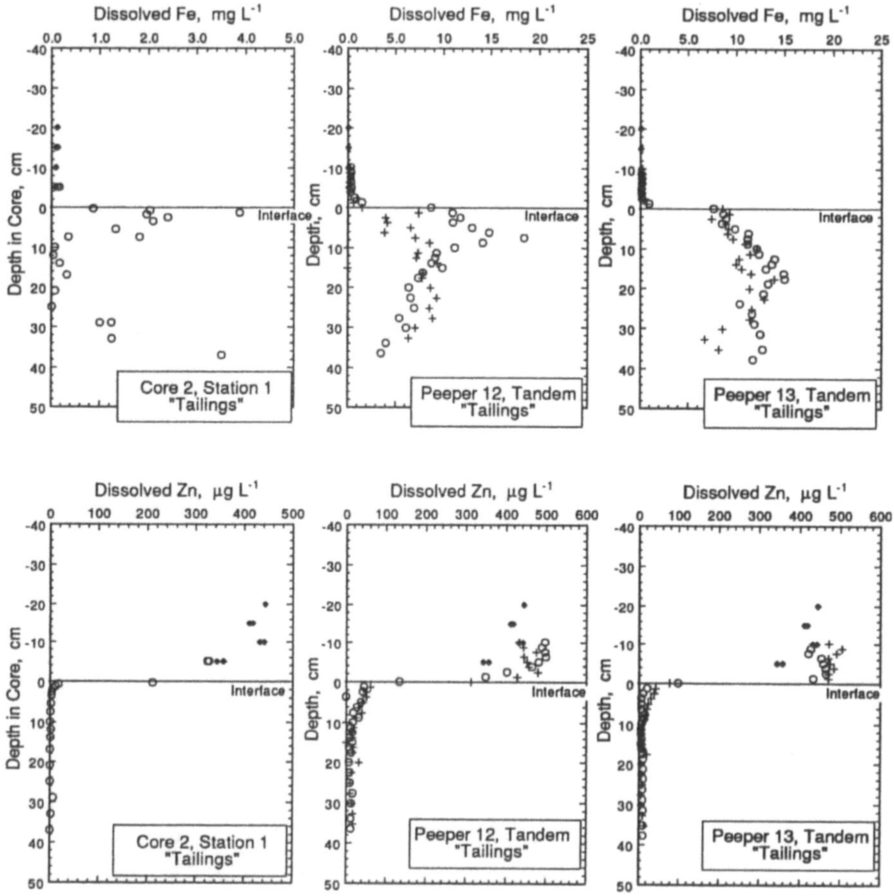

Fig. 6.10. Dissolved iron and zinc distributions in tandem peeper and core pore waters at the Tailings site in August 1993. *Open circles* on peeper plots represent samples from the left cell bank; *solid diamonds* represent samples collected from the water column by peristaltic pumping

sistent with previous results and with the entrainment hypothesis because the iron-rich surface material smeared downward during insertion of the core barrel is carefully removed during extrusion before core slices are centrifuged.

Dissolved Zn, Cu and Pb distributions are shown in Figs. 6.10 and 6.11. Without exception, the profiles show very steep downward declines in the concentrations of these three metals at or immediately below the sediment–water interface. The profiles from the core agree well with those from the peepers, with the exception that the average concentrations at

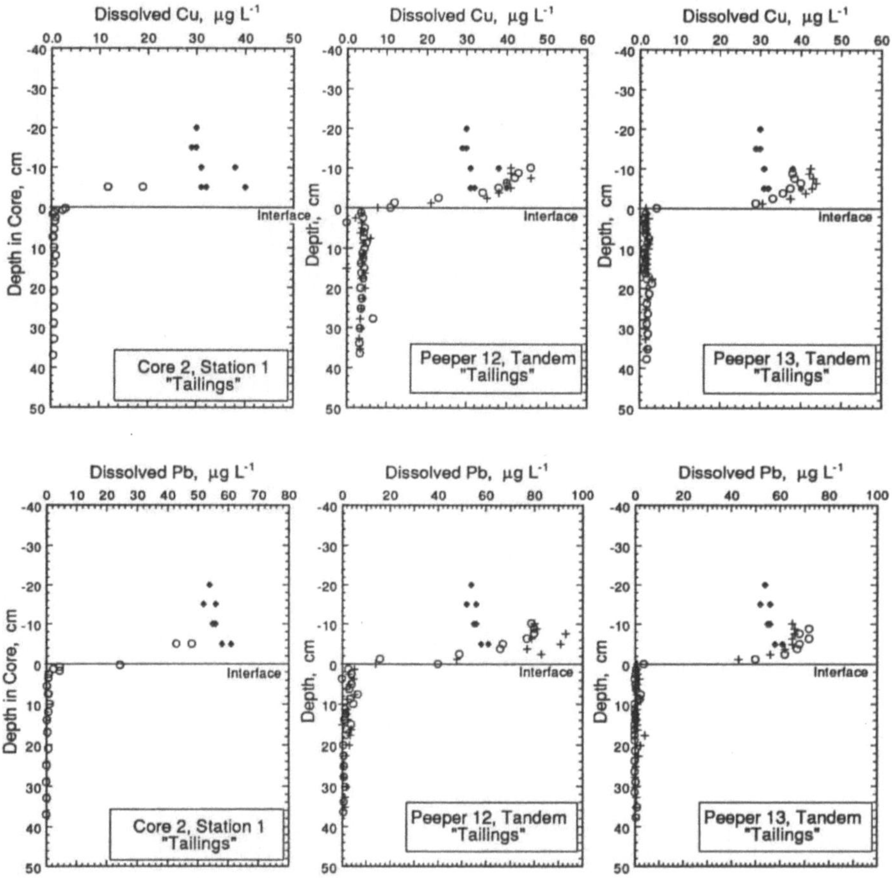

Fig. 6.11. Dissolved copper and lead distributions in tandem peeper and core pore waters at the Tailings site in August 1993. *Symbols* as in Fig. 6.10

depth are lowest in the core samples. Concentrations measured in core-top water and in the peeper samples above the interface agree reasonably well with levels observed in near-bottom samples collected by peristaltic pumping. Dissolved sulphide was not detected in any of the samples analysed at this location.

Natural Sediments Site. The analyses revealed that the peeper emplacement at the Natural Sediments site was unsuccessful, probably due to the unconsolidated nature of the deposits which may have allowed the arrays

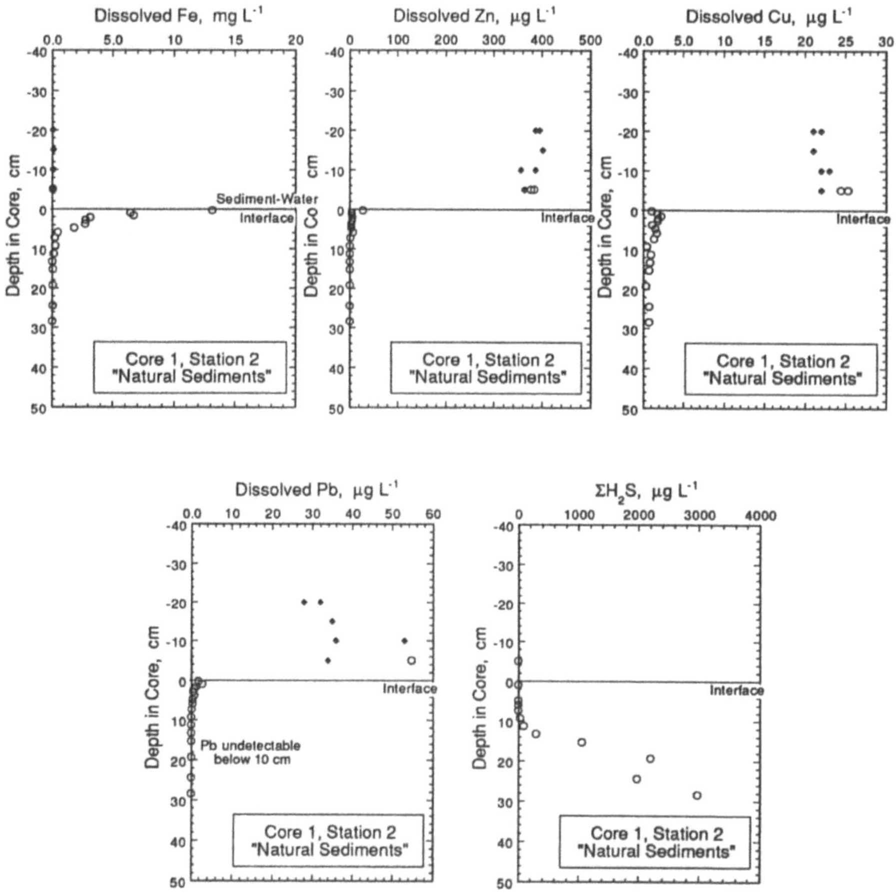

Fig. 6.12. Dissolved iron, zinc, copper, lead and H_2S distributions in core pore waters at the Natural Sediments site in August 1993. *Symbols* as in Fig 6.10

to fall over during their 2-week emplacement. The peeper data from this site are therefore not used here. The core profile for dissolved iron at this location exhibits a sharp maximum (~ 13 mg l^{-1}) just below the interface and a smooth decrease with depth to very low values below 6 cm (Fig. 6.12). Dissolved Zn, Cu and Pb distributions in the core conform with the pattern established by the iron profile (Fig. 6.12). The profiles illustrate the sharp decreases with depth immediately below the interface that have been seen previously (Pedersen et al. 1993). Dissolved sulphide occurs in the core pore water below 8-cm depth (Fig. 6.12), which matches observations based on odour during core extrusion. Sulphide was undetectable (< ~ 17 µg l^{-1}) above 8-cm depth in the core.

6.4.2
Buttle Lake

6.4.2.1
Water Column

Dissolved oxygen and temperature measurements indicated that the water column was moderately stratified and reasonably well oxygenated (~ 5 – 10 mg l^{-1} O_2) during the sampling period in October 1993. The water is relatively soft, with the hardness averaging about 33 mg l^{-1} (as $CaCO_3$). The waters in Anderson Lake are much harder, hosting major ion concentrations 40- to 50-fold higher (100-fold higher in the case of sulphate).

The slightly higher dissolved zinc concentrations seen in the deeper water at the three locations (24 µg l^{-1} at 35-m depth compared with ~ 5 µg l^{-1} at the surface) may be the result of three phenomena. First, the creek that runs past the mine site continues to contribute dissolved metals to the lake at levels higher than would be expected in a pristine watershed; concentrations measured in the creek water throughout 1993 ranged from 41 to 254 µg l^{-1}. The creek water is often very cold and therefore relatively dense, and it usually sinks and spreads laterally at a considerable depth in the south basin. The effect of this input is to raise the dissolved zinc concentration artificially in the deeper waters. Second, zinc is an essential element for phytoplankton growth and is actively sequestered by cells in the euphotic zone and released at depth or to sedimentary pore waters when the cells are subsequently degraded (Reynolds and Hamilton-Taylor 1992). However, Buttle Lake is oligotrophic, and its waters have a relatively short residence time, so such biological vectoring is unlikely to have more than a minor influence on the zinc distribution in the water column. Third, benthic recycling at or just below the sediment surface, for example via the reductive dissolution of oxide phases, could theoretically support an efflux of dissolved zinc that might augment the concentration of the metal in bottom waters, despite the short residence time. This possibility will be discussed further in Section 6.5.1.

6.4.2.2
Sediments

The natural sediments collected from site 7 on the southeastern margin of the deep central basin of the lake (the Natural Sediments site) consist of relatively organic-rich detritus. Organic carbon concentrations range from ~ 4.5 wt% in surface deposits to a maximum of ~ 7 wt% at 20-cm depth (Fig. 6.13). Solid-phase zinc, copper and lead profiles in core 7A

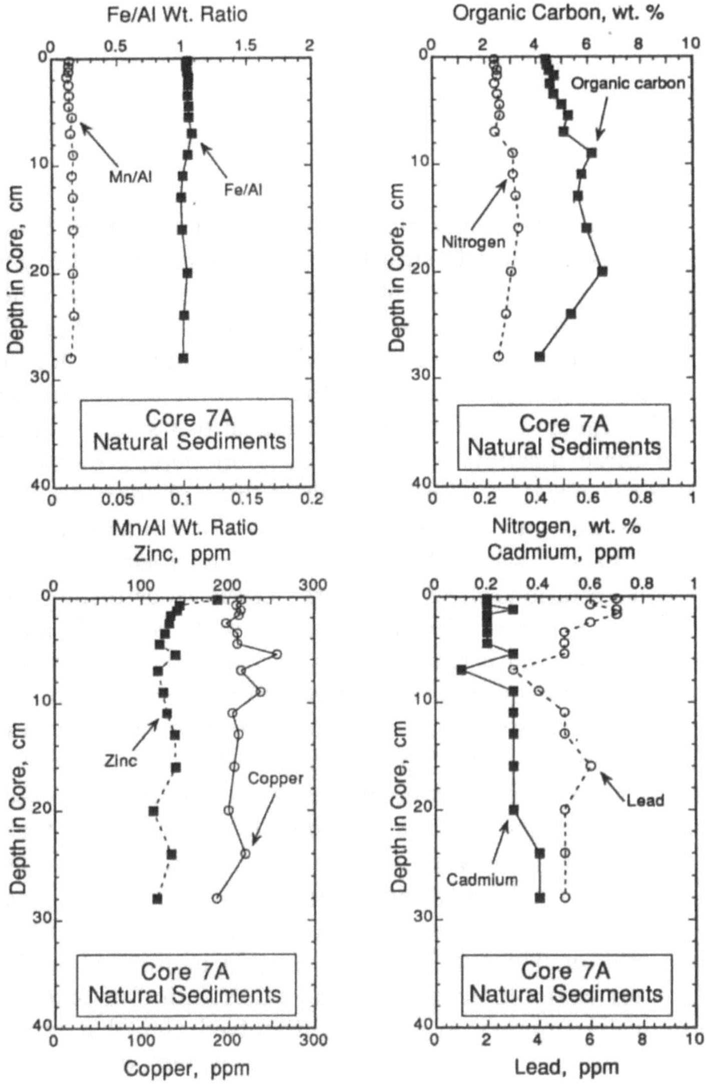

Fig. 6.13. Fe/Al and Mn/Al weight ratio profiles and organic carbon, nitrogen, zinc, copper, cadmium and lead distributions at the Natural Sediments site in the central basin of Buttle Lake

suggest that there has been insignificant accumulation of tailings at this site, with the possible exception of the upper two centimetres where the data are equivocal, indicating slight enrichments of Zn and Pb, but not Cu. Fe/Al and Mn/Al weight ratio distributions in the core (Fig. 6.13) indicate that near-surface oxide enrichments, which are commonly observed in sediment profiles in lakes (Davison 1993), are essentially absent at site 7.

The deposits collected from station 6 very near the site of the historical tailings outfall in the south basin consist of about 4 cm of organic-rich, mostly natural sediments overlying homogeneous and organic-poor tailings. Organic carbon (C_{org}) concentrations range from ~5 wt% in surface deposits to barely detectable in the Pure Tailings site below 5 cm (Fig. 6.14). Solid-phase zinc, copper and lead profiles in core 6A (Fig. 6.14) define the metal-sulphide-rich tailings chemically below 5-cm depth where, for example, Zn concentrations exceed 10,000 $\mu g\ g^{-1}$ (1 wt%). High Fe/Al ratios below 5-cm depth mark the high pyrite content that is characteristic of the tailings. In contrast, high relative Mn contents occur only in the upper 3 cm of the core, where concentrations reach nearly 1 wt%. Such high values are attributed to the diagenetic accumulation of Mn oxides in the aerobic zone just below the sediment–water interface, as discussed in Section 6.5.2. Profiles of dissolved iron in interstitial waters indicate that there must also be an enrichment of Fe oxides in the upper few centimetres, but the high Fe content attributable to other phases, particularly pyrite, makes the oxide fraction difficult to distinguish. Visual observations in concert with the Zn, Cu and Pb data indicate that the surface sediments at this site still contain as much as 15% tailings by weight (somewhat less by volume) which may reflect ongoing bioturbation.

The deposits collected from station 4 (the Distal site) about 1.5 km northeast of the old outfall consist of about 4 cm of organic-rich, mostly natural sediments overlying about 20 cm of relatively homogeneous and organic-poor tailings which were deposited onto pre-existing organic-rich sediments. Organic carbon (C_{org}) concentrations range from ~6–7 wt% at the surface and in the underlying natural sediments in the lower portion of the core to 0.5–1 wt% in the intervening tailings between ~5 and 20-cm depth (Fig. 6.15). Solid-phase zinc and copper profiles in cores 4A and 4B define the tailings chemically between 5 and 20-cm depth. Zn concentrations in the tailings at this Distal site reach as high as 1 wt% and sulphur contents as high as 8 wt%, substantially higher than in the natural deposits. Despite the presence of abundant pyrite in the tailings at this location, the Fe/Al ratio profiles in the two cores show lower values in the tailings stratum than in the natural deposits that

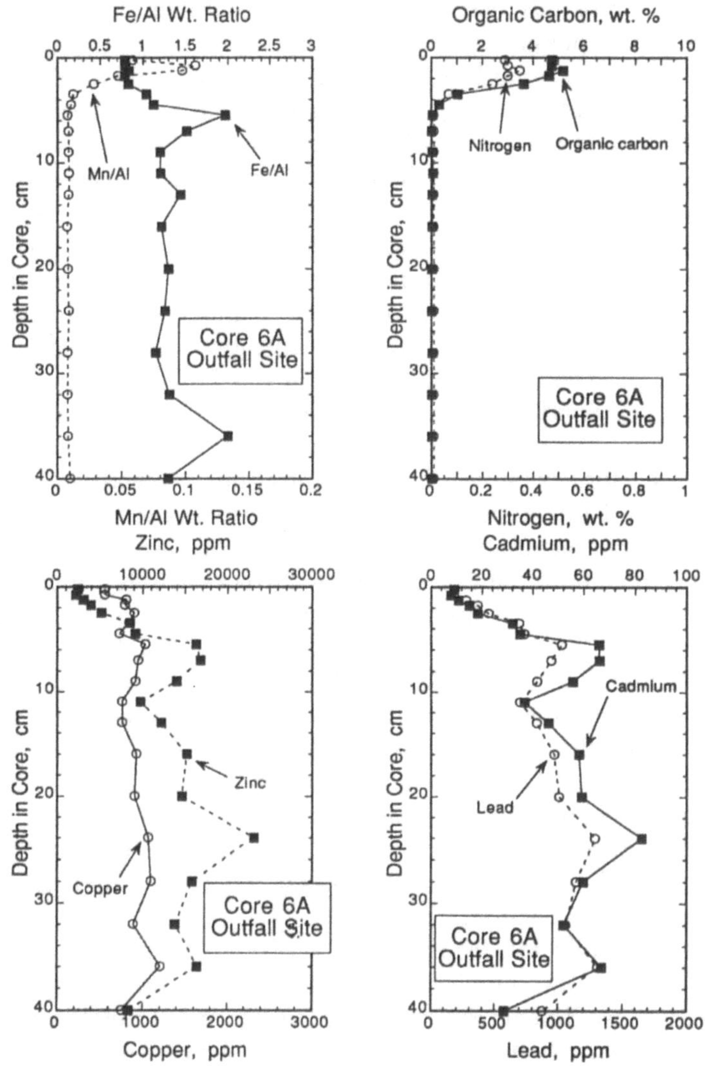

Fig. 6.14. Fe/Al and Mn/Al weight ratio profiles and organic carbon, nitrogen, zinc, copper, cadmium and lead distributions at the Outfall site in the south basin of Buttle Lake

sandwich it (Fig. 6.15), indicating either that Fe oxides are abundant both above and below the buried tailings, or that the natural sediments contain a relatively high concentration of iron-bearing detrital minerals. As discussed in Section 6.5.2, pore water iron distributions suggest that the former explanation is the most likely. Manganese concentrations in the near-surface sediments from both cores are very high, ranging up to

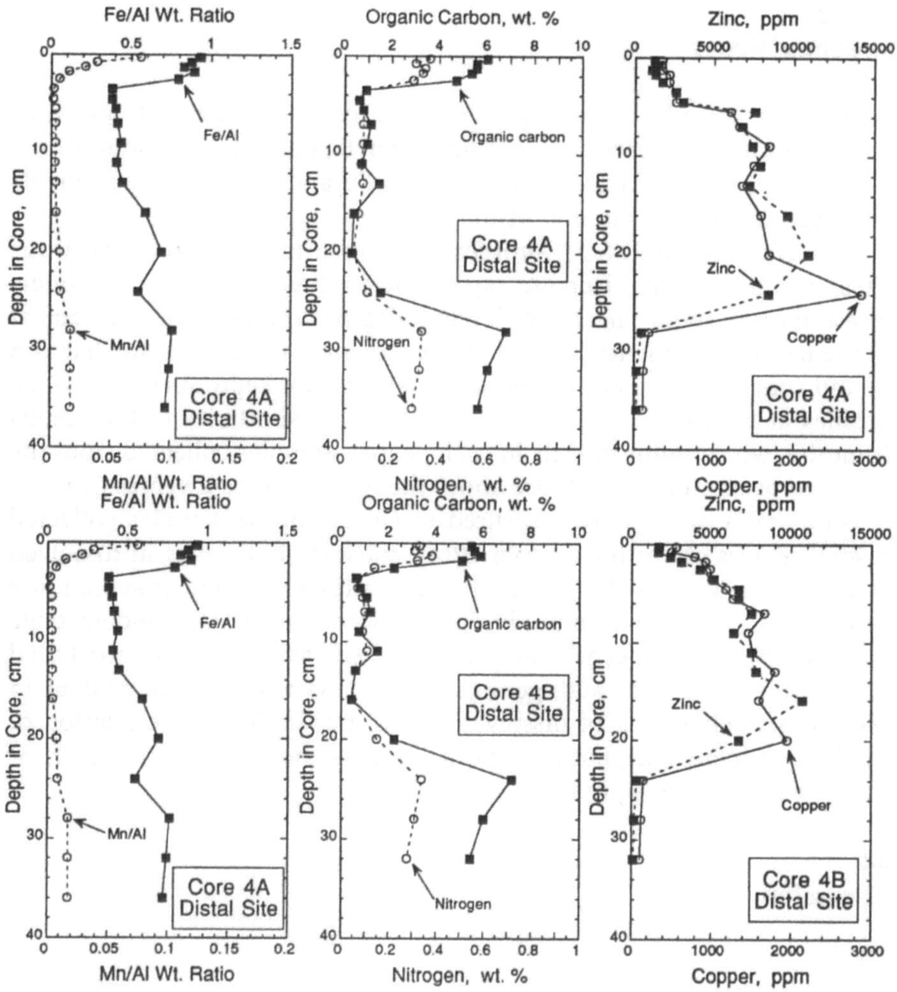

Fig. 6.15. Fe/Al and Mn/Al weight ratio profiles and organic carbon, nitrogen, zinc, copper, cadmium and lead distributions at the Distal site in the south basin of Buttle Lake

1.3 wt% in the upper 5 mm of core 4B. These high values are matched by high Mn/Al ratios (Fig. 6.15), and indicate that manganese oxides are abundant in the deposits now accumulating on top of the tailings at this site. Slight increases in the Mn/Al ratio and higher absolute manganese levels below the tailings, i.e. below ~20 cm, suggest that a manganese oxide phase is present in these buried natural deposits.

6.4.2.3
Interstitial Waters

Pore water samples from the dialysis arrays emplaced at all three sampling locations were analysed for sulphate, nitrate, ΣH_2S and a suite of dissolved metals, including Fe, Mn, Zn, Cu and Pb. pH measurements were also made on a number of the samples.

Dissolved NO_3^- and SO_4^{2-} profiles are shown in Fig. 6.16. At all three sites, nitrate contents decrease rapidly below the sediment–water interface and reach zero or near-zero levels at depths ranging from ~10–20 cm in the natural sediments (station 7), ~5–10 cm at the Outfall site (station 6) and roughly 10 cm at the Distal site (station 4). Reproducibility of the profiles at each site is very good, based on the similarity of the results from the two columns of cells in each peeper. Samples collected from the emerged peeper cells (i.e. those above the interface) yield values generally comparable with those measured in Go-Flo bottle samples collected several metres above the bottom. Steep subsurface declines in dissolved SO_4^{2-} are also witnessed at the south basin sites, in particular at station 4 (peeper 14). Dissolved sulphide species were universally undetectable ($<40\ \mu g\ l^{-1}$) in the peeper samples analysed, which can be attributed to the precipitation of authigenic sulphides. As will become evident in Section 6.5 below, lack of dissolved iron is not limiting precipitation of sulphide minerals in Buttle Lake sediments, but the relative lack of sulphate almost certainly is.

6.5
Discussion

6.5.1
Anderson Lake

Pedersen et al. (1993) described the distribution of dissolved metals in pore waters collected in summer 1990 at three sites in Anderson Lake. Their principal conclusion was that the submerged deposited tailings in Anderson Lake were not releasing metals to the overlying water column at that time. Indeed, the opposite appeared to be true: sharp decreases with depth in the concentrations of dissolved Zn, Cu, Pb and Cd in the upper few centimetres of the deposits were interpreted as representing consumption of metals from the contaminated overlying lake water. Precipitation of authigenic sulphide phases at shallow depths in anoxic tailings and natural sediments was suggested as an explanation for the very low metals concentrations in the pore waters.

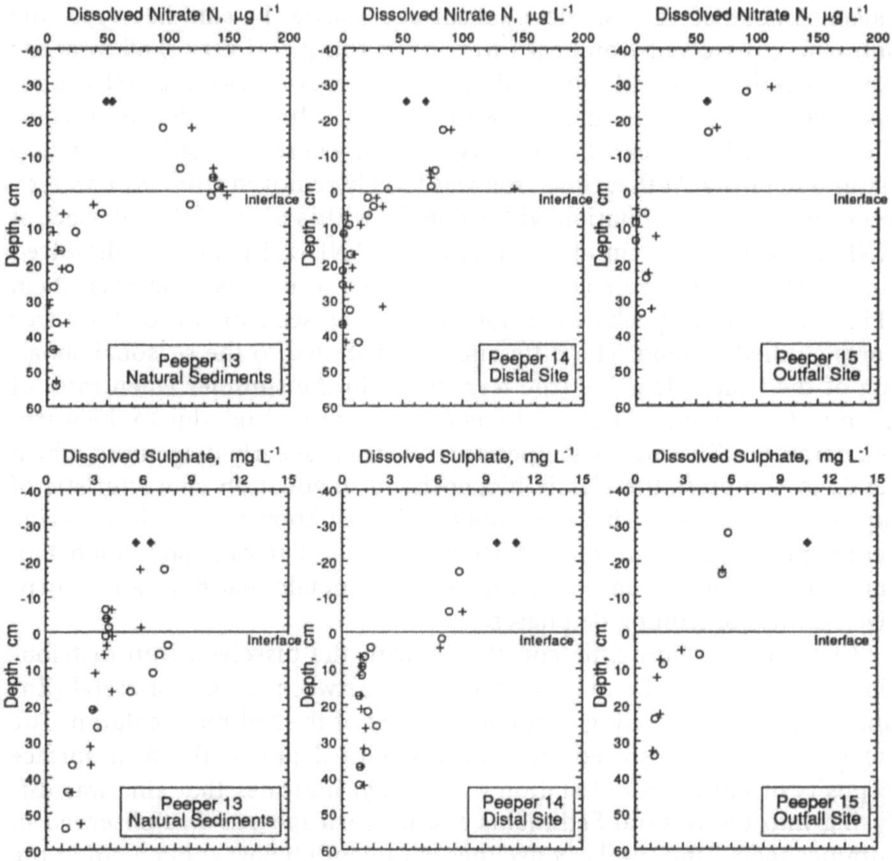

Fig. 6.16. Dissolved nitrate and sulphate in interstitial waters at the three sites sampled in this study. Left and right columns of cells in tandem peepers are represented respectively by *circles* and *crosses*. *Filled diamonds* in each panel represent NO_3^- concentration measured in the deepest samples collected from the water column at the three locations

The data collected in the Summer and Winter Surveys in 1993 show similar behaviour to that observed in summer 1990, with some exceptions. The oxidation of organic matter during early diagenesis in sediments typically proceeds via the progressive exhaustion of a suite of oxidants including, in general order, O_2, NO_3^-, Mn oxides, Fe oxyhydroxides and sulphate. The dissolved iron distributions in winter in both cores at both sites show relatively shallow maxima which are interpreted as representing reductive dissolution of oxyhydroxides upon burial, and the establishment of very shallow subsurface anoxia. In summer 1990, the subsurface dissolved-iron peaks were quite sharp and very shallow at the Natural Sediments and Pure Tailings sites (cores AND-1 and AND-2, respectively; Pedersen et al. 1993), and somewhat broader and more complex at a third site near the discharge outfall where bulk tailings sat atop natural deposits. At that time, iron was largely removed from pore waters, presumably by precipitation with sulphide, within 2 cm of the interface at AND-1 and within 4 cm of the interface at AND-2. In contrast, dissolved iron persisted in pore solution in April 1993 to depths of at least 8 cm (Fig. 6.5), and the peaks were not as sharp as seen in two of the three cores studied in 1990. The difference is attributed to the seasonal shoaling of the iron redoxcline (the ferricline?) in the summer when rates of chemical reaction and bacterial metabolic rates are high due to the warm temperatures. The higher temperatures accelerate early diagenesis, which tends to compress the classic biogeochemical zonation characteristic of aquatic sediments. Such a phenomenon is well known in shallow marine sediments (e.g. Klump and Martens, 1981), and it can have important implications for seasonal remobilization of metals that have an adsorptive affinity for iron oxide phases.

Zinc profiles show little correlation with the dissolved iron distributions, implying that there is no association between the minor metal (Zn) and oxyhydroxide cycling in the upper part of the sediment column. The steep decline in dissolved zinc content with depth in the near-surface deposits of both cores at station A (Fig. 6.5) indicates that zinc was diffusing into the Natural Sediments site in April 1993, as was observed in summer 1990. Zinc similarly declines with depth below about 2 cm in the tailings near the outfall (station B, Fig. 6.5), but the profiles hint that there may have been a slight efflux of the metal at the time of sampling; the maximum in core B2 pore waters at 1- to 2-cm depth would support diffusion upward and downward from this interval. The profile in core B1 is slightly different and suggests that no efflux was occurring. This contrast is problematic. Although a slight diffusive efflux of Zn from the tailings at the site of core B2 cannot be explicitly ruled out, it is possible that the near-surface Zn maximum in the core is artificial, since the tailings were overlain by oxygen-poor water in April 1993. Under such conditions, no Zn should be released (for example, via oxidation of the surfaces of ZnS particles). Indeed, water column dissolved Zn data at station B imply uptake of the metal from bottom waters in winter, rather than addition from the underlying tailings (Pedersen et al., unpubl. data). An historical non-steady state perturbation might explain the Zn efflux implied by the dissolved Zn data in the upper centimetre of core B2. This suggestion implies that temporally varying chemical reactions and hydrographic processes in the overlying contaminated water column may play a role in dictating the distributions of dissolved metals in the near-surface pore waters.

Copper and lead concentrations in pore waters from the Tailings site in winter were so low as to be essentially undetectable (Fig. 6.6). Thus, no conclusions can be drawn about their chemical or diagenetic behaviour other than to suggest that the tailings were not releasing these metals to the overlying water column at that time. Lead contents in the pore waters at station A were similarly very low, being at or below detection limits (Fig. 6.6). These data imply that there was no significant lead efflux from the Natural Sediments site nor influx at the time of sampling. In contrast, dissolved copper concentrations increased immediately below the sediment surface in core A2, reaching a minor maximum ~3 μg l^{-1} at about 1.5-cm depth. This implies upward diffusion from the maximum, and consumption at depth. However, water column data (Pedersen et al., unpubl.) indicate removal of copper from the near bottom waters, which presumably can be attributed to active precipitation of authigenic sulphide minerals at or near the sediment surface. Thus, the (small) efflux of dissolved copper from the natural deposits implied by the profile in core A2 is at odds with the water column data and with the overall body of pore water data. A possible explanation is that the two samples that define the subsurface maximum may have been slightly contaminated with copper.

The Zn, Cu and Pb Summer-Survey pore water profiles reported here show precipitous declines in the concentrations of these metals immediately below the sediment–water interface at all three locations (Figs. 6.8–6.12). This result is consistent with those seen in the previous surveys, and the implication is clear: both the natural sediments *and* the tailings on the floor of Anderson Lake appear to be a sink for these metals rather than a source. The leading candidate for the implied sequestration of dissolved metals by both sediments and tailings is the precipitation in situ of sulphide minerals. Framboidal pyrite is a common trace constituent in the natural sediments, but was not directly observed in samples from the Pure Tailings site cores. Concave-downward decreases in sulphate with depth in the sediments at station A seen in both winter and summer indicate that sulphate reduction occurs at shallow depths, and this observation is consistent with the presence of framboids in the top 5 mm in core A1, and with the presence of dissolved sulphide in the pore waters. Dissolved sulphide was not detected in the tailings, but neither was nitrate: in both winter and summer, NO_3^- concentrations in pore waters were typically not measurable (< ~5 μg l^{-1}). This implies that nitrate is quantitatively reduced throughout the deposits, and that anoxia is established at very shallow depths. The fact that sulphate is being added to the pore waters at depth in the tailings, presumably via dissolution of sulphate salts (Pedersen et al. 1993), does not obviate simultaneous reduction of SO_4^{2-}. Sulphide production could easily be masked by reaction with dissolved iron to form FeS or FeS_2.

One specific objective of the Summer Survey was to assess whether or not metal release might be occurring at very shallow locations that are probably more frequently overlain by oxic bottom water. The well-oxygenated, Shallow Tailings site appears to have provided an answer to this question: triplicate profiles for each metal consistently indicate that the flux of dissolved Zn, Cu and Pb at this site was into and not from the deposited tailings. The texture of the deposits indicated vigorous reworking (i.e. resuspension, redeposition and translation on the bottom), but this does not seem to have promoted oxidation of the tailings: there was no indication in any of the visual observations or the analytical data that the tailings at this site were even marginally oxidized. The very sharp increase in the pore water iron content within the top 2 cm of the deposits at this location (Fig. 6.8) strongly implies that anoxia occurred at very shallow depths, despite the reworking. Upward diffusion of iron along that sharp gradient and oxidation at the interface must be contributing to the ferruginous veneer observed on the sediment surface.

6.5.2
Buttle Lake

A principal objective of the Buttle Lake study was to assess the role of post-depositional cycling of oxides in influencing the mobility of metals such as Zn, Cu and Pb. The Fe and Mn results from the peepers (Fig. 6.17), in conjunction with the nitrate and sulphate data (Fig. 6.16), show that the aerobic zone was thin at all three stations sampled in the 1993 survey, and that accumulation of oxides must have been confined to the upper 2 cm at both the Outfall and the Distal sites. Manganese oxides are reduced at about the same redox potential as nitrate during early diagenesis in sediments, but at a higher potential and thus a shallower depth than iron oxide phases (Froelich et al. 1979). This behaviour is illustrated by the slightly shallower manganese concentration maxima in the profiles, relative to dissolved iron, at both of the south basin sites (Fig. 6.17). These results confirm that the south basin sediments are dysaerobic (essentially oxygen-free) below about 2-cm depth. Solid-phase oxides are actively cycled in these deposits; solution during burial clearly supports upward diffusion of dissolved Mn and Fe, and, upon encounter-

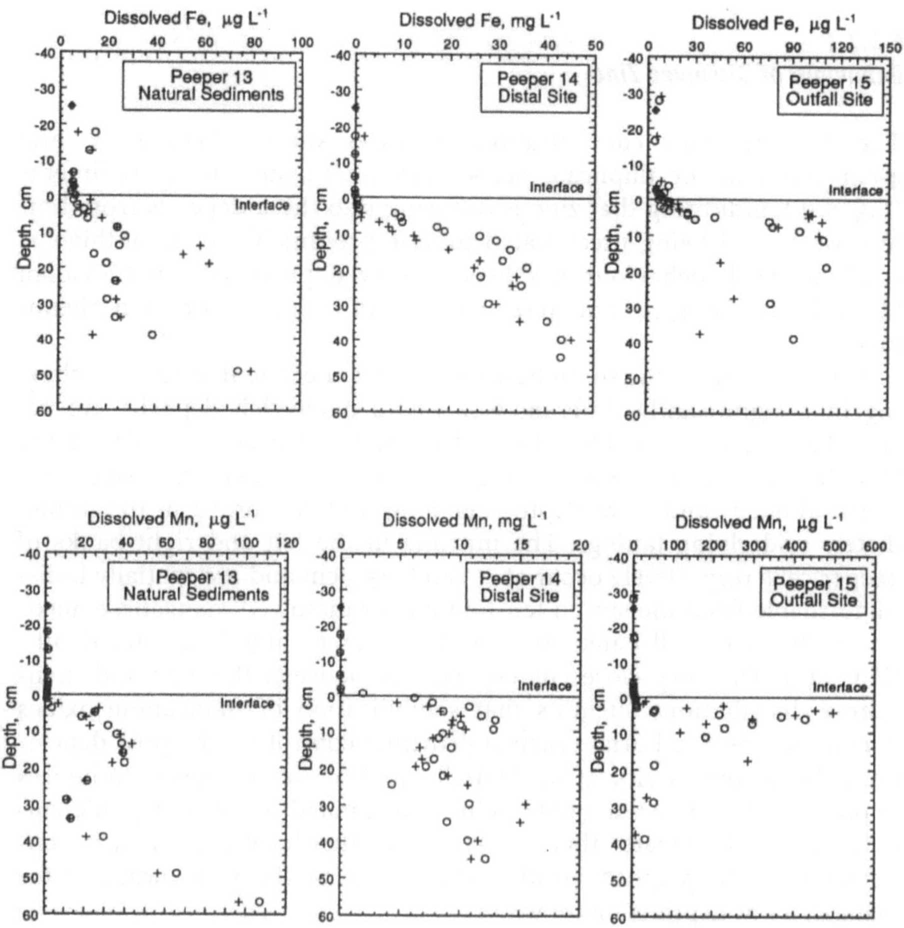

Fig. 6.17. Dissolved iron and manganese profiles in interstitial waters from peepers emplaced at the three sites in Buttle Lake. *Symbols* as in Fig. 6.16. Note the different scales on abscissae

ing molecular O_2, these species reprecipitate near the sediment-water interface, as illustrated by the high solid-phase Mn/Al ratios in the upper 1-1.5 cm of the south basin cores (Figs. 6.14 and 6.15). Although there is little indication in the Fe/Al and Mn/Al ratio solid-phase data that significant concentrations of oxides are accumulating in the surface deposits at the Natural Sediments site, visual observations (e.g. orange-brown coloration) suggest that an oxide fraction, albeit minor, is present in the upper few millimetres or so.

6.5.2.1 *Behaviour of Dissolved Zinc*

The dissolved zinc concentration decreases sharply between 3- and 4-cm depth in the duplicate peeper profiles in the natural sediments (Fig. 6.18), indicating that zinc is diffusing into these deposits from bottom water and being precipitated in situ, presumably as an authigenic sulphide. Such behaviour has been observed previously in Canadian Shield lakes (Carignan and Tessier 1985) and may be a common phenomenon.

The results from the south basin sediments indicate that zinc is behaving in a significantly different way where oxide-rich deposits are accumulating on top of the tailings. The duplicate peeper profiles at the Outfall site show that zinc is being released to the interstitial waters between about 1- and 5-cm depth (Fig. 6.18) and consumed in the immediately underlying tailings. The maxima in the left and right banks of peeper cells respectively occur at ~3 and ~5.5 cm and are spatially indistinguishable from the six- to ten-fold higher dissolved manganese maxima in the same cell banks. Iron peaks occur at slightly greater depths (Fig. 6.17). The very close correspondence between the zinc and manganese distributions suggests that solubilization of manganese oxides during progressive burial exerts a principal control on the post-depositional behaviour of zinc at the Outfall site. Iron oxide appears to be less important. Zinc is presumably being precipitated as an authigenic sulphide at depths greater than ~8–10 cm at this location, which is consistent with the sulphate profile. The nature of the sink indicated for manganese at depth is unknown.

Dissolved zinc profiles from the Distal site (station 4) have a fundamentally different character (Fig. 6.18); concentrations in peeper 14 increase generally with depth, reaching ~300 $\mu g\, l^{-1}$ below ~30 cm, several-fold higher than the maxima seen at the Outfall station. Unlike the other two sites, some scatter is evident in the concentrations measured just above the interface. Similar scatter is seen in the Cu and Pb results (discussed below), which may reflect slight contamination by particles introduced during sampling of the peeper cells. The scatter makes it difficult to define the near-interface behaviour of zinc. There is no clear evidence in the top 2 cm of an efflux, although this cannot be conclusively ruled out; nor is there evidence for consumption of Zn from pore water onto, for example, freshly precipitating oxides. However, Cu and Pb show small deficits in the pore waters in the top few centimetres at this site relative to bottom waters and/or slightly deeper pore waters (Figs. 6.18 and 6.19), implying consumption from solution. By analogy to these

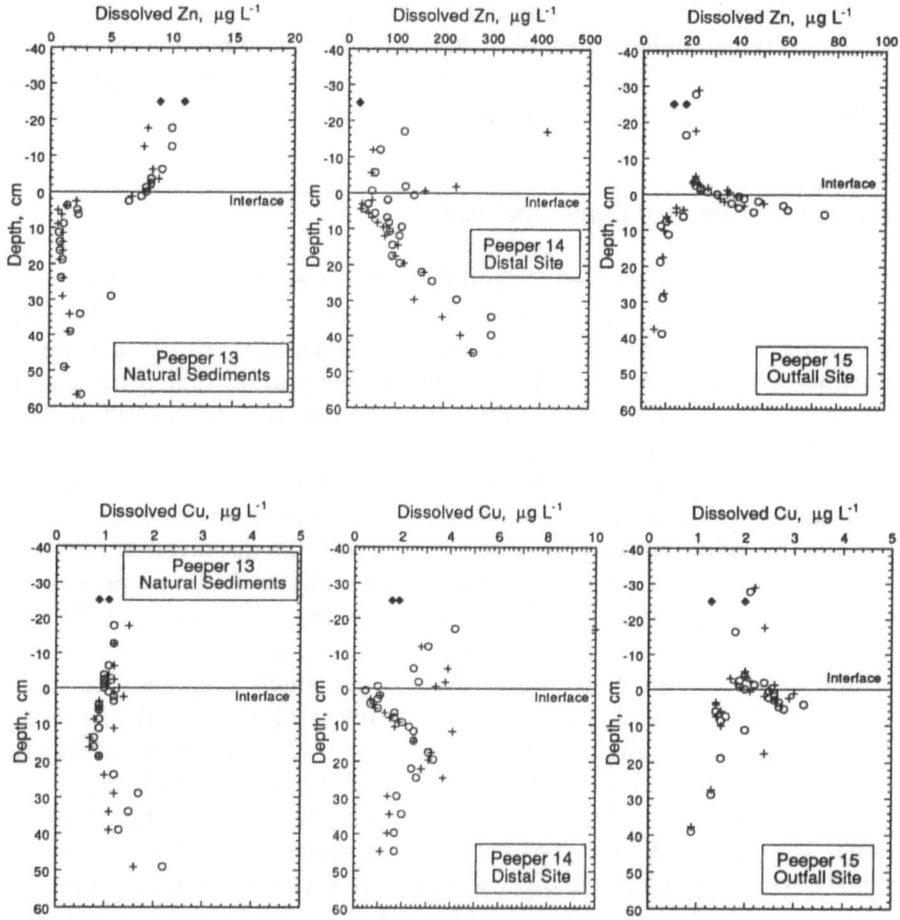

Fig. 6.18. Dissolved zinc and copper profiles in interstitial waters from peepers emplaced at the three sites in Buttle Lake. *Symbols* as in Fig. 6.16. Note the different scales on abscissae

metals, the scatter in the Zn data near the interface in peeper 14 may be disguising minor uptake. There is no ready explanation for the high dissolved Zn levels seen below 20 cm in both the peeper profiles. If the peeper did not penetrate natural sediments, then the high Zn values might represent Zn release from the more deeply buried tailings. No such release is seen in the peeper profile in the rapidly deposited Pure Tailings site at the Outfall site, nor was such an addition to pore water seen in the rapidly accumulating tailings in previous work (e.g. Pedersen 1983). However, release of Zn to pore solution in the slowly accumulating distal tailings facies remains a possibility. Alternatively, some of the high concentrations could represent "relict" Zn that was buried contemporaneously with the tailings in metal-rich lake water during the 1970s and 1980 s, and not removed by subsequent precipitation of authigenic ZnS.

6.5.2.2
Behaviour of Dissolved Copper and Lead

Concentrations of dissolved Cu and Pb in the peepers from all three sites are low, typically $< 5\ \mu g\ l^{-1}$ (Figs. 6.18 and 6.19). At the Natural Sediments site, little variation in Cu or Pb contents with depth is seen in the upper decimetre; there is no indication in the peeper profiles that either Cu or Pb are diffusing from bottom water into the deposits. Cu and Pb distributions in the duplicate pore water profiles from the Outfall site (station 6) are similar to Zn. Maxima occur in the upper several centimetres, and consumption at depth is indicated by the general downward concavity seen below about 5 cm. Comparison with the dissolved Mn and Fe distributions (Fig. 6.17) suggests that Cu and Pb are cycling with oxides during early diagenesis in the natural sediments that now overly the tailings at this location. The Cu and Pb profiles correspond more closely to the dissolved Mn distribution than to Fe, which suggests that solution of MnO_2 below about 2-cm depth is releasing the adsorbed metals to the pore waters. Slight upward concavity in the profiles immediately below the interface implies that fractions of the metals are being readsorbed by Mn (or Fe?) oxides that reprecipitate in the thin aerobic zone. The concentration gradients also extend into the waters immediately above the interface, suggesting that the readsorption is not quantitative, and that small effluxes into south basin bottom water exist for these elements.

Dissolved Cu and Pb distributions in the peeper at the Distal site (station 4) are similar in form and magnitude, with slight minima being evident immediately below the interface (Figs. 6.18 and 6.19). These minor depletions imply scavenging of the metals from solution at about 3-cm depth, possibly by freshly precipitated oxides. This is consistent with the dissolved Fe profile (Fig. 6.17) which indicates precipitation at that same horizon.

6.5.2.3
Implications of Diffusive Effluxes

The data discussed above in Section 6.5.2 confirm that dysaerobic or anoxic conditions prevail at shallow sub-bottom depths in all sediment-

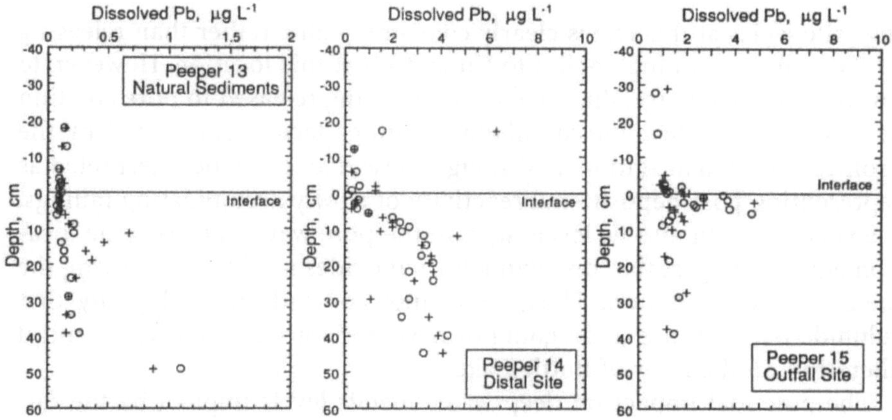

Fig. 6.19. Dissolved lead profiles in interstitial waters from peepers emplaced at the three sites in Buttle Lake. *Symbols* as in Fig. 6.16. Note the different scales on abscissae

ary facies sampled in Buttle Lake, and this is clearly a welcome condition for the storage in perpetuity of sulphide-bearing tailings. The natural sedimentation rate indicated by the current depth of the Tailings–Natural Sediment boundary is about 4 mm $year^{-1}$, which agrees very well with previous work (Rescan 1990). This implies that the tailings now buried below the aerobic zone were exposed to molecular oxygen (at progressively decreasing concentrations) for 7 or 8 years following initial deposition, and that those tailings particles that have been mixed upward into the near-surface natural deposits will have an oxygen exposure time perhaps twice as long. Has this nearly decade-long exposure of sulphide-particle surfaces to oxygen promoted significant release of metals to Buttle Lake bottom waters? Although this question can be addressed from a semi-quantitative standpoint by calculating diffusive fluxes, as shown below, the magnitude of sulphide oxidation cannot be clearly determined with the data at hand. This is because Buttle Lake was contaminated with acid-rock drainage during and after the 15 years of tailings discharge to the lake, so that the metal cycling seen in the pore water profiles is not just an indication of tailings reactivity. Instead, the profiles reflect the accumulated chemical history associated with scavenging by oxide phases of metals from the contaminated water column. Thus, some or all of the zinc that appears to be actively cycling in the upper few centimetres at the Outfall site, for example, could be zinc that has been derived by progressive adsorption from the water column since ~1970 – it might not be related to tailings reactivity at all. Indeed, the actual buried Pure Tailings site facies at this site is clearly consuming zinc rather than releasing it. The same reasoning applies to Cu and Pb at this location. However, to put this in perspective, Zn, Cu and Pb *are* being released to pore solution at depth in the "slowly accumulating" tailings facies represented by the profiles collected at station 4. Although these data could be interpreted as representing post-depositional reactivity of slowly accumulating tailings, the very high Mn and Fe levels seen in the pore waters at the same location equally suggest that dissolving buried oxides might be supplying the metals to pore solution. Thus, we cannot conclude unequivocally that sulphide particles are undergoing or have undergone oxidative chemical alteration on the floor of Buttle Lake.

The potential impact on deep-water metals levels implied by the diffusive effluxes indicated by the peeper data at the Outfall and Distal sites is assessed using Fick's First Law of diffusion:

$$J = -\phi(K_c/F)(\delta C/\delta x),$$

where ϕ is the porosity (assumed here to be 1; the true porosity is slightly less so fluxes will be slightly overestimated by the adoption of 100% porosity). K_c is the diffusion coefficient for each metal (from Li and Gregory 1974) corrected for estimated in situ temperature (6 °C); the metals are presumed to occur largely as divalent ions – no allowance is made for complexation in the calculations. The coefficients are thus 4×10^{-6} cm^2 s^{-1} for Zn, Cu and Cd and 5.5×10^{-6} cm^2 s^{-1} for Pb. $\delta C/\delta x$ is the concentration gradient across the interface indicated by the pore water profiles. F is the formation factor (Manheim 1970) which takes into account the tortuous diffusion path of an ion in wet sediments. Given the high water content of the uppermost sediments at each site, we estimate the formation factor to be only slightly greater than unity; thus F is taken to be 1.1. No corrections are applied here for the possible electrical coupling of the divalent metal ions to the fluxes of major ions (see Lasaga 1979), since the latter are unknown. In eastern Canadian lakes, Carignan and Tessier (1985) found that correcting for the coupling effect required a relatively small adjustment in the calculated flux for Zn^{2+} (about + 10% in one lake and – 7% in another). Since we have very limited major ion data, we have chosen to ignore this small potential effect on our calculated fluxes.

The estimates for zinc effluxes range from 0.6–3.5 µg cm^{-2} year^{-1} at the Outfall site to 8 µg cm^{-2} year^{-1} at the Distal site. For copper the estimates at the Outfall site range from 0 to 0.1 µg cm^{-2} year^{-1} (an influx is indicated at the Distal site), while for Pb, 0.2 – 0.7 µg cm^{-2} year^{-1} diffuse into the overlying bottom water at the Outfall site and no efflux is indicated at the

Distal location. These values represent "worst case" effluxes. To put into perspective the potential impact of such fluxes on dissolved metal concentrations in Buttle Lake deep waters, we have integrated the maximum flux for each element indicated above over the entire area of the tailings deposit on the bottom of the south basin, which we estimate to be 2 km^2 (4×0.5 km). In 1 year (approximately the residence time of water in the south basin), the total release of Zn, Cu and Pb to bottom waters would then be about 160, 2 and 14 kg, respectively. These are of the same order as estimates computed in a previous survey (Rescan 1990). Adding these masses to the lower 50 m of south basin water (a volume of about 0.3×10^{12} l) would increase the dissolved Zn, Cu and Pb concentrations respectively by 0.5, 0.01 and 0.05 μg l^{-1}. These estimates are both very crude and very generous, and almost certainly represent a net overestimation of any augmentation of the metal inventories in the lake deep waters. We conclude that the benthic effluxes of Zn, Cu and Pb implied by the pore water profiles at sites 4 and 6 are so small as to have an immeasurable impact on water quality.

6.6 Conclusions

The new results from Anderson Lake confirm previous findings – metal release from fresh, sulphide-rich tailings is limited by submerged storage, in contrast to the behaviour of sulphide-bearing tailings deposits stored subaerially. There is no indication from either the chemical data at hand or microscopic inspection of submerged tailings samples that sulphide phases in the tailings are chemically reacting on the lake floor. The fact that this conclusion also applies to very shallowly submerged tailings that may be frequently resuspended into well oxygenated bottom water is of particular importance, as it suggests that tailings in permanently flooded ponds may be prevented from oxidizing. Further research is required to confirm this.

The recent data from the Buttle Lake project indicate that the pyrite-rich tailings on the floor of the lake, abandoned more than a decade ago, are similarly not chemically reacting. There is no indication that direct oxidation of the tails is occurring. Slight releases of Zn and Pb to the overlying water appear to stem from early diagenetic recycling of authigenic naturally occurring iron and manganese oxide phases which host adsorbed zinc and lead. Prior contamination of Buttle Lake with acid rock drainage derived from a waste dump on land may have contributed to the uptake of dissolved metals by the oxides and the ongoing cycling.

In any event, the quantum of metal possibly being released to the deep waters in the lake from the submerged deposits appears to be so small as to be not measurable.

Taken collectively, the new information from Anderson and Buttle Lakes strongly supports the hypothesis that unaltered sulphide-rich mining waste will remain chemically benign if stored permanently under a water cover.

Acknowledgments. This work was supported by the Mine Environmental Neutral Drainage program, which is funded jointly by government and industry in Canada, and the Natural Sciences and Engineering Research Council of Canada. We are indebted to our many colleagues who provided field, laboratory and logistic assistance during the course of this research.

References

Carignan R (1984) Interstitial water sampling by dialysis: methodological notes. Limnol Oceanogr 29:667–670

Carignan R, Tessier A (1985) Zinc deposition in acid lakes: the role of diffusion. Science 228:1524–1526

Carignan R, Pierre SS, Gachter R (1994) Use of diffusion samplers in oligotrophic lake sediments: effects of free oxygen in sampler material. Limnol Oceanogr 39:468–474

Cline JD (1969) Spectrophotometric determination of hydrogen sulphide in natural waters. Limnol Oceanogr 14:454–458

Davison W (1993) Iron and manganese in lakes. Earth Sci Rev 34:119–163

Deniseger J, Erickson LJ, Austin A, Roch M, Clark MJR (1990) The effects of decreasing heavy metal concentrations on the biota of Buttle Lake, Vancouver Island, British Columbia. Water Res 24:403–416

Froelich PN, Klinkhammer GP, Bender ML, Luedtke NA, Heath GR, Cullen D, Dauphin P, Hammond D, Hartman B, Maynard V (1979) Early oxidation of organic matter in pelagic sediments of the eastern equatorial Atlantic: suboxic diagenesis. Geochim Cosmochim Acta 43:1075–1090

Klump JV, Martens CS (1981) Biogeochemical cycling in an organic-rich coastal marine basin. II. Nutrient sediment-water exchange processes. Geochim Cosmochim Acta 45:101–121

Lasaga AC (1979) The treatment of multi-component diffusion and ion pairs in diagenetic fluxes. Am J Sci 279:324–346

Li YH, Gregory S (1974) Diffusion of ions in sea water and in deep-sea sediments. Geochim Cosmochim Acta 38:703–714

Manheim FT (1970) The diffusion of ions in unconsolidated sediments. Earth Planet Sci Lett 9:307–309

Pedersen TF (1983) Dissolved heavy metals in a lacustrine mine tailings deposit – Buttle Lake, British Columbia. Mar Pollut Bull 14:249–254

Pedersen TF (1985) Early diagenesis of copper and molybdenum in mine tailings and natural sediments in Rupert and Holberg Inlets, British Columbia. Can J Earth Sci 22:1474–1484

Pedersen TF, Mueller B, McNee JJ, Pelletier CA (1993) The early diagenesis of submerged sulphide-rich mine tailings in Anderson Lake, Manitoba. Can J Earth Sci 30:1099–1109

Rescan (1990) Geochemical assessment of subaqueous tailings disposal in Buttle Lake, British Columbia. Unpublished report, British Columbia Acid Mine Drainage Task Force

Reynolds GL, Hamilton-Taylor J (1992) The role of planktonic algae in the cycling of Zn and Cu in a productive soft-water lake. Limnol Oceanogr 37:1759–1769

7 Phytoplankton Composition and Biomass Spectra Created by Flow Cytometry and Zooplankton Composition in Mining Lakes of Different States of Acidification

C.E.W. Steinberg[1], *H. Schäfer*[2], *J. Tittel*[3] *and W. Beisker*[2]

[1] Institute for Gewässerökologie and Binnenfischerei, Postfach 19, 12561 Berlin, Germany

[2] Arbeitsgruppe Durchflußzytometrie, GSF-Forschungszentrum für Umwelt und Gesundheit GmbH, Neuherberg, Postfach 1129, 85758 Oberschleißheim, Germany

[3] UFZ, Umweltforschungszentrum Leipzig-Halle GmbH, Sektion Gewässerforschung, Am Biederitzer Busch 12, 39114 Magdeburg, Germany

7.1 Introduction

In most aquatic systems, phytoplankton organisms are the most important primary producers. Thus, any alterations in the phytoplankton community have strong effects on the whole system. Apart from the fulfillment of basic requirements such as temperature, dissolved oxygen, ion composition and pH, zooplankton abundance and community structure depend mainly on the amount and usability of phytoplankton as the main source of nutrition. Conversely, effects of zooplankton on the phytoplankton community can be found – feeding types, selection (cell size and shape) and release of nutrients – which can strongly alter phytoplankton abundance and diversity.

Since most enzymes and physiological functions only work effectively within small pH ranges in order to maintain the viability of organisms, intracellular pH regulation is necessary for all organisms to be independent from the pH of the surrounding water. The better the external pH fits the internal pH optimum, the less energy is needed for pH regulation processes (e.g. ion pumps). Only few aquatic organisms are adapted to live in extreme pH regions, for instance owing to special enzymatic equipment, very effective regulation mechanisms or highly selective and resistant cell/body covers. Further effects of pH include influence on photosynthesis, effectiveness of phytoplankton organisms (availability of CO_2, HCO_3^- or CO_3^{2-}) and, especially of low pH values, release of toxic aluminium ions from silicates.

Chemical data about the mining lakes investigated in this study for zooplankton are presented by Geller et al. (this Vol.). A detailed description of the examined Lusatian mining lakes is given in Nixdorf (this Vol.). Lake B-Loch is a non-acidic (pH 7.0–8.5) mining lake while F-Loch still remains in a stage of high acidification (pH 2.2–2.9). Both lakes are located in the same lignite field at Schlabendorf. A higher share of marl in the surrounding fields of B-Loch caused its change to a neutral calcium sulphate lake (Klapper and Schultze 1995). Since the flooding of B-Loch started in 1964 and that of F-Loch in 1974, the latter is not much younger than the former. Although there are some relatively old and very acidic lakes and young non-acidic lakes, and vice versa, Nixdorf (this Vol.) could not find a relationship between age and the degree of acidification. According to her results, pH gives more indication of plankton diversity of a lake than its age. Decreasing diversity and stepwise exclusion of some plankton groups with decreasing pH, a well-known phenomenon in lakes affected by acid deposition (Nilssen 1980; Dillon et al. 1984; Havens 1993; Stenson et al. 1993), appear to persist at lower pHs in geogenically acidified lakes (cf. Müller 1961) as well.

Phytoplankton examinations with respect to productivity estimations or evaluation of human impacts are laborious and time-consuming if done by visual microscopy. Quite often no distinct taxonomic classification of the organisms is possible (e.g. picoplanktonic cyanophytes) or necessary; cell density and volume or other biomass parameters such as protein content may be sufficient. In this case, flow cytometry can facilitate and accelerate sample evaluation, especially for water samples with low cell density, high detritus concentration or a large amount of very small organisms such as the picoplanktonic cyanophytes (Campbell and Yentsch 1989 a, b; Campbell et al. 1989; Legendre and Yentsch 1989; Phinney and Cucci 1989; Troussellier et al. 1993; Li 1994). The high degree of automation allows, for instance, the continuous monitoring of drinking water reservoirs or the control of eutrophication or acidification processes.

Biomass spectra are a method of determining plankton system integrity based on allometric relations, i.e. cell/body volume/mass or organic carbon content of plankton organisms (Platt 1985; Sprules and Munawar 1986; Gaedke 1992). Many of the investigations of the structure of planktonic communities indicate a homogeneous distribution of biomass over all classes of body mass, from bacteria to large zooplanktonic predators in undisturbed stable water bodies such as the open ocean and large lakes. Abundance decreases with increasing biomass, and a linear relationship between biomass and abundance, as shown for Lake Constance by Gaedke (1992) in Fig. 7.1, can be evaluated. As an

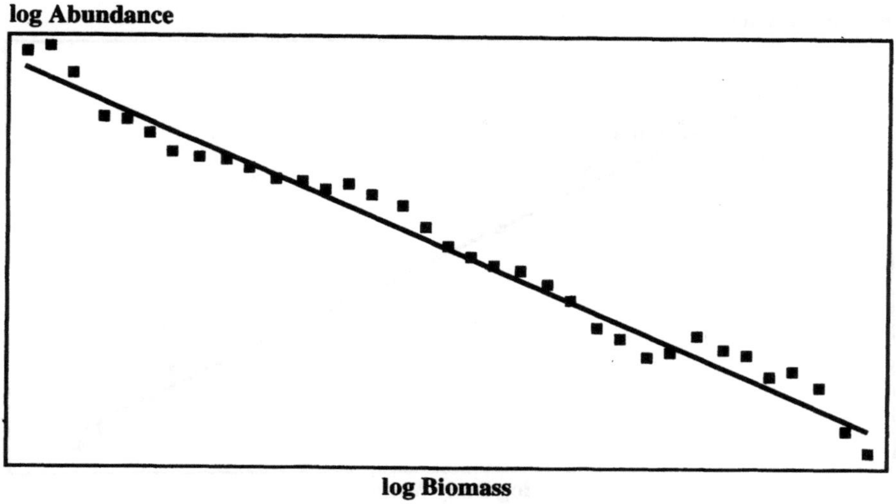

Fig. 7.1. Biomass spectrum of Lake Constance. Log biomass is plotted against log abundance; each point represents a biomass (size) class. This biomass spectrum is an example of a stable, undisturbed pelagic system. (Modified from Gaedke 1992)

example of irregularities in a biomass spectrum caused by wastewater and xenobiotic pollution, a biomass spectrum of a contaminated bay of Lake Superior is shown in Fig. 7.2 (from Sprules and Munawar 1986). Apart from a large gap resulting from the absence of organisms in these size classes, the biomass distribution is less regular if compared with Lake Constance.

Instead of measuring cell volume by microscopy or a Coulter Counter as described in many papers, fluorescein isothiocyanate (FITC) protein staining can be used as biomass descriptor for flow cytometry (Freeman and Crissman 1975; Darzynkiewicz et al. 1985).

Chlorophyll a (CHLa) fluorescence of the phytoplankton cells can be recorded as an identification parameter to discriminate between detritus and cells. Frequent field surveys and observations using ecologically relevant endpoints, such as changes in species (genetic) diversity, spatial and temporal distribution of species and alterations in energy flow, are indispensable for the detection and quantification of ecosystem alterations caused by man (Steinberg et al. 1994).

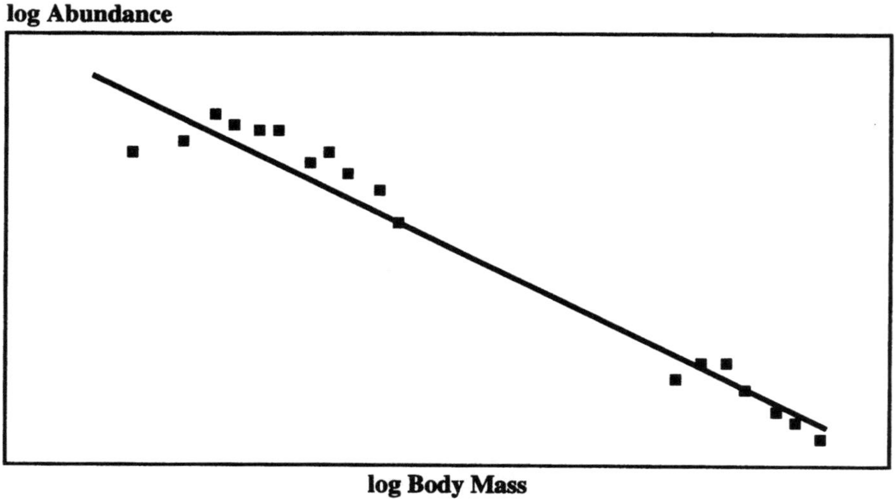

Fig 7.2. Biomass spectrum of a contaminated bay of Lake Superior. Log body mass is plotted against log abundance; each point represents a biomass class. Notable is the large gap between biomass classes, indicating a lack of organisms in those biomass classes, which is a hint for a pelagic system disturbed, e.g., by wastewater or xenobiotica. (Modified from Sprules and Munawar 1986)

7.2 Materials and Methods

7.2.1 Sampling

Phytoplankton samples of the mining lakes B-Loch and F-Loch were taken over the whole water column of both lakes. Sampling depths were generally from 0 to 9 m, apart from some at 1 m in B-Loch and 5 m in F-Loch; the sampling date was 4 July 1995. Immediately after sampling, the water samples were fixed using 0.2% glutaraldehyde and kept in the dark at 4 °C for a maximum of 2 weeks until flow cytometric analysis.

Zooplankton samples where taken by vertical net hauls over the whole water column and sugar-formol was added up to a final concentration of 4% (Haney and Hall 1973). A net with 55-µm mesh size was used. If not denoted otherwise, all samples were taken during the growing season. Some lakes were sampled only once between 19 July and 23 July 1993.

7.2.2
Phytoplankton Evaluation by Flow Cytometry

A dual laser flow cytometer (FACStar^{Plus}, Becton Dickinson; Fig. 7.3 shows a simplified construction scheme) equipped with an Ar and HeNe ion laser was used for all flow cytometric measurements. Particles up to a diameter of about 40 µm were measured. Data were stored in list-mode files and transferred for further analysis using the FASTFILE program (Becton Dickinson) to an IBM-compatible computer. The Data Analysis System (DAS version 4.29) was used for all data analyses and also graphics display (Beisker 1994).

For pigment composition analyses the excitation wavelengths (Ex) were set to 528 nm (Ar laser) and 632 nm (HeNe laser) with a laser power of 100 mW for 528 nm and 50 mW for 632 nm. The fluorescence emission (Em) of chlorophyll was measured at wavelengths > 665 nm. The amplifier

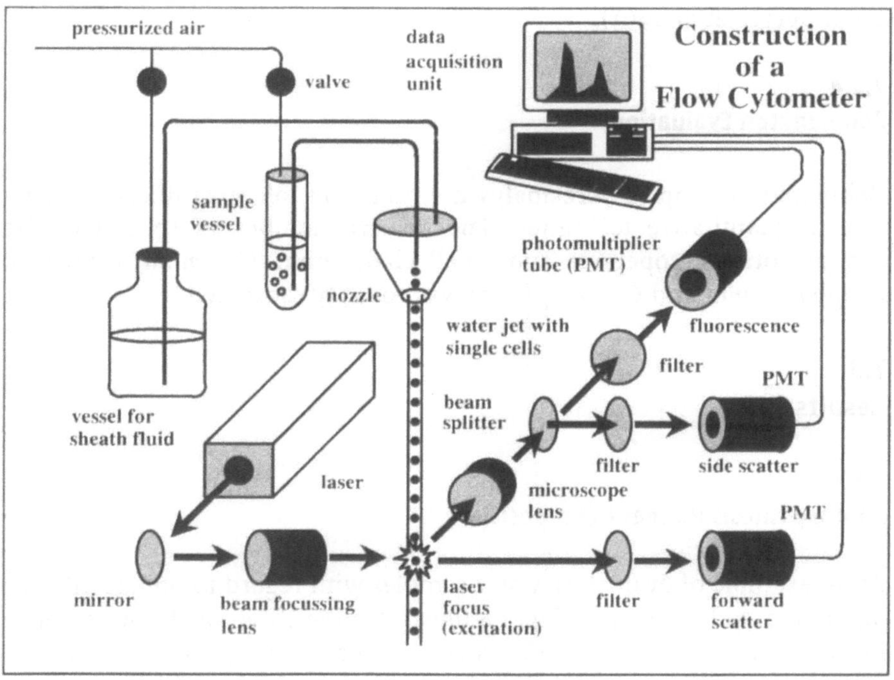

Fig. 7.3. Simplified construction scheme of a flow cytometer. The cells which are to be measured are transported to a fine nozzle where they are focussed hydrodynamically in a fine water jet. When they pass the laser focus, their optical properties (scattering, fluorescence light emission) can be detected and used to characterize the cells

gains of both chlorophyll fluorescence channels were adjusted to get a chlorophyll fluorescence ratio (CFR, CHLa Ex 528 nm/CHLa Ex 632 nm) of about one for chlorophytes. Phycoerythrin was excited at 528 nm and emission was recorded from 570 to 620 nm; phycocyanin (Ex 632 nm) emission was measured together with CHLa (Yentsch and Phinney 1985; Beeler SooHoo et al. 1986; McMurter and Pick 1994). FITC was excited at 488 nm (500 mW) and measured with a 530–15 nm bandpass filter in separate runs.

7.2.3 Protein Staining with FITC for Flow Cytometry

All cells were fixed using 0.2% glutaraldehyde, spun down, washed with 0.5-M $NaHCO_3$ solution (pH 8) and stained with 200 µg/ml FITC (isomer I, Sigma F-7250, dissolved in DMSO, final DMSO concentration 0.1%) in the $NaHCO_3$ solution (4°C, 72 h). Before flow analysis, the cells were washed in the $NaHCO_3$ solution to remove unbound FITC (see also Stoscheck 1990; Berges et al. 1993).

7.2.4 Zooplankton Evaluation

Zooplankton samples were analysed at 30 to 100-fold magnification using a semi-quantitative technique. The relative number of species of the groups rotifers, copepods and small cladocerans (*Bosmina*, *Chydorus*, *Diaphanosoma* and *Ceriodaphnia*) was roughly estimated.

7.3 Results

7.3.1 Photosynthesis Pigment Composition

As an example of evaluating water samples with regard to photosynthesis pigment autofluorescence, Fig. 7.4 illustrates the chlorophyll fluorescence ratio (CFR) analysis to differentiate between groups containing few carotenoids and groups rich in carotenoids as accessory pigments. Carotenoids are excited effectively around 530 nm and pass the absorbed energy immediately to CHLa which then releases excessive energy as dark-red fluorescence light emission (> 660 nm). Since the two absorption maxima of CHLa are around 450 and 650 nm, CHLa shows only

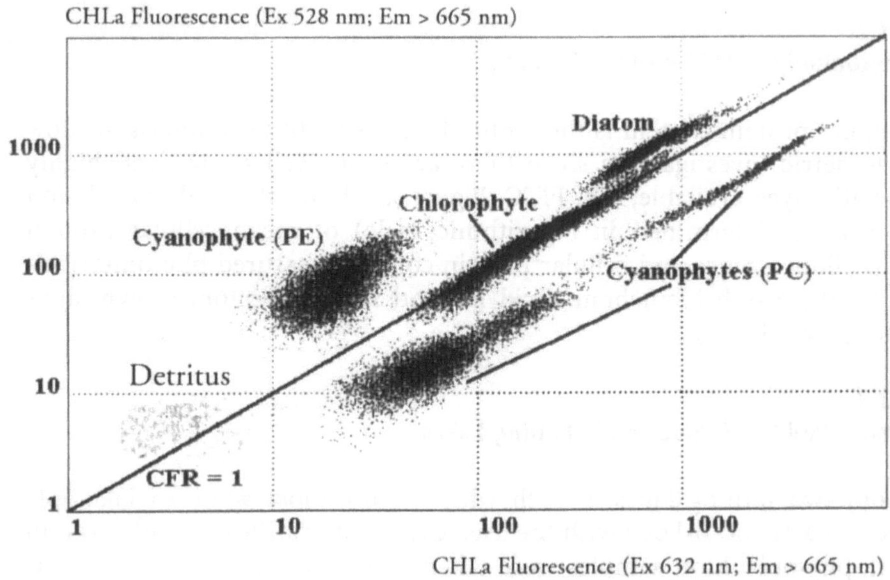

Fig. 7.4. Flow cytometric CFR analysis of phytoplankton organisms. CHLa fluorescence (Em >665 nm) excited with 632 nm is plotted against CHLa fluorescence excited with 528 nm. The CFR value of the chlorophyte with low amounts of accessory carotenoids is set to 1; the CFR of the carotenoid-rich diatom and also of the phycoerythrin(PE)-containing cyanophyte is >1, while the phycocyanin (PC)-containing cyanophytes show CFR values of <1

weak absorption in the green (around 530 nm); algae with minor carotenoid content exhibit strong CHLa fluorescence if illuminated in the blue and red (632 nm) and weak fluorescence if excited with green light. However, cells rich in carotenoids fluoresce strongly if illuminated with blue, red or green light.

If the ratio of CHLa-fluorescence excited with 528 nm and of 632 nm excitation (CFR) is calculated, algae containing minor amounts of carotenoids, such as chlorophytes, show CFR values around 1 (the diagonal in Fig. 7.4), whereas cells rich in carotenoids exhibit higher CFR values (>1.5; e.g. the diatom in Fig. 7.4). Phycoerythrin (PE)-containing cyanophytes and cryptophytes which can be detected by their PE autofluorescence (Ex 528 nm, Em 570–620 nm) also show high CFR values due to the strong excitation of PE around 530 nm and energy transfer to CHLa, whereas for phycocyanin (PC)-containing cyanophytes and cryptophytes the CFR values are <1 due to the superposition of PC and CHLa fluorescence and energy transfer from PC to CHLa.

7.3.2
Phytoplankton FITC Protein Staining

Cellular protein content is thought to be the best fit for biomass for flow cytometric investigations since there are some well-known and highly specific dyes available, e.g. FITC. Figure 7.5 shows the high correlation ($r = 0.979$, determined in logarithmic scale) of mean cellular protein FITC fluorescence and cellular protein content measured photometrically by the Smith bicinchonic acid method with laboratory-grown algae (Berges et al. 1993).

7.3.3
Phytoplankton Evaluation in Mining Lakes

From two mining lakes, F-Loch which is the most acidified lake (pH around 2.7) and B-Loch with the least degree of acidification (pH around 8.3), phytoplankton samples were taken and measured using flow cytometry and epifluorescence microscopy. Figures 7.6 and 7.7 show vertical profiles of phytoplankton abundance and pigment group composition for these lakes (Fig. 7.6, F-Loch; Fig. 7.7, B-Loch). While in B-Loch a rich phytoplankton community (picoplanktonic cyanophytes, centric diatoms, chrysophytes and a few chlorophytes and dinophytes) with high abundance (up to 40,000 cells/ml, mostly picoplanktonic cyanophytes) was found, F-Loch only showed two populations of cyanophyte-like organisms with low abundance (maximum 1000 cells/ml). In further microscope investigations, both of these organisms were found to be filamentous, with a thickness of <1 μm. Cell walls were not visible, and the lengths of the filaments were between about 10 and 30 μm. The filaments of one species were mostly randomly conglomerate with short regular helical sections resembling *Spirulina* sp. (Pascher 1925); the other species was of an *Oscillatoria*-like shape (Pascher 1925). If examined by epifluorescence microscopy, both organisms showed distinct CHLa fluorescence and the fluorescence properties in the flow cytometric measurements gave strong evidence for the occurrence of PC (CFR <1).

The dependence of species diversity and abundance on pH could also be seen in the zooplankton investigations, but some phytoplankton organisms seem to be able to cope even with pH values of <2.9 – too low for zooplankters. Abundance versus cellular protein content as a descriptor for body mass ("size") is plotted in Figs. 7.8 and 7.9 to create a kind of biomass spectrum. Since cellular protein content had to be arranged into classes, a factor of 3 between the protein classes was found to be suitable for lake B-Loch and was also applied to lake F-Loch. The counts of the

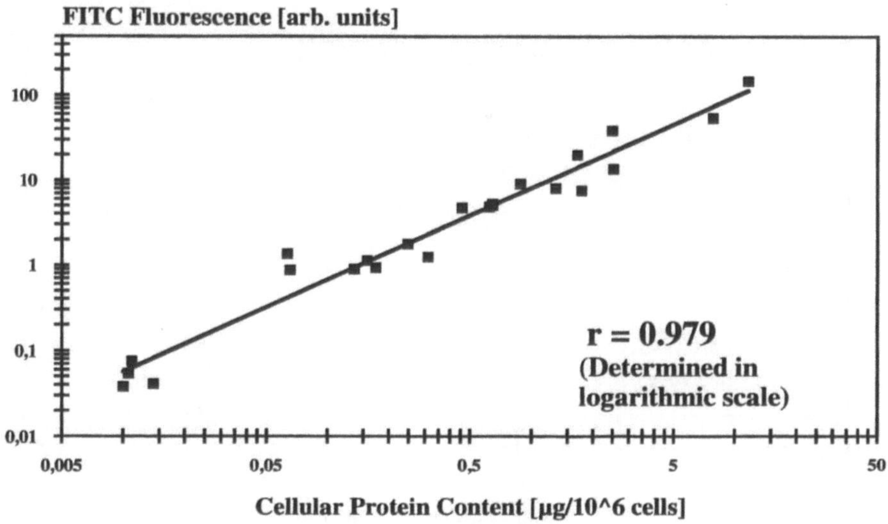

Fig. 7.5. FITC protein staining of laboratory-grown algae. The correlation ($r = 0.979$, determined in logarithmic scale) of the cellular protein content determined photometrically by the Smith bicinchonic acid method with the FITC protein fluorescence of 23 different strains of algae from 6 taxonomic groups is depicted

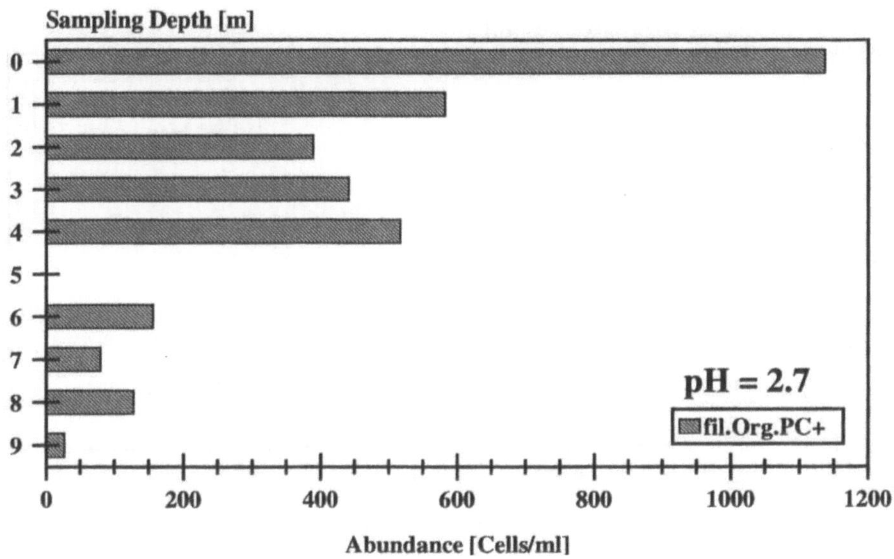

Fig. 7.6. Phytoplankton abundance and vertical distribution in lake F-Loch (pH 2.7). Only phycocyanin (PC)-containing filamentous cyanophyte-like organisms (*fil. Org. PC+*) were found

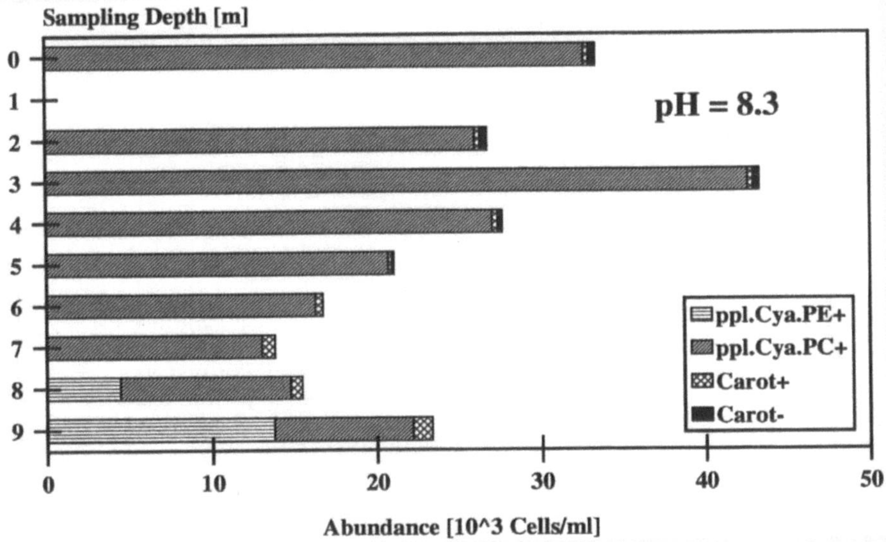

Fig. 7.7. Phytoplankton abundance and vertical distribution in lake B-Loch (pH 8.3). Picoplanktonic phycoerythrin (PE)- and phycocyanin (PC)-containing cyanophytes (*ppl.Cya.PE+/PC+*), carotenoid-rich species, mainly diatoms (*Carot+*) and chlorophytes with low amounts of carotenoids (*Carot−*) were found

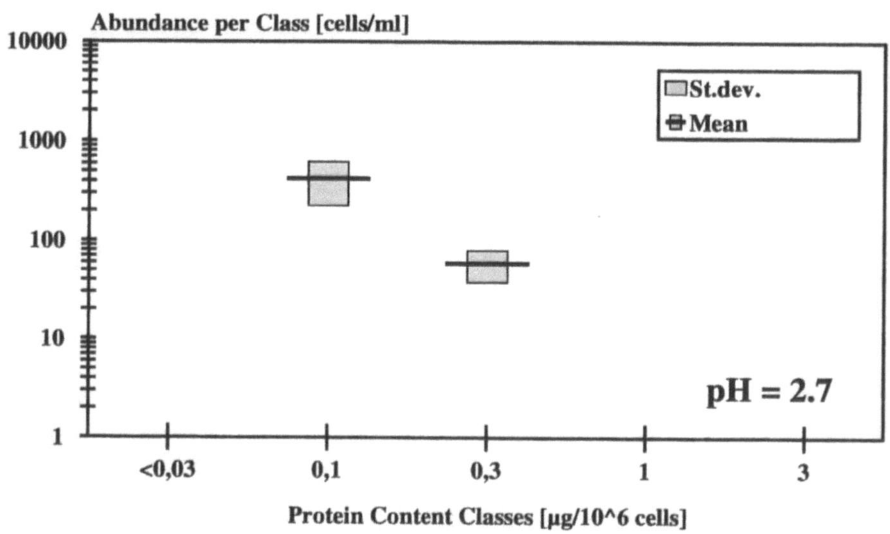

Fig. 7.8. Phytoplankton biomass distribution of lake F-Loch. Protein content as an estimate of biomass is plotted against abundance. Results of the vertical samples were added and the mean value and standard deviation, which is an estimate of the changes in abundance with sampling depth, were calculated. As in Fig. 7.6, only two phytoplankton populations were found

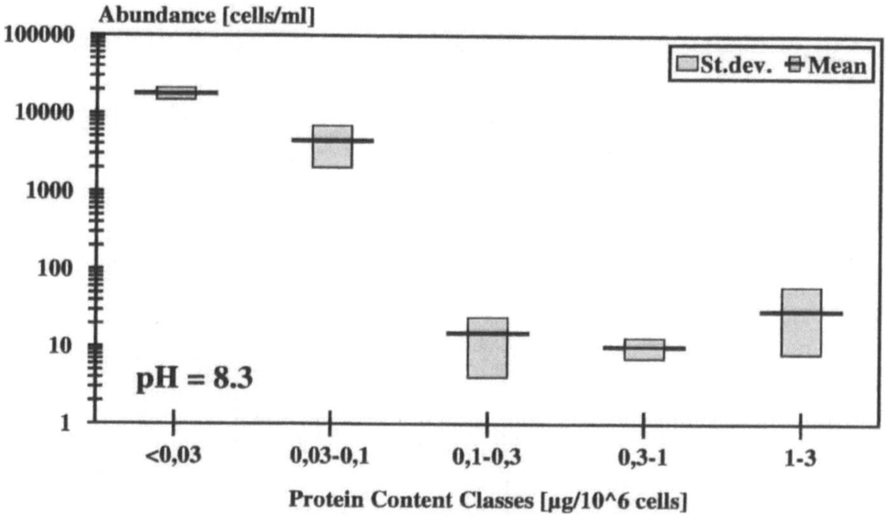

Fig. 7.9. Phytoplankton biomass distribution of lake B-Loch. Protein content as an estimate of biomass is plotted against abundance. Results of the vertical samples were added and the mean value and standard deviation, as an estimate of the changes in abundance with sampling depth, were calculated. In the not acidified lake B-Loch, a rich phytoplankton community can be found (see also Fig. 7.7) compared with the highly acidified lake F-Loch

nine vertical samples in each particular class were added; the standard deviation gives a clue for the vertical distribution of the populations within the size class. The differences between the acidified lake F-Loch (Fig. 7.8) and lake B-Loch (Fig. 7.9) are evident. In all five size classes, organisms were found in lake B-Loch, indicating a more structured phytoplankton community compared with lake F-Loch with only two size classes of organisms. Smaller algae displayed a higher abundance than larger ones in both lakes.

7.3.4 Zooplankton Evaluation in Mining Lakes

In Fig. 7.10 the occurrence of different zooplankton groups in mining lakes sorted by ascending pH value is depicted. Rotifers occurred in nearly all samples of lakes with pH above 2.9. Below this value neither rotifers nor crustaceans were found. In all lakes with pH values between 2.9 and 4 the rotifer *Brachionus urceolaris* was present, and in the two most acidic lakes of this range (Lake 117 and Lake Halbendorf, pH 2.9 and 3.0) *B. urceolaris* was the only rotifer. Species of the "small cladocerans" group

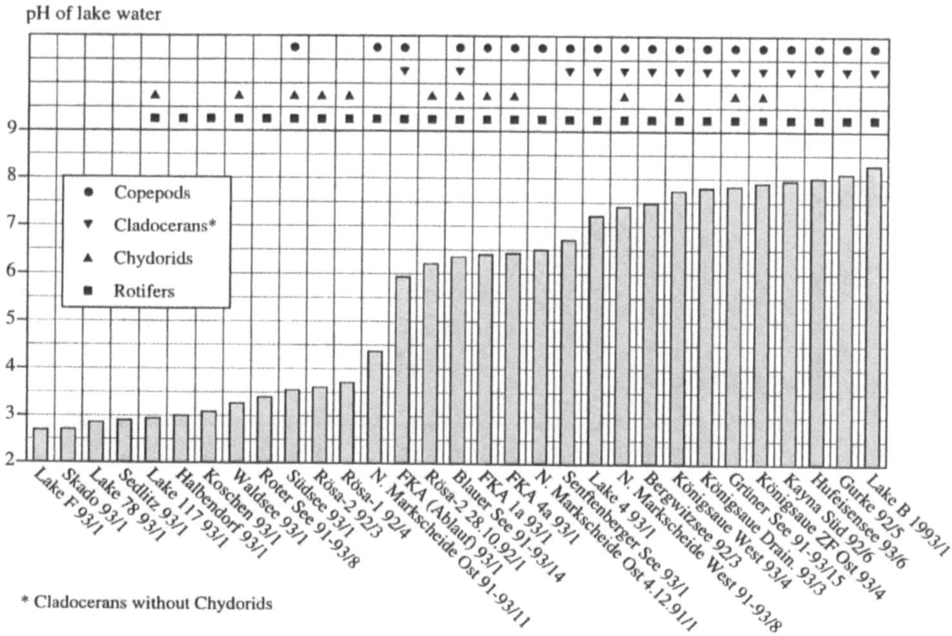

Fig. 7.10. Occurrence of rotifers, chydorids, other cladocerans and copepods in all investigated mining lakes sorted by ascending pH of lake water. Rotifers were found in all lakes with a pH of >2.9; chydorids were present only in some lakes and almost independent from pH; the other cladocerans and copepods, respectively, were found frequently in all lakes of pH > 6.5 and 6.3

occurred at pH 5.5–6.0 and more, except for the species *Chydorus sphaericus* which was also found in more acidic lakes down to pH 2.9. Daphnids were present only in non-acidic or slightly acidic lakes with average pH values of 6 and more. In two samples below pH 6 from Lake Blauer See, daphnids were found, but all other samplings of this lake showed higher pH values. Copepods were also abundant in non-acidic or slightly acidic lakes, and in lower densities they even occurred in more acidic lakes; these were *Cyclops strenuus* in Lake Nördliche Markscheide Ost (pH 4.4) and *Diacyclops languidus* in Lake Südsee (pH 3.5).

Figures 7.11–7.14 show the occurrence of different zooplankton groups in mining lakes depending on the season and pH value. The data plotted are from single net samples from the investigated mining lakes as presented in Fig. 7.10 and summarized from 1991 to 1993. All groups show their typical seasonal pattern with a summer maximum within the tolerated pH ranges. Daphnids generally occurred in lower densities then copepods.

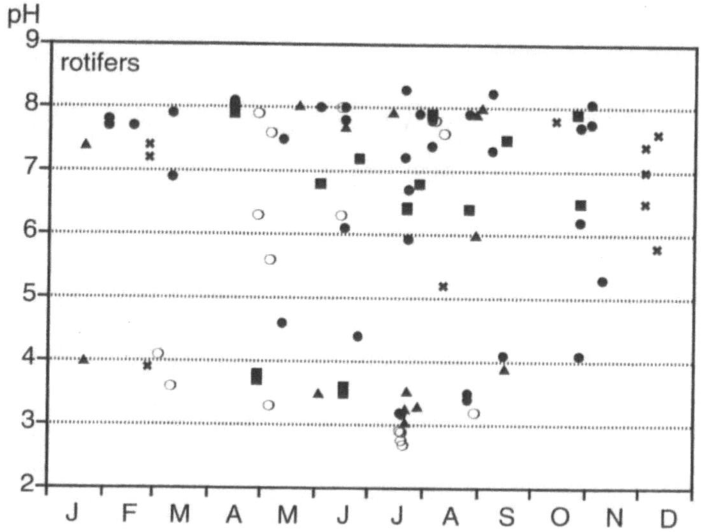

Fig. 7.11. Occurrence of rotifers in the investigated mining lakes (see Fig. 7.10) depending on season (months on x-axis) and pH of lake water. Data are from single net samples and summarized from 1991 to 1993. Abundance is depicted with the following *symbols*: *open circles* absent; *triangles* seldom; *closed circles* frequent; *squares* very frequent; *crosses* present, frequency not estimated

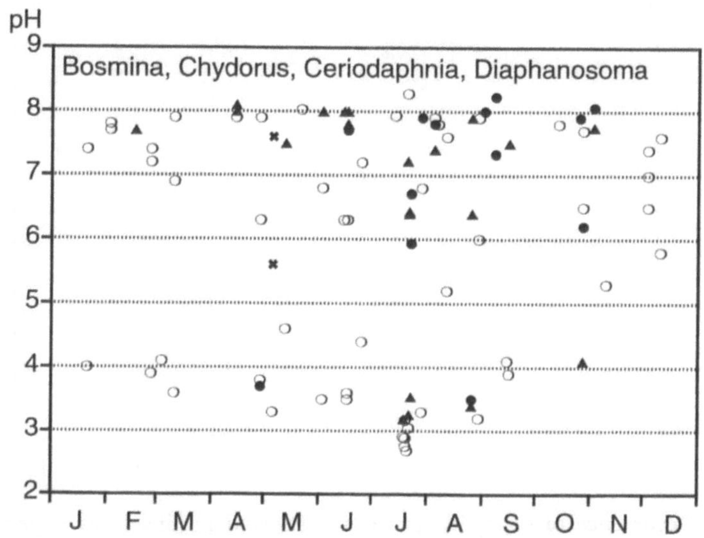

Fig. 7.12. Occurrence of small cladocerans in the investigated mining lakes depending on season (months on x-axis) and pH of lake water. For database and *symbols* see Fig. 7.11

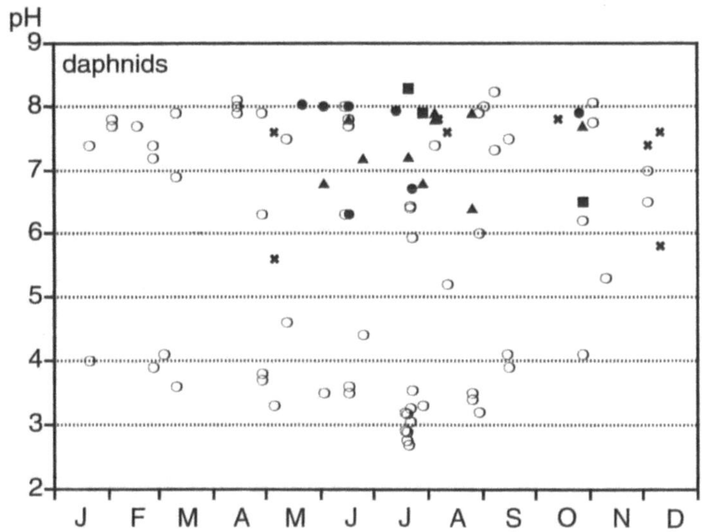

Fig. 7.13. Occurrence of *Daphnia* in the investigated mining lakes depending on season (months on x-axis) and pH of lake water. For database and *symbols* see Fig. 7.11

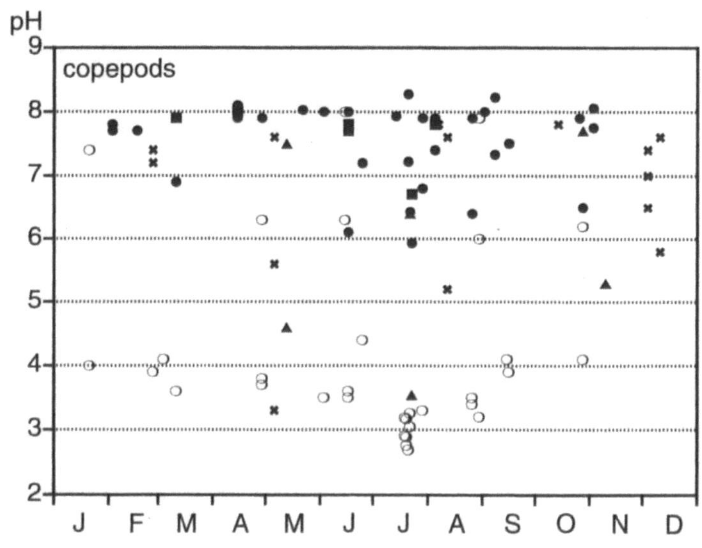

Fig. 7.14. Occurrence of copepods in the investigated mining lakes depending on season (months on x-axis) and pH of lake water. For database and *symbols* see Fig. 7.11

7.4 Discussion

Acidified mining lakes are peculiar ecosystems. Their only natural counterparts are thought to be acidic crater lakes (Geller et al., this Vol.). Life in these lakes appears to be imbalanced: acidified mining lakes may house, if any, only very short trophic chains and no or poor macrophyte stands. Knowledge of these lakes is also relatively poor. There is almost nothing known about the ontogeny of acidic mining lakes with particular respect to natural gains of acid-neutralizing-capacity in time and space. The study of Bramkamp et al. (this Vol.) tries to correlate the age of a series of Bavarian acidified mining lakes with the pH status. However, although knowledge of acidified mining lakes appears to be little developed, public opinion is very strong. The public, driven by (non-scientist) decision-makers, requests "restored" ecosystems (it would be better to say "somewhat treated", since naturally acidic lakes are by no means degraded, i.e. acidified ecosystems that need to be restored), in the form of normal circumneutral freshwater recreation lakes with scenic surroundings, known from holiday brochures of landscapes abroad. Due to the lack of detailed ecological knowledge, one major aim in treating acidic mining is to raise the pH. In our opinion, this aim is necessary but not sufficient enough to meet all ecological demands.

Provided that the essentials of the biomass spectra (e.g. Figs. 7.8 and 7.9) and the occurrence of zooplankton dependent on pH (Fig. 7.10) also hold true on an annual basis, we are able to establish ecological aims for the treatment of acidified mining lakes. *Bracheonus urceolaris* is cosmopolitan and known to be a typical rotifer in highly acidified lakes (Vallin 1952; Ohle 1981; Havas and Hutchinson 1983). Koste (1978) quoted a pH value of 2.8 for the lower tolerance limit while the lowest pH reported is 2.6 (McConathy and Stahl 1982). Nixdorf (this Vol.) found *B. urceolaris* in the southern enclosed part of lake F (lake F-Süd) in which the pH was slightly higher than in the northern basin. It is assumed that at the transition between pH 2.6 and 2.9, *B. urceolaris* occurs in very low numbers or only episodically, in relation to fluctuating pH conditions. However, there are also rotifers in lake F-Loch (*Cephalodella*; Nixdorf, this Vol.) which can only be detected using finer mesh sizes and by a live counting technique.

The occurrence of cyanophytes in lakes with low pH has not been reported much to date (Almer et al. 1974; Kwiatkowski and Roff 1976). However, the fluorescence properties as well as the microscope examination support the assumption that the two phytoplankton species found in lake F-Loch are cyanophytes. Photosynthesis pigment decomposition in acidic samples causing a shift in the CFR is not very likely, since in a study of Bavarian mining lakes no significant changes in the fluorescence properties in samples of different pH between 2.8 and 4.5 were observed (Schäfer et al., in prep.) when stored under dark and cool conditions for up to 2 weeks. Picoplanktonic cyanophytes (*Synechococcus* sp., *Synechocystis* sp.), however, were found only in a lake with a pH of 4.5 and not in the more acidic lakes.

Provided that life in acidified mining lakes always starts in the form of prokaryotic consortia, the constitution of food chains is the first step in the ontogeny of these lakes. The next step is the occurrence of eukaryotic phytoplankters and of small-bodied cladocera. The plankton community becomes more and more diverse, and the biomass spectra become more complete. The ultimate aim in the treatment of acidic mining lakes is the establishment of a fish community, completing the trophic chain within the open water of the lake.

The emerging picture is the absence of some groups (crustaceans) in the very acidic lakes and the reduction in the diversity of functional groups (primary producers, consumers) to a few acid-tolerant species. Such a drastically reduced community will depict an irregular plankton size distribution, characterized by large gaps, i.e. unoccupied size classes. Because the size of an organism is the basis for its physiological and ecological properties, an unoccupied size class can be seen as an unoccupied ecological niche in the system. The real existence of gaps or discontinuities in the plankton size distribution is questionable (see Gaedke 1992) and seems to be true only for other extreme limnic environments, as shown for the high mountain lake La Caldera (Echevarria et al. 1990). Thus, the construction of plankton size spectra provides the opportunity to gather quantifiable information about the stage of succession of an acidic mining lake. Alternatively, an approach based on species diversity is time-consuming, needs the work of taxonomic specialists and is mostly restricted to one taxonomic group (MacIsaac et al. 1987; Siegfried et al. 1989). In addition, the most important organisms in these young and often extreme environments (bacteria, autotrophic picoplankton and possibly protozoa) are taxonomically difficult to treat.

By means of flow cytometry, ecologists are able to semi-automatically create biomass spectra. However, these spectra include only a small section of the total range of body size of pelagic organisms. With our method, organisms up to 40–50 μm (bacteria, cyanobacteria, algae and small-bodied protozoa and rotifers) can be measured and counted. Using larger capillaries, our approach can be applied also to crustacean zooplankters (Kachel et al. 1992).

Comparing Figs. 7.8 and 7.9, the conclusion may be derived that, even in the small section of the biomass spectrum studied with our methods, a clear increase in completeness takes place. Future studies will show how to apply the semi-automatically created biomass spectra as an estimate of integrity or disturbance of planktonic systems.

7.5 Conclusions

Both phyto- and zooplankton investigations depicted a clear difference in community structure and abundance between the acidified and the circumneutral mining lakes. While in non-acidified lakes the diversity of both phyto- and zooplankton communities was quite high, in lakes with very low pH values only a few insensitive species were found. In the case of the extremely acidic lakes, results are based on a single sampling. So far, there are only a few observations of the zooplankton in such extreme environments and other investigations on components of the microbial food web (e.g. heterotrophic flagellates, ciliates and other protozoa) are in an initial stage at present. Therefore further research is needed for validation of the hypothesis described here. Of course, the phytoplankton investigations were only a snapshot of the situation on the sampling day in these two lakes and firm conclusions about phytoplankton integrity should not be drawn. The aim was to show the feasibility of flow cytometric methods for preparing biomass spectra of the phytoplankton community. For estimation of phytoplankton integrity, biomass spectra based on repeated sampling during at least one vegetation period are necessary. A solid database gained from repeated measurements over some years is needed for the detection of long-term alterations in the planktonic community and for the prediction of trends in aquatic systems.

References

Almer B, Dickson W, Ekström C, Hörnström E, Miller U (1974) Effects of acidification on Swedish lakes. Ambio 4:30–36

Beeler SooHoo JB, Kiefer DA, Collins DJ, McDermid IS (1986) In vivo fluorescence excitation and absorption spectra of marine phytoplankton: I. Taxonomic characteristics and responses to photoadaptation. J Plankton Res 8:197–214

Beisker W (1994) A new combined integral-light and slit-scan data analysis system (DAS) for flow cytometry. Comp Meth Prog Biomed 42:15–26

Berges JA, Fisher AE, Harrison PJ (1993) A comparison of Lowry, Bradford and Smith protein assays using different standards and protein isolated from the marine diatom *Thalassiosira pseudonana*. Mar Biol 115:187–193

Campbell JW, Yentsch CM (1989a) Variance within homogeneous phytoplankton populations, I: theoretical framework for interpreting histograms. Cytometry 10:587–595
Campbell JW, Yentsch CM (1989b) Variance within homogeneous phytoplankton populations. II. Analysis of clonal cultures. Cytometry 10:596–604
Campbell JW, Yentsch CM, Cucci TL (1989) Variance within homogeneous phytoplankton populations. III. Analysis of natural populations. Cytometry 10: 605–611
Darzynkiewicz HA, Crissman Z, Tobey RA, Steinkamp JA (1985) Correlated measurements of DNA, RNA, and protein in individual cells by flow cytometry. Science 228:1321–1324
Dillon PJN, Yan D, Harvey HH (1984) Acidic deposition: effects on aquatic ecosystems. CRC Crit Rev Environ Contr 13:167–194
Echevarria F, Carrillo P, Jimenez F, Sanchez-Castillo P, Cruz-Pizarro L, Rodriguez J (1990) The size-abundance distribution and taxonomic composition of plankton in an oligotrophic, high mountain lake (La Caldera, Sierra Nevada, Spain). J Plankton Res 12:415–422
Freeman DA, Crissman HA (1975) Evaluation of six fluorescent protein stains for use in flow microfluorometry. Stain Technol 50:279–284
Gaedke U (1992) The size distribution of plankton biomass in a large lake and its seasonal variability. Limnol Oceanogr 37:1202–1220
Haney JF, Hall DJ (1973) Sugar-coated *Daphnia*: a preservation technique for *Cladocera*. Limnol Oceanogr 18:331–333
Havas M, Hutchinson TC (1983) The Smoking Hills: natural acidification of an aquatic ecosystem. Nature 301:23–27
Havens KE (1993) Pelagic food web structure in Adirondack Mountain, USA, lakes of varying acidity. Can J Fish Aquat Sci 50:149–155
Kachel V, Hüller R, Glossner G, Burkhill P, Tarran G (1992) Optical and electrical methods for the analysis of single aquatic particles and organisms. Proceedings of the conference: optics within life sciences II, Münster 1992
Klapper H, Schultze M (1995) Geogenically acidified mining lakes – living conditions and possibilities of restoration. Int Rev Ges Hydrobiol 80:639–653
Koste W (1978) Rotatoria, vol 1, 2nd edn. Borntraeger, Berlin
Kwiatkowski RE, Roff JC (1976) Effects of acidity on the phytoplankton and primary productivity of selected northern Ontario lakes. Can J Bot 54:2546–2561
Legendre L, Yentsch CM (1989) Overview of flow cytometry and image analysis in biological oceanography and limnology. Cytometry 10:501–510
Li WKW (1994) Primary production of prochlorophytes, cyanobacteria, and eukaryotic ultraplankton: measurements from flow cytometric sorting. Limnol Oceanogr 39:169–174
MacIsaac HJ, Hutchinson TC, Keller W (1987) Analysis of planktonic rotifer assemblages from Sudbury, Ontario, area lakes of varying chemical composition. Can J Fish Aquat Sci 44:1692–1701
McConathy JR, Stahl JB (1982) Rotifera in the plankton and among filamentous algal clumps in 16 acid strip mine lakes. Trans Ill Acad Sci 75:85–90
McMurter HJG, Pick FR (1994) Fluorescence characteristics of a natural assemblage of freshwater picocyanobacteria. J Plankton Res 16:911–925
Müller H (1961) Zur Limnologie der Restgewässer des Braunkohlenbergbaus. Verh Int Ver Limnol 14:850–854

Nilssen JP (1980) Acidification of a small watershed in southern Norway and some characteristics of acidic aquatic environments. Int Rev Ges Hydrobiol 65: 177–207

Ohle W (1981) Photosynthesis and chemistry of an extremely acidic bathing pond in Germany. Verh Int Ver Limnol 21:1172–1177

Pascher A (1925) Die Süßwasserflora Deutschlands, Österreichs und der Schweiz, Heft 12: Cyanophyceae. Fischer, Jena

Phinney DA, Cucci TL (1989) Flow cytometry and phytoplankton. Cytometry 10:511–521

Platt T (1985) Structure of the marine ecosystem: its allometric basis. Can Bull Fish Aquat Sci 213:55–64

Siegfried C, Bloomfield JA, Sutherland JW (1989) Planktonic rotifer community structure in Adirondack, New York, USA, lakes in relation to acidity, trophic status and related water quality characteristics. Hydrobiologia 175:33–48

Sprules WG, Munawar M (1986) Plankton size spectra in relation to ecosystem productivity, size, and perturbation. Can J Fish Aquat Sci 43:1789–1794

Steinberg CEW, Geyer HJ, Kettrup AF (1994) Evaluation of xenobiotic effects by ecological techniques. Chemosphere 28:357–374

Stenson JAE, Svensson JE, Cronberg G (1993) Changes and interactions in the pelagic community in acidified lakes in Sweden. Ambio 22:277–282

Stoscheck CM (1990) Quantitation of protein. Methods Enzymol 182:50–68

Troussellier M, Courties C, Vaquer A (1993) Recent applications of flow cytometry in aquatic microbial ecology. Biol Cell 78:111–121

Vallin S (1952) Zwei acidotrophe Seen im Küstengebiet von Nordschweden. Rept Inst of Freshwater Research, Drottningholm 39:167–189

Yentsch CS, Phinney DA (1985) Spectral fluorescence: an ataxonomic tool for studying the structure of phytoplankton populations. J Plankton Res 7:617–632

8 Ecological Potentials for Planktonic Development and Food Web Interactions in Extremely Acidic Mining Lakes in Lusatia

B. Nixdorf, K. Wollmann and R. Deneke

Brandenburg Technical University of Cottbus, Faculty of Environmental Sciences and Process Engineering, Chair of Water Conservation, Seestr. 45, 15526 Bad Saarow, Germany

8.1 Introduction

Lakes can be acidified naturally (crater lakes or other volcanic waters) or by anthropogenic impact (acid rain and mine tailing). Contrary to the very well investigated physicochemical mechanisms and ecological consequences of the atmospheric acidification of lakes (Steinberg and Wright 1994), knowledge about the limnology in geogenically acidified lakes is limited (Geller et al., this Vol.). Normally it is expected that, except for specialized bacteria and fungi, only a few organisms are able to survive in lake waters with $pH < 3$.

Since 1993 we have been investigating the pelagic and littoral colonization (algae, bacteria, zooplankton and further invertebrates) in a number of geogenically acidified Lusatian lakes of different age and maturity to find out the status of primary succession and food web relations and control as a base for possible biological mechanisms of neutralization.

Some hypotheses regarding the planktonic colonization and shifts in moderate acidic lakes (pH mostly between 4 and 6) will be tested:

1. The phytoplankton biomass in acidic lakes is generally less than 1 mg fresh weight (fw)/l (Almer et al. 1974; Kwiatkowski and Roff 1976; Yan 1979).
2. The situation in acidic lakes is comparable with neutral/non-acidic oligotrophic waters concerning the high transparency of the water and the low nutrient concentration.
3. Therefore, the predominant planktonic algae in acidified lakes are mainly cryptophytes, dinoflagellates and diatoms, whereas blue-greens are rare (Duty and Ostrofsky 1974).
4. The dominance of the primary producers can be observed in the littoral zone rather than in the pelagial.

It is known of pelagic primary producers in acidic lakes that their species richness, diversity and biomass are very low. Blouin (1989) found that the pH tended to affect plankton diversity more than abundance or standing crop. Unfortunately, these observations are confined to pH-ranges between 4 and 5.5, not to extreme pH (lower than 3) as often measured in the geogenically acidified lakes in Lusatia (former East Germany). The aim of this chapter is, therefore, to answer at least two questions: What kind of lakes have been arising in the Lusatian area concerning morphometry, hydrography and mixing as well as chemistry? What is the "status quo" of planktonic colonization and what are the presuppositions for primary production in these extremely acidic lakes (nutrients, light, mixing, inhibition, grazing pressure, mineralization) as a base for effectiveness of neutralization measures due to lake-internal biological processes?

8.2 Study Sites

Lusatia comprises the south-eastern part of Brandenburg and the north-eastern part of Saxony. The standing waters in the region have their origin in mining activities. The Lusatian region (Brandenburg) around Cottbus has been one of the most important open-cast lignite mining areas in Germany since the end of the nineteenth century. As a consequence of the intensive exploitation and the drastic reduction in brown coal production after 1990, a number of holes were filled with water (mainly re-rising groundwater and water from rivers) and are expected to become residual lakes in the near future. Because of the geogenic potential for acidification due to pyrite oxidation, most of these lakes are extremely acidic ($pH < 3.5$). The main coal mining areas are illustrated in Fig. 8.1. Up till now, lakes in the following regions have been investigated: Schlabendorfer Felder (2), Muskauer Faltenbogen (6), Grünewalde/Plessa (3), around Hoyerswerda (7) and Senftenberg (4), around Drebkau/Vetschau (5) near Cottbus, and outside Lusatia near Frankfurt/Oder (1) (Helenesee and Katjasee; Table 8.1 and Fig. 8.1).

8.3 Methods

Water samples for the chemical and biological analysis (planktonic composition, chlorophyll, oxygen demand) were taken monthly or irregularly as mixed samples from the surface to the bottom during mixing and as vertically integrated samples from the epi- and hypolimnion with a 2.3-l LIMNOS sampler considering the different mixing regimes of the lakes. The sampling stations were above the deepest points and as a consequence of horizontal inhomogeneities also at other points if necessary

Fig. 8.1. Map of the investigated region in eastern Brandenburg, Germany

(Schultze et al. 1994) where the Secchi depth was also estimated. Depth profiles of temperature, oxygen (absolute and relative as saturation), pH, redox potential and conductivity were measured at 0.5-m intervals by means of a HYDROLAB 20 probe and a field computer (HUSKY Hunter). The attenuation was calculated from light measurements every 50 cm underwater with two spherical quantum sensors (LICOR) with a distance of 39 cm between them to reduce the disturbances of the incident light by clouds and waves. Qualitative and quantitative net samples (25 µm mesh) were taken for the identification and the calculation of zooplankton abundance and biomass.

The different fractions of anions (among them the nutrients as phosphate, nitrate and silicate) and cations, dissolved inorganic carbon (DIC), dissolved organic carbon (DOC) and metals were determined according to the recommendations given in the "Limnological methods for investigation of mining lakes" (Schultze et al. 1994) and to the standard methods (Deutsche Einheitsverfahren 1986–1996). For the estimation of the acid neutralization capacity (ANC), samples were titrated with 0.02 mol/l NaOH.

Table 8.1. Morphometric characteristics according to intended state after flooding, range of pH (1995–1996) and age of some mining lakes in Brandenburg and Saxonia

Region and lakes	Range of pH	Z_{max} (m)	Area (ha)	Volume (10^6 m^3)	Start of flooding
Schlabendorfer Felder (2)					
Hindenburger See (A-Loch)	7–8	4	44		1962
Stöbritzer See (B-Loch)	7.1–8.2	12	6		1964
Stoßdorfer See (C-Loch)	7–8	8	167		1966
Lichtenauer See (F-Loch)	2.6–3.0	32.5	230	25.2	1974
Drehnasee (RL 12)	3–3.5	44.5	180	17.2	
Bergen-Niederhof (RL 13)	2.8–3.1	26.8	48	5.2	
Beesdau (RL 14/15)	2.6–3.2	33	570	47.5	1968
Muskauer Faltenbogen (6)					
Felixsee	3.8–4.0	18	11		1921
Waldsee (epilimnion)	2.9–3.5	5	0.1		1920
Lerche (Lohnteich)	7–9	10			1926
Grünewalde/Plessa (3)					
Plessa	2.5–2.7	10	10		1965
Bergheide		53.0	290	38.1	
Heidesee (RL 129)	2.9–3.0	10.0	ca. 13	0.7	
Heidesee (RL 130)	3.1–3.4	47.6	500	90	
Heidesee (RL 131)	2.7–2.8	31.0	ca. 73	8.9	
Restloch 107	2.2–2.4	4	13		
Hoyerswerda (7)					
Dreiweibern	2.8–3.3	37.9	286	34.8	
Burghammer	6.7–9.0	49.4	390	38	
Lohsa II	2.5–2.8	56	960	99	
Spreetal NE	2.6–2.8	58	314	96.9	
Spreetal (Partwitz)	2.7–2.9	34.7	347	64	
Bluno	2.7–2.9	43	446	54	
Bärwalde	2.6–2.8	59.0	1500	148.4	
Scheibe	2.8–3.8	51.5	712	133.3	
Senftenberg (4)					
Sedlitz	2.7–3.0	12	1170	202	
Koschen	2.7–3.1	25	620	38.4	
Skado	2.5–2.8	46.4	989	130.5	
Meuro (Ilse-See)		56.5	750	133	
Drebkau/Vetschau (5)					
Greifenhain	5.1–6.9	70	970	330	
Gräbendorf	3.2–3.4	47.6	500	90	1996
Frankfurt (Oder) (1)					
Helenesee	7.5–8.6	55	250	90	1959–1970
Katjasee	8–8.5	12	50		1970

Z_{max}, maximum depth. Numbers in parentheses refer to areas on Fig. 8.1.

The phytoplankton was analysed after fixation with Lugol's solution using the conventional Utermöhl technique (Utermöhl 1958). The biovolume was calculated according to Nixdorf and Hoeg (1993). Chlorophyll a and chemical parameters were determined according to the standard methods, with some modifications concerning the peculiarities of acidic water samples.

Bacterial cell numbers and their shape were estimated using the DAPI staining method (Porter and Feig 1980) and epifluorescence microscopy (magnification, 1250). From the same sample the autotrophic picoplankton (APP) was counted by means of epifluorescence technique using the autofluorescence of cells containing chlorophyll. Zooplankton samples were analysed using transmission (magnification, 400) and stereomicroscope (20–100). Corixids were captured with underwater light traps, fixed in ethanol and analysed by a stereo microscope (20–100).

8.4 Results and Discussion

8.4.1 Diversity in Morphometry and Chemistry of Acidic Mining Lakes in the Lusatian Region

The area, shape, depth, exposure to wind, retention time and further abiotic structures of a lake determine the potential for biological conversion of production resources. As can be seen from Table 8.1, there is great diversity of age and morphometry of lakes as a result of the flooding of the mining holes. After flooding, Lusatia will be rich in great lakes compared with the natural lakes in Brandenburg. For example, Lake Müggelsee, the largest lake in Berlin, has a volume of about 36×10^6 m^3. The Scharmützelsee in the Bad Saarow region is, up till now, the greatest lake in Brandenburg with a volume of 121×10^6 m^3. Some of the mining lakes arose more than 70 years ago; others are still in the state of flooding with groundwater and/or surface water. The lakes are quite different in size and morphometry; as a consequence some are well stratified and others are polymictic water bodies (Mischke et al. 1995; Nixdorf et al. 1995). As far as the acidity of the lakes is concerned, there are also pointed differences between them: there are young lakes with neutral water and old lakes with very acidic water, and vice versa. The acid neutralization capacity (ANC) is in a wide range (Fig. 8.2), indicating the different amounts of the neutralization potential necessary for the restoration of the lakes.

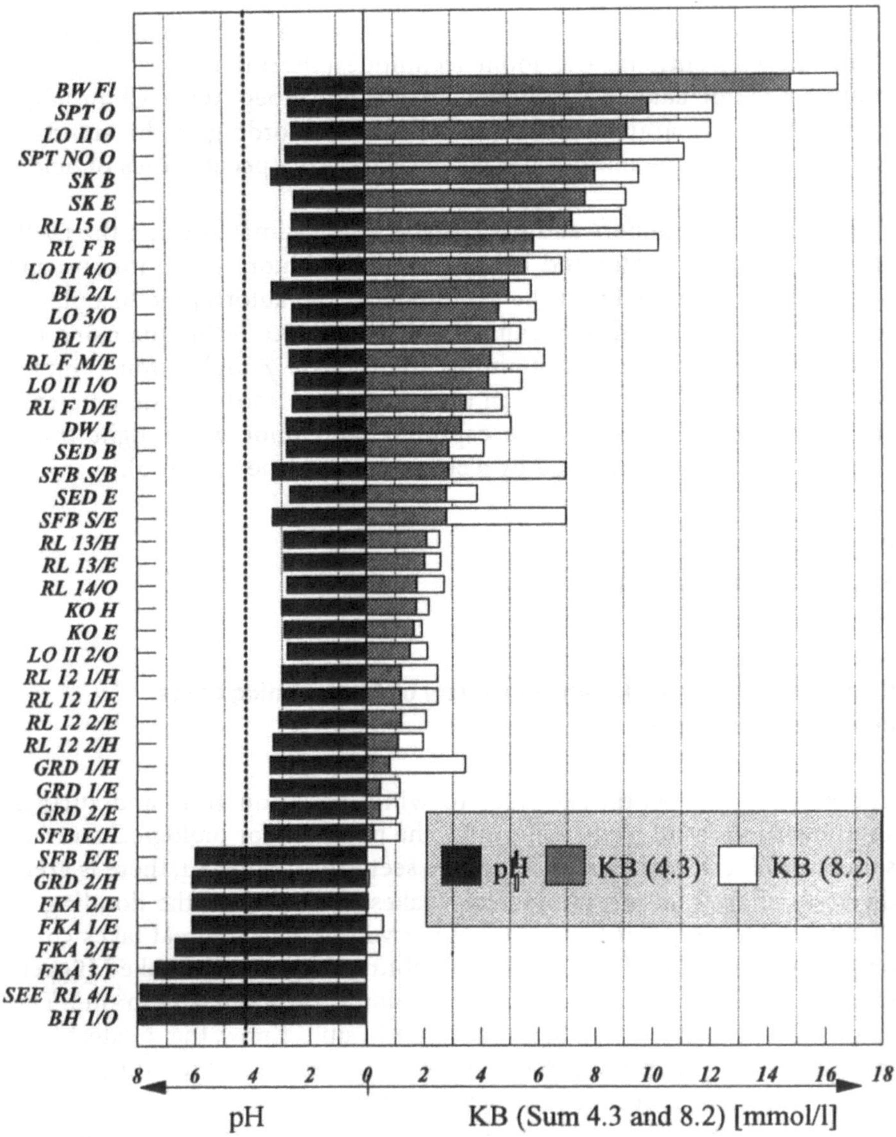

Fig. 8.2. Acid neutralization capacity (ANC) expressed by KB values and pH of selected lakes in Lusatia

Morphometric presupposition for the formation of trophic states according to the approach of Klapper (TGL 27885/01, 1982) using the ratio between epi- and hypolimnion volume is very favourable for the large, deep lakes (e.g. Gräbendorf, Bärwalde, Scheibe, Burghammer). From the morphometric point of view, they are potentially oligotrophic or mesotrophic, like the Helenesee, a 40-year-old non-acidic mining lake near Frankfurt/Oder which has meanwhile become one of the most popular lakes in Brandenburg.

Summarizing the situation at this point of limnological investigations, no regular patterns can be derived for the classification of lakes according to their age, their acidity or the local geological conditions. These lakes represent only a small part of the number and variation of man-made lakes in the Lusatian region. On the basis of a "normal" food web and its main components, attention is first paid to the conditions and inhibitions of primary production in extremely acidic mining lakes.

8.4.2
Presuppositions for Primary Production

Factors and processes influencing primary production in acidic mining lakes are schematically summarized (Fig. 8.3) according to their availability and, furthermore, to their inhibiting or promoting effects on algal biomass development. These aspects are key questions in order to find out whether primary production in acidic lakes is limited by the availability of resources or controlled by the acidity of the standing waters.

Dissolved inorganic carbon is one prerequisite to maintain a certain level of autotrophic production that is the basis for the development of a more or less complex food web structure in balanced systems. In naturally neutral lakes, the primary producers are mainly photoautotrophic organisms dominated by phytoplankton or macrophytes, depending on the trophic state. Chemoautotrophy is negligible or only important at boundary layers in meromictic lakes.

8.4.2.1
Dissolved Inorganic Carbon (DIC)

The concentration of DIC in the epilimnion of extremely acidic lakes ($pH < 3.5$) is very low and has generally been below the level of detection (< 0.5 mg C/l). Therefore, a limitation of primary production by inorganic carbon in most of the extremely acidic lakes is assumed (Schindler and Holmgren 1971; Goldman et al. 1974). The studies of Ohle (1981) in a man-made acidic pond gave evidence that original carbon dioxide limits the

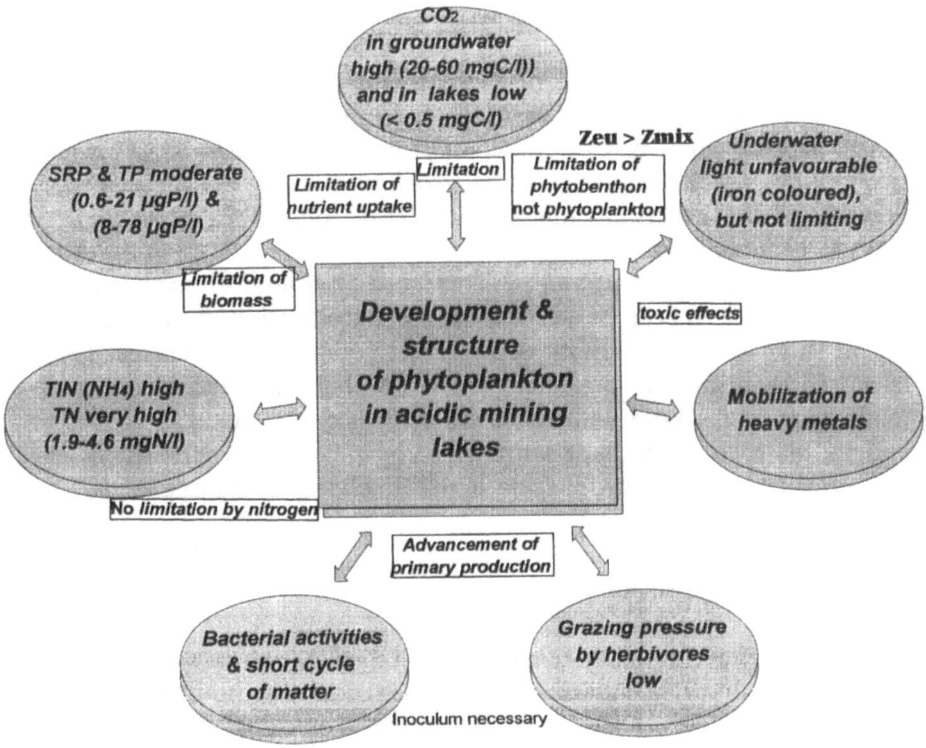

Fig. 8.3. Schematic presentation of presuppositions for planktonic primary production in acidic mining lakes. *Zeu* Euphotic depth; *Zmix* mixing depth; *SRP* soluble reactive phosphorus; *TP* and *TN* total phosphorus and nitrogen, respectively; *TIN* total inorganic nitrogen

natural intensity of photosynthesis and that 2 or 3 mg/l additional CO_2 support the activity vigorously. This is in contrast to the results of Pietsch (this Vol.) who calculated considerable amounts of carbonic acid in the lakes of the Lusatian region. On the other hand, the DIC concentration in the hypolimnetic water may be high as a consequence of DIC-rich groundwater flooding or a high intensity of the decay of organic matter. There are indications that CO_2-concentrations in hypolimnetic waters are higher because of the more intensive decomposition and, therefore, a more intensive respiration in these layers. During the extremely cold winter of 1995/1996, there were also higher concentrations of inorganic carbon under ice because of the mineralization processes and prevention of gas exchange.

Contrary to the pelagic water, the concentration of inorganic carbon in the groundwater in some regions of Lusatia is high (Schlabendorf, near lake F-Loch, between 60 and 80 mg C/l). These samples are taken from the anoxic groundwater which is normally neutral and allows the establishment of a lime–carbonic acid equilibrium. After aeration and pyrite oxidation, it becomes acidic, but it is unclear when and where these great amounts of CO_2 are diminished or lost while flooding the mining holes. Further tests to detect metabolic pathways of inorganic carbon in these waters are still being conducted.

Occasionally, there must be considerable amounts of inorganic carbon in the investigated lakes because of the phytoplankton peaks, even in extremely acidic lakes (F-Loch, Plessa) or moderately acidic lakes (Felixsee; Mischke et al. 1995).

8.4.2.2
Nutrients (Phosphorus, Nitrogen and Silica)

In Table 8.2 the range of fractions of the most important nutrients is shown for the period from the middle of 1993 to early summer 1995. Total phosphorus (TP) as well as soluble reactive phosphorus (SRP), which are considered to be the most important limiting factors for primary production in natural lakes, are at a low level in all lakes, but are sufficient to allow a synthesis of algal biomass at a mesotrophic level or even at a eutrophic level in F-Loch. This is in contrast to the recent hypothesis that most acid-sensitive lakes are phosphorus-limited. Even small changes in phosphorus would have a major impact on productivity. However, since we do not have long-term data about the nutrient dynamics in geogenically acidified lakes, the problem of the phosphorus limitation of primary production will remain unsolved.

Total nitrogen (TN) is considerably higher compared with TP. The main reason is the high concentration of ammonia because of inhibition

Table 8.2. Soluble reactive phosphorus (SRP) and total phosphorus (TP), sum of nitrate and nitrite nitrogen (NO_t), total nitrogen (TN), dissolved inorganic silica (DSi) and dissolved organic carbon (DOC) in acidic lakes and neutral lake B-loch in the Lusatian region from 1993 to 1995

Lake	SRP (µg P/l)	TP (µg P/l)	NO_t (µg N/l)	NH_4 (mg N/l)	TN (mg N/l)	DSi (mg Si/l)	DOC (mg C/l)
Felixsee	0.6–12.4	9–17	430–880	0.99–1.24	1.9–2.5	0.2–2.2	0.9–3.3
Waldsee (epi)	1.9–20.4	18–25	20–58	1.01–6.50	0.8–6.5	0.4–19.5	0.9–1.4
F-Loch	6.6–21.5	27–78	100–260	1.20–3.20	2.2–2.3	14.7–21.7	1.7–4.1
Plessa	4.9–8.0	<8	200–900	3.90–4.20	1.8–4.6	0.8–19	<1–1.7
B-Loch	0.6–3.5	8–9	3–29	0.08–0.72	1.2–1.3	0.08–1.46	4.7–6.4

of nitrification processes in acidic waters. Nitrate as the dominant fraction of the NO_t component is low in the neutral mesotrophic lake B-Loch and in the epilimnion of Waldsee, whereas it is at a moderate or high level in the remaining acidic lakes (Table 8.2). Apart from the latter examples, no limitation of primary production by nitrogen is expected.

There also appears to be enough silica available for the development of diatoms and chrysophytes in all lakes. Extremely high amounts were detected in the very acidic waters. With regard to nutrient limitation, the data permit the conclusion that primary production in a mesotrophic range would be possible. Only in the lake B-Loch could a limitation due to phosphorus deficiency be expected.

8.4.2.3
Energy Sources for Primary Production

8.4.2.3.1
Underwater Light

A rough estimation of the turbidity of waters is the measurement of the Secchi depth. In all mining lakes with $pH > 4$, including non-acidic lakes, we found clear water with Secchi depths between 5 and 10 m (Felixsee, Katjasee, Helenesee, B-Loch; Nixdorf and Kühne 1998). These lakes are well supplied with light and the euphotic depth often reaches the bottom of the lakes. Depending on the depth of thermocline, this underwater light climate allows algal growth in the metalimnion or hypolimnion (Nixdorf et al. 1995). Extremely acidic lakes (F-Loch, Plessa) are influenced by the light attenuation of iron species. Secchi depths range between 1 and 3 m and the corresponding euphotic depths estimated by the attenuation of the photon flux density are the same or greater during the vegetation period. Considering these results, there will be no or a negligible limitation of primary production due to light resources in acidic water. Up till now, we have only found two lakes of meromictic character that may be colonized by pigmented phototrophic bacteria (Chlorobineae, Chromaticeae) because of the sufficient light supply and H_2S at the chemocline.

8.4.2.3.2
Chemical Compounds Favouring Chemotrophy

The conditions for the growth of chemolitoautotrophic bacteria are relatively favourable in acidic mining lakes as far as availability of Fe- and S-compounds is concerned. Different species of *Thiobacillus* were reported to be responsible for or to propagate the pyrite oxidation. This process is of minor importance in the pelagial of the acidic lakes. Whether further chemoautotrophic organisms are involved in primary production in the lakes will be investigated in the near future.

8.4.3 Colonization with Organisms

8.4.3.1 *Phytoplankton*

It is apparent from our investigations that phytoplankton biomass is at a low level, mostly between 1 and 3 mg fw/l, which is higher than reported by the authors in the first hypothesis (see Sect. 8.1). In some cases we found considerably higher amounts of phytoplankton, exceeding 30 µg Chl a/l, especially in the hypolimnion (Kapfer et al 1997; Nixdorf et al. 1998), comparable with the biomass development in eutrophic lakes. These phenomena indicate the relatively high potential for primary production even in extremely acidic lakes such as F-Loch, Plessa and also in the moderately acidic Felixsee. In addition to the reported shift in algal communities due to acidification (see Sect. 8.1, hypotheses 2 and 3), our investigations include primary colonization by the taxa listed in Table 8.3.

Besides the importance of littoral primary producers (Kapfer, submitted), a distinct difference in the phytoplankton distribution was observed in several mining lakes: Phytoplankton tends to concentrate in certain layers of the water body (above the sediment in Lake Plessa; in or below the thermocline in the Felixsee and Helenesee). The relevance of this habitat has to be aggregated to hypothesis 4. Dixit and Smol (1989) also observed the development of *Chlorella* below the thermocline in an acidified lake. From these results we have to conclude that sampling only epilimnic water may lead to deficient results concerning algal colonization.

The relationship between pH and phytoplankton development and the regulation of primary production is still controversial (Dixit and Smol 1989), but it became evident that the pH effect is not direct. The limitation of primary production by availability and concentration of inorganic carbon and phosphorus and underwater light climate in iron-coloured waters appears to be more probable than the direct inhibition of processes by low pH. The process of P-precipitation due to high Al- and Fe-concentrations may be the reason for low SRP values in the lakes. One consequence is the high phosphatase activity of algae in acidic lakes (Jannson 1981). Another possibility to influence phytoplankton growth is

Table 8.3. Dominant phytoplankton taxa and species found in literature and in selected acidic lakes in the Lusatian region as a percentage of total phytoplankton biomass

Taxa and species of phytoplankton in acidic lakes reviewed by Dixit and Smol (1989)	Taxa and species found in the Lusatian region	Biomass (%)			
		B	F	Fs	Fx
Cyanophyta					
Merismopedia					
Aphanothece	*Aphanothece* (?)				
Pyrrophyta					
Dinophyceae:					
Gymnodinium uberrium	*Gymnodinium* sp.	14			27
Peridinium inconspicuum	*Peridinium umbonatum*				
	Ceratium hirundinella				
	Gyrodinium sp.				
Cryptophyta					
Chrysochromulina breviturrita	*Cryptomonas erosa*				
Euglenophyta	*Euglena mutabilis*				
Euglena sp.	*Trachelomonas* sp.				
Chrysophyta					
Chrysophyceae:			99	65	5
Chrysomonadales	*Ochromonas* sp.				
Dinobryon	*Synura sphagnicola*				
	Chromulina sp.				
Bacillariophyceae:		82			< 1
Asterionella ralfsii	*Achnanthes affinis*				
(var. *americanica* Korn – two	*Achnanthes bottnica*				
forms of this taxon;	*Fragilaria crotonensis*				
see Anderson et al. 1993)	*Frustulia rhomboides*				
Eunotia spp.	*Nitzschia* sp.				
	Stephanodiscus sp.				
	Synedra acus				
	Synedra affinis				
	Synedra nana				
	Synedra ulna				
	Synedra pulchella				
	Thalassiosira fluviatilis				
	Eunotia sp.				
Chlorophyta					
Volvocales:	*Chlamydomonas* sp.	3		33	65
Chlamydomonas sp.	*Carteria* sp.				
C. acidophila	*Scourfieldia cordiformis*				
C. sphagnophila	*Chlorogonium* sp.				
Chlorococcales:					
Chlorella sp.	*Chlorella* sp.				
	Schroederia sp.				
	Coelastrum microporum				
	Stichococcus sp.				

B, B-Loch; F, F-Loch; Fs, F-Loch south basin; Fx, Felixsee.

the leaching of trace metals in acidified waters (Graneli and Haraldson 1993). This influence can be stimulating or inhibiting, depending on the concentration.

8.4.3.2
Picoplankton

Autotrophic picoplankton, the algal fraction which plays an important role in the microbial food webs (Weisse 1993), decreases with decreasing pH. Compared with non-acidic lakes, bacterial cell numbers in acidic mining lakes are also lower. They are in a range from 0.5 to 2×10^6 cells per ml (Fig. 8.4). Reports about the occurrence of bacteria reflect the discrepancies in estimating the bacterial potential in acidic lakes. Scheider and Dillon (1976) found significant differences in bacterial numbers between acidified and non-acidified lakes in Canada, whereas Boylen et al. (1983) did not find any differences. Lenhard and Steinberg (1984) gave a complex view of the occurrence of organisms in acidic waters. The authors suggested, among other things, that methodological differences are one reason for these discrepancies. The range of the reported bacterial cell numbers is from 0.5 to 4×10^6 cells/ml. Our investigations in the Lusatian region supply the following results:

1. The number of bacteria is lower in extremely acidified lakes (see Fig. 8.4).
2. The shape of the bacteria in acidic lakes differs from that in natural eutrophic lakes in the Brandenburg/Berlin area (Köcher and Nixdorf 1994). The bacteria are larger and dominated by rods; filamentous forms are also common. Whether this phenomenon is induced by the grazing pressure of mixotrophic flagellates consuming the small bacteria or by the lack of zooplankton functioning as a grazer of larger particles or by other reasons will be investigated in the near future. Even in very acidic waters a relatively high bacterial production is possible (Nixdorf et al. 1995), similar to that in eutrophic lakes in Berlin/Brandenburg.

Mills et al. (1989) concluded from their inhibition and recovery experiments on microbial communities which were grown in acidic water (pH 2.9) and in neutral lakes (pH 6.5) and suspended in the opposite station that bacteria from extreme environments tend to be more often generalists compared with organisms from mesic (circumneutral) environments.

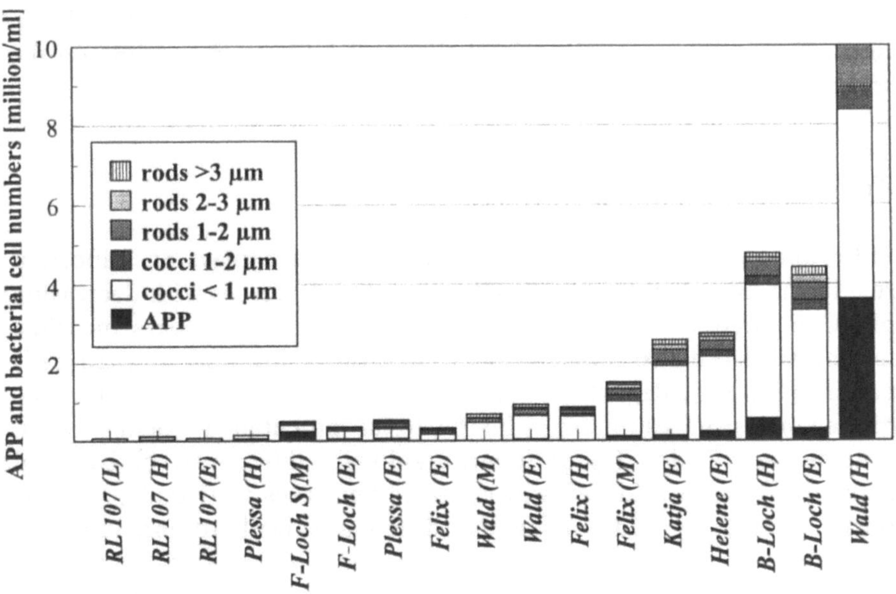

Fig. 8.4. Autotrophic picoplankton (*APP*) and heterotrophic bacteria in mining lakes of Lusatia. *L* Littoral; *E* epilimnion; *M* metalimnion; *H* hypolimnion. Lakes are arranged according to pH

8.4.3.3
Zooplankton and Insects

Up till now, we have found consumers in all of the investigated mining lakes. Even in lake RL 107, the most acidic lake (pH 2.3), there were a few pioneer species of ciliates, rotifers, chironomids and corixids (Table 8.4). The zooplankton community showed a very early stage of colonization (Table 8.5). Two other acidic mining lakes of higher pH [Waldsee (epilimnion): pH 3; Felixsee: pH 4] have already been colonized by crustaceans. All species found in the pelagic environment of acidic mining lakes belong to the group of ubiquitous benthic or littoral zooplankton. Characteristic rotifer and crustacean species are *Cephalodella hoodi*, *Elosa worallii*, *Brachionus sericus* and *Chydorus sphaericus*. All lakes investigated were inhabited by the omnivore corixid *Sigara nigrolineata*, which seems to be a typical species of mining lakes of the Lusatian region. Generally, species richness of corixids and rotifers seems to be similar in acidic mining lakes. The zooplankton of neutral mining lakes of low productivity (B-Loch, Helenesee, Katjasee) in comparison with acidic ones is completely different and much more diverse (Table 8.4).

Table 8.4. Zooplankton and insects found in different lakes of the Lusatian region

Lakes	Characteristic taxa and species		
	Rotifers	Crustaceans	Others
RL 107	*Cephalodella hoodi*	–	Ciliates, Chironomidae (plumosus group) *Sigara nigrolineata*
Plessa	*Cephalodella hoodi*, *Elosa worallii*	–	Chironomidae (plumosus group) *Sigara nigrolineata*, *Glaenocorisa p. propinqua*, *Callicorixa praeusta*, *Sigara striata*, *Corixa dentipes*, Hydracarina
F-Loch	*Cephalodella hoodi*, *Cephalodella gibba*, *Elosa worallii*, *Brachionus sericus*	– –	Ciliates, Chironomidae *Arctocorisa germari*, *Sigara nigrolineata*
Waldsee	*Cephalodella hoodi*, *Cephalodella gibba*, *Elosa worallii*, *Brachionus sericus*, Bdelloidae	*Chydorus sphaericus*	Ciliates, gastrotrichs, *Sigara nigrolineata*, *Sigara semistriata*, *Sigara distincta*, *Callicorixa praeusta*, *Glaenocorisa p. propinqua*, *Corixa dentipes*, *Iliocoris cimicoides*, *Bezzia* sp., Dytiscidae
Felixsee	*Cephalodella hoodi*, *Cephalodella gibba*, *Elosa worallii*, *Brachionus sericus*, Bdelloidae	*Chydorus sphaericus*, *Diacyclops* sp.	Ciliates, *Arctocorisa germari*, *Glaenocorisa p. propinqua*, *Callicorixa praeusta*, *Sigara nigrolineata*, *Sigara distincta*, *Corixa dentipes*, Micronectinae, Dytiscidae, Zygoptera larvae, Hydracarina
B-Loch	*Keratella cochlearis*, *Filinia terminalis*, *Kellicottia longispina*, *Keratella quadrata*, *Polyarthra* spp.	*Thermocyclops oithonoides*, *Ceriodaphnia* spp., *Daphnia* spp., *Bosmina longirostris*, *Cyclops* sp.,	Ciliates, *Chaoborus flavicans*, *Sigara striata*, *Sigara lateralis*, *Sigara nigrolineata*, *Corixa punctata*, Micronectinae,

Table 8.4 (continued)

Lakes	Characteristic taxa and species		
	Rotifers	Crustaceans	Others
B-Loch		*Eudiaptomus gracilis*	*Notonecta glauca, Iliocoris cimicoides,* Hydracarina, Gammarus spec., Chironomidae, Dytiscidae
Katjasee	*Keratella cochlearis, Conochilus unicornis, Kellicottia longispina, Gastropus stylifer, Polyarthra* spp., *Filinia terminalis*	*Eudiaptomus gracilis, Bosmina longirostris, Daphnia cucullata, Thermocyclops oithonoides, Polyphemus pediculus*	Chironomidae
Helenesee	*Keratella cochlearis, Polyarthra* spp., *Pompholyx sulcata, Keratella quadrata, Kellicottia longispina, Gastropus stylifer*	*Ceriodaphnia* spp., *Bosmina longirostris, Daphnia cucullata, Eudiaptomus gracilis, Thermocyclops oithonoides*	Not investigated

Table 8.5. Stages and pH thresholds for the colonization of the pelagic zone by zooplankton and insect larvae

Stage	Zooplankton groups	pH
I	Ciliates, rotifers, chironomids, corixids	2–3
II	Colonization by small cladocerans	>3
III	Colonization by copepods	>4
IV	Colonization by daphnids; fish possible	>5
V	Normal species assemblage	>6

Our results show that zooplankton and insect larvae are present at a pH of about 2. With rising pH spieces richness increases and very simple food webs develop and become more complex. Stages in the development of the zooplankton community are proposed referring to threshold values for the occurrence of particular taxonomic groups (Table 8.5). An important step is the first occurrence of crustaceans (*Chydorus sphaericus*) at a pH of about 3.

The colonization of the littoral environment may differ markedly from the pelagic environment. Here, littoral or benthic consumers such as chironomids and corixids dominate rather than acid-adapted specialists. The reasons we found were a high tolerance of acid, the lack of competitors and possibilities of colonization from other lakes in the region. The lack of fish at a pH below 5 is another biological factor for a shift in food web interactions. Compared with natural lakes, the importance of predator–prey relations in acidic lakes are not only changed, but also reduced to a low level (Eriksson et al. 1980). We found the corixid *Glaenocorisa p. propinqua* in several lakes (Table 8.4). These predatory invertebrates are able to control planktonic food webs as top predators in acidic lakes (Henrikson and Oscarson 1981). The change from fish to insect predators was observed in atmospherically acidified lakes where this shift brought about a change in zooplankton community towards smaller species. While *G. p. propinqua* is known to be an efficient predator of all kinds of zooplankton, the main source of food in the investigated lakes seemed to be the chironomid larvae, which were detected in considerable amounts. Only in Lake Felixsee were crustaceans part of the diet of predatory corixids. It is important to investigate the consumption basis of chironomids and the share of omnivore corixids in their function as detritus consumers because of the generally low autochthonous production level and, consequently, the low detritus formation.

8.4.4
Acidified Lakes, Neutralization Capacities and Eutrophication

Different biological reactions increase pH or alkalinity (primary production, reduction of nitrate, manganese, iron and sulphate, fermentation of amino acids, methanogenesis; Mills et al. 1989; Hermann 1994). It is obvious that a number of biochemical and/or biological processes can lead to a remarkable increase in alkalinity and pH, respectively. In this chapter, attention is paid to one of the in-lake restoration processes: primary production as a natural potential for the neutralization of acidic lakes. How fast an acidified lake can respond to neutralization measures depends on the base capacity, estimated by $K_{B\,4.3}$ (Fig. 8.2). High neutralization capacity is needed for a number of acidic lakes (Bärwalde, Skado, Lohsa), whereas some lakes will respond after the addition of only a few quantities of alkaline substances (Senftenberger See, Gräbendorf, Koschen). It is known of Senftenberger See and the river treatment plant "Laubusch" that the load of organic and inorganic substances due to water from the river Schwarze Elster or wastewater was sufficient to induce neutral conditions in this geogenically acidified lake because of lake internal processes (Benndorf 1994; Klapper et al., this Vol.). The important role of nutrient additions while restoring acidified lakes in Sweden was shown by Blomquist et al. (1993) who found that nutrient additions may be necessary in addition to liming. Comparing the data from Davison (1987) concerning the actually introduced amount of phosphate and the appropriate amount of lime for the neutralization of an acidic lake, it becomes evident that in-lake measures have to be applied. Phosphorus addition has been shown to be beneficial to the restoration of acidified waters by promoting the biological production, altering the balance of carbon dioxide metabolism and invasion and generating hydroxyl ions (Reynolds 1992). This ecotechnology is one of the in-situ measures to support maturation mechanisms in the lake by the addition of nutrients and the increase in primary production by controlled eutrophication. Thus, one has to take care that eutrophication mechanisms and consequences are regulated regarding the later use of the lakes.

One topic of the International Workshop on Abatement of Geogenic Acidification in Mining Lakes, 4–6 September 1995, Magdeburg, Germany, (Group No. 1) was the discussion about classification and natural succession or natural recovery of acidified lakes. Should or should we not wait? Treatment of acidified lakes is possible by a number of measures (Klapper et al., this Vol.) including addition of neutralizing chemicals, phosphate or establishing anaerobic conditions using organic substrates. Most measures may be successful in neutralizing processes, but they involve a risk of eutrophication. Because ferric iron precipitates phosphate in lakes with a $pH < 3$, this risk of eutrophication becomes evident in lakes with a higher pH. As was shown in our investigations, there is a high potential for primary production even in the extremely acidic lakes (Nixdorf et al. 1997). Therefore, one aim of all neutralization measures which use nutrients and organic substrates to force primary production should be the control of the eutrophication process. Regarding the high variability of phytoplankton development depending on primary production resources, eutrophication has to be kept at a low level. What amount of available TP resources is converted into algal biomass depends on a complex of abiotic and biotic factors and processes (Nixdorf and Deneke 1997). Compared with the natural lakes in eastern Brandenburg, the mining lakes in Lusatia have a good chance of avoiding the effective conversion of resources. The mining lakes are deep, with a great hypolimnion. This means favourable presuppositions for lower intensity of primary production at given nutrient loads. A further reason for a better control of eutrophication in mining lakes is the characteristic of the catchment area where diffuse and point nutrient loadings are low.

8.5 Conclusions

The lakes which have been arising in Lusatia are quite different in morphometry, area, chemistry and further abiotic characteristics. No regular patterns can be derived for the classification according to their age, their acidity and the local geological conditions. In contrast to the results of Almer et al. (1974), Kwiatkowski and Roff (1976) and Yan (1979), phytoplankton biomass can reach up to 3 mg fw/l and considerably more in certain layers or under certain conditions with a favourable nutrient supply. These phenomena indicate that the limnetic ecosystems are very sensitive and may respond to drastic changes in abiotic influences with intensive biotic developments. A moderate promotion of biological processes, including a higher intensity of food web relations, by ecotechnological measures to produce more alkalinity is recommended.

Acknowledgements. We would like to thank Gudrun Lippert who was responsible for the hydrochemical program. Furthermore, we wish to thank Ingo Henschke and Wolfgang Terlinden for their careful work in the field.

References

Almer B, Dickson W, Ekström C, Hörnström E, Miller U (1974) Effects of acidification on Swedish lakes. Ambio 3 (1) : 30 – 36

Anderson DS, Davis RB, Ford MS (1993) Relationship between sedimented diatom species (Bacillariophyceae) to environmental gradients in dilute northern New England lakes. J Phycol 29 (3) : 264 – 277

Benndorf J (1994) Sanierungsmaßnahmen in Binnengewässern: Auswirkungen auf die trophische Struktur. Limnologica 24 (2) : 121 – 135

Blomquist P, Bell RT, Olofsson H, Stensdotter U, Vrede K (1993) Pelagic ecosystem responses to nutrient additions in acidified and limed lakes in Sweden. Ambio 22 (5) : 283 – 289

Blouin AC (1989) Patterns of plankton species, pH and associated water chemistry, in Nova Scotia lakes. Water Air Soil Pollut 46 (1 – 4) : 343 – 358

Boylen CW, Shick MO, Roberts DA, Singer R (1983) Microbiological survey of Adirondack lakes with various pH values. Appl Environ Microbiol 45 : 1538 – 1544

Davison W (1987) Internal element cycles affecting the long-term alkalinity status of lakes: implications for lake restoration. Schweiz Z Hydrol 49 : 186 – 201

Deutsche Einheitsverfahren zur Wasser-, Abwasser- und Schlammuntersuchung (1986 – 1996). Verlag Chemie, Weinheim

Dixit SS, Smol JP (1989) Algal assemblages in acid-stressed lakes with particular emphasis on diatoms and chrysophytes. In: Rao SS (ed) Acid stress and aquatic microbial interactions. CRC Press, Boca Raton, pp 91 – 114

Duty HC, Ostrofsky ML (1974) Plankton chemistry, and physics of lakes in the Churchill Falls region of Labrador. J Fish Res Board Can 31 : 1105

Eriksson MOG, Henrikson L, Nilsson BI, Nyman G, Oscarson HG, Stenson AE (1980) Predator-prey relations, important for biotic changes in acidified lakes. Ambio 9(5):248–249

Goldman JC, Oswald WJ, Jenkins D (1974) The kinetics of inorganic carbon-limited algal growth. J Water Pollut Control Fed 46(3):554–574

Graneli E, Haraldson C (1993) Can increased leaching of trace metals from acidified areas influence phytoplankton growth in coastal waters? Ambio 22(5):308–311

Henrikson L, Oscarson HG (1981) Corixids (Hemiptera-Heteroptera), the new top predators in acidified lakes. Verh Int Ver Limnol 21:1616–1620

Hermann R (1994) Die Versauerung von Oberflächengewässern. Limnologica 24(2): 105–120

Jannson M (1981) Induction of high phosphatase activity by aluminium in acid lakes. Arch Hydrobiol 93(1):32–44

Kapfer M. Colonisation and primary production of microphytobenthos in the littoral of acid mining lakes in Lusatia (Germany). Water, Air & Soil Pollution (submitted)

Kapfer M, Mischke U, Wollmann K, Krumbeck H (1997) Erste Ergebnisse zur Primärproduktion in extrem sauren Tagebauseen der Lausitz. In: Deneke R, Nixdorf B (Hrsg) Gewässerreport (Teil III). BTU Cottbus Aktuelle Reihe 5:31–40

Köcher B, Nixdorf B (1994) Bakterien und autotrophes Picoplankton in natürlichen und künstlichen Seen der Region Berlin/Brandenburg. Deutsche Gesellschaft für Limnologie, Erweiterte Zusammenfassungen, Coburg, pp 284–288

Kwiatkowski RE, Roff JC (1976) Effects of acidity on the phytoplankton and primary productivity of selected northern Ontario lakes. Can J Bot 54:2546–2561

Lenhard B, Steinberg C (1984) Limnochemische und limnobiologische Auswirkungen der Versauerung von kalkarmen Oberflächengewässern. Inform Bayer Landesamt Wasserwirtschaft 4/84

Mills AL, Bell PE, Herlihy AT (1989) Microbes, sediments, and acidified water: the importance of biological buffering in acid stress and aquatic microbial interactions. In: Rao SS (ed) Acid stress and aquatic microbial interactions. CRC Press, Boca Raton, pp 1–20

Mischke U, Rücker J, Kapfer M, Nixdorf B (1995) Besiedlungsstruktur und Interaktionen im Plankton geogen versauerter Tagebaurestseen der Lausitz. Deutsche Gesellschaft für Limnologie, Erweiterte Zusammenfassungen, Hamburg, pp 700–704

Nixdorf B, Deneke R (1997) Why very shallow lakes are more successful opposing reduced nutrient loads. Hydrobiologia 342/343:269–284

Nixdorf B, Hoeg S (1993) Phytoplankton-community structure, succession and chlorophyll content in Lake Müggelsee from 1979 to 1990. Int Revue Ges Hydrobiol 78(3):359–377

Nixdorf B, Leßmann D, Grünewald U, Uhlmann W (1997) Limnology of extremely acidic mining lakes in Lusatia (Eastern Germany) and their fate between acidity and eutrophication. Proc Conf on Acid rock drainage, Canada 1997, Vol IV, pp 1745–1760

Nixdorf B, Mischke U, Leßmann D (1998) Chrysophyta and Chlorophyta – pioneers of planktonic succession in extremely acidic mining lakes in Lusatia. Hydrobiologia (in press)

Nixdorf B, Kühne M (1998) Besonderheiten im Stoffhaushalt künstlicher Klarwasserseen Südostbrandenburgs (Tagebauseen der Lausitz) – ein Überblick. Beiträge zur Gewässerökologie Norddeutschlands. Sonderheft Klarwasserseen (submitted)

Nixdorf B, Rücker J, Köcher B, Deneke R (1995) Erste Ergebnisse zur Limnologie von Tagebaurestseen in Brandenburg unter besonderer Berücksichtigung der Besiedlung im Pelagial. In: Geller W, Packroff G (eds) Abgrabungsseen - Risiken und Chancen. Limnologie Aktuell, no 7. Fischer, Jena, pp 39-52

Ohle W (1981) Photosynthesis and chemistry of an extremely acidic bathing pond in Germany. Verh Int Ver Limnol 21:1172-1177

Porter KG, Feig YS (1980) The use of DAPI for identifying and counting microflora. Limnol Oceanogr 25(5):943-948

Reynolds CS (1992) Eutrophication and the management of planktonic algae: what Vollenweider couldn't tell us. In: Sutcliffe DW, Jones JG (eds) Research and application to water supply. Freshwater Biol Assoc, Ambleside, pp 5-29

Scheider W, Dillon P (1976) Neutralization and fertilization of acidified lakes near Sudbury, Ontario. Water Pollut Res Canada 11:93-100

Schindler DW, Holmgren SK (1971) Primary production and phytoplankton in the Experimental Lakes Area, northwestern Ontario, and other low carbonate waters, and a liquid scintillation method for determining C14 activity in photosynthesis. J Fish Res Board Can 28:189

Schultze M, Klapper H, Nixdorf B, Mischke U, Grünewald U (1994) Methodik zur limnologischen Untersuchung und Bewertung von Bergbaurestseen. Bund-Länder Arbeitsgruppe Wasserwirtschaftliche Planung, Berlin

Steinberg CEW, Wright RF (eds) (1994) Acidification of freshwater ecosystems - implications for the future. Dahlem Workshop Report, ESR 14. Wiley, Chichester

TGL 27885/01 (1982) Fachbereichstandard-Nutzung und Schutz der Gewässer-Stehende Binnengewässer, Klassifizierung, DDR. Berlin

Utermöhl H (1958) Zur Vervollkommnung der quantitativen Phytoplanktonmethodik. Mitt Int Verein Limnol 9:1-38

Weisse T (1993) Dynamics of autotrophic picoplankton in marine and freshwater ecosystems. In: Jones JG (ed) Advances in microbial ecology, vol 13. Plenum, New York, pp 327-370

Yan ND (1979) Phytoplankton community of an acidified, heavy metal-contaminated lake near Sudbury, Ontario: 1973-1977. Water Air Soil Pollut 11:43-55

9 Colonization and Development of Vegetation in Mining Lakes of the Lusatian Lignite Area Depending on Water Genesis

W. Pietsch

Brandenburg Technical University of Cottbus, Faculty of Environmental Sciences, Chair of Soil Protection and Recultivation (Special Recultivation), P.O. Box 101344, 03013 Cottbus, Germany

9.1 Introduction

Before mining activities started in the Lusatian lowlands, this area was regarded as a region where Atlantic–sub-Atlantic aquatic plants of the class Littorelletea uniflorae were commonly and frequently distributed at their eastern dispersal border (Barber 1893). Representatives of the oceanic-acidophile littoral associations are particularly common in the shallow heath and bog waters and in fish ponds of the Lusatian region. They include the following species: *Apium inundatum, Deschampsia setacea, Eleocharis multicaulis, Pilularia globulifera, Juncus bulbosus, Luronium natans, Potamogeton polygonifolius, Eleogiton fluitans, Hypericum elodes* and *Utricularia ochroleuca.*

The eutrophic old branches of the River Schwarze Elster and fish ponds with subacidic to alkaline water conditions used to be typical sites of true aquatic plants of the classes Potametea and Lemnetea. Younger stages were characterized by *Trapa natans* and older waters in the stage of silting-up by floating mats of *Stratiotes aloides* and *Hydrocharis morsus-ranae*. Species of acidophile Atlantic Littorelletea associations, particularly of the order Juncetalia bulbosi, did not occur in these waters.

Due to intensive mining activities between 1920 and 1980, the central parts of this former area – located in the districts of Senftenberg, Hoyerswerda and Weisswasser – were completely destroyed. The Atlantic aquatic plant species could only survive in the peripheral areas of the former Lusatian region, especially in the fish and heath ponds and in the heath bogs of the western Senftenberg district, particularly in the surroundings of Ruhland. The sites of true aquatic plants were completely destroyed.

For a long time, the mining lakes and residual waters of the Lusatian lignite mining district have been regarded as dead, bare waters of wasteland having a growth-hostile character. While data on the colonization of terrestrial areas, dumps and mine spoils of the post-mining landscape and the occurrence of various plant species were available, information on the colonization of water areas was completely missing.

First comprehensive studies on the ecological and hydrochemical quality of 158, and later 219, residual waters of the Lusatian lignite mining district were made by Pietsch (1970, 1973, 1979a, b), who also recorded floristic structures of the aquatic macrophyte vegetation. For the first time, a close correlation was found between the development of vegetation and the regular metamorphic processes of physico-chemical conditions taking place in the water bodies and at the bottoms of mining lakes and residual waters. First data on the distribution of aquatic plants and the development of vegetation in residual waters of the Lusatian lignite mining district originate from Pietsch (1965, 1970, 1973) and Heym (1971).

9.2 Materials and Methods

9.2.1 Location of the Study Area

The area of the Lusatian lignite mining district is situated in the extended physical units of the Lower Lusatian basin and heathland and the Upper Lusatian heathland. This strip of a plain extends between the Lusatian Frontier Wall and the marginal edge of the low mountain rift which differs from the adjacent Elbe–Mulde lowlands by a lower extent of clearance. In addition to the plains of the River Schwarze Elster, wide sandy areas in the valleys, followed by hilly, sandy regions are predominant. Especially the central part is characterized by the glacial valley of the River Schwarze Elster with its wide sand terraces. The area is surrounded by the towns Senftenberg, Luckau, Cottbus, Hoyerswerda, Spremberg and Weisswasser. The northern part belongs to the State of Brandenburg, the southern part of Saxony.

9.2.2 Methods of Water Analysis

For hydrochemical analyses, surface water was taken from a depth of 1 m in various seasons. The following parameters were analysed in the laboratory: pH value, alkalinity, total and carbonate hardness, free dissolved carbonic acid (CO_2), total dry residue, electrolyte content and the main cations and anions (sodium, potassium, calcium, magnesium, iron, manganese, aluminium, sulphate, chloride, silicate and hydrogen carbonate). In addition, ammonium, nitrate, nitrite, phosphate and the content of brown matter ($KMnO_4$ mg/l), oxygen and oxygen saturation were determined. For quantitative analyses, the usual laboratory methods were applied.

9.2.3 Nomenclature of Aquatic Macrophyte Species and Aquatic Plant Associations

Macrophyte species of mining lakes and residual waters were recorded and specified according to the nomenclature of *Rothmaler's Exkursionsflora*, Vol. 4 (1994). Plant stands and associations of the acid residual waters were termed after Pietsch (1973). The Braun–Blanquet method was used to record the conditions of vegetation.

9.3 Criteria of Colonization

In residual mining waters, the geological conditions of dumped and naturally occurring soil substrates are of major importance. They are decisive for the hydrochemical quality of the water bodies. Morphometric conditions, water depth and total volume determine the natural aging of waters, including time sequence and the pattern of natural developmental processes. The colonization by aquatic, marsh and reed plants is based on the slope angle of the developing littoral and shallow water areas.

The formation of water types and the colonization by macrophytes and their associations depend on the following criteria which are important for the specification and development of water types:

- Quality and geological origin of dumped substrates (marcasite and pyrite);
- Chemical quality of water bodies;
- Depth of residual waters and mining lakes;
- Slope angle of littoral zones dependent on hydraulic fills and dumps created by spreaders;
- Leaching and alluviation processes of surrounding mine spoil substrates, especially the high sulphuric acid potentials of adjacent mine spoils;
- Effect of various interfering factors, particularly erosion phenomena;
- Content of free dissolved carbonic acid (CO_2) and combined carbonic acid (HCO_3) in the water body which is subject to pH-dependent solubility distributions.

9.4 Range of Species in Mining Lakes and Residual Waters

Since 1963 we have studied 234 mining lakes and residual waters in the Lusatian lignite mining district. Beyond the determination of hydrochemical water body conditions, the development of vegetation on virgin soil substrates was recorded, where a total of 36 aquatic plant species and 16 species of reeds and high sedges were found. Table 9.1 summarizes the total of 52 aquatic macrophytes in the order of occurrence. Moreover, the percentage of occurrence in the total number of 234 waters is given.

Juncus bulbosus is the most common species and of main importance for the colonization of mining and residual lakes. This species always initiates the plant colonization in mining waters and forms extensive floating and submersed mats (Pietsch 1965). The distribution of *Juncus bulbosus* is nearly equivalent to the frequency distribution of hydrochemical factors since this species is present in 213 of the 234 waters included in the study (91%). Of these 234 waters, 218 are colonized by aquatic macrophytes. Sixteen residual waters are free of macrophytes. These are, on the one hand, very young waters of the initial stage which developed only recently and did not allow any plant colonization because of their extreme water qualities. On the other hand, this group includes older mining lakes of adverse geomorphological conditions, especially very steep banks, making the development of vegetation impossible.

The reed species *Phragmites australis* and *Typha latifolia* occur in 75% of the waters and often form extensive stands. Table 9.1 summarizes 52 aquatic macrophyte species which are allocated to the following classes of European aquatic plant associations:

– Littorelletea associations	12
– Associations of Utricularietea intermedio-minoris	8
– Potamogetonetea associations	10
– Lemnetea layers	2
– Charetea associations	2
– Phragmitetea associations	16
– Associations of Scheuchzerio-Caricetea nigrae	2

9.5 Developmental Process and Plant Colonization

Data on the ecological water situation and knowledge of the physicochemical quality variations in water bodies and sediments as a result of metamorphic processes provide the best basis to understand the colonization of mining lakes and residual waters by aquatic macrophytes.

Table 9.1. Occurrence of 52 aquatic macrophyte species in 234 mining lakes and residual waters of the Lusatian lignite mining district

Plant species	Total occurrence in 234 waters	In %
Aquatic plant species:		
Juncus bulbosus	213	91.0
Potamogeton natans	102	43.6
Nymphaea alba	88	37.7
Utricularia minor	64	27.4
Pilularia globulifera	62	26.5
Sparganium minimum	53	22.6
Utricularia ochroleuca	48	20.5
Utricularia australis	46	19.7
Eleocharis acicularis	42	17.9
Utricularia intermedia	41	17.5
Myriophyllum heterophyllum	36	15.4
Sphagnum cuspidatum	32	13.7
Lemna minor	31	13.2
Potamogeton polygonifolius	29	12.4
Sphagnum inundatum	29	12.4
Myriophyllum spicatum	29	12.4
Deschampsia setacea	22	9.4
Polygonum amphibium	21	9.0
Ranunculus flammula	18	7.7
Hydrocotyle vulgaris	18	7.7
Potamogeton crispus	17	7.3
Hottonia palustris	14	6.0
Potamogeton gramineus	14	6.0
Luronium natans	12	5.1
Elatine hydropiper	11	4.7
Littorella uniflora	9	3.8
Potamogeton berchtoldii	9	3.8
Eleocharis multicaulis	9	3.8
Elodea canadensis	8	3.4
Potamogeton alpinus	8	3.4
Carex oederi	6	2.6
Chara foetida	6	2.6
Spirodela polyrrhiza	5	2.1
Eleogiton fluitans	4	1.7
Potamogeton lucens	3	1.3
Chara hispida	2	0.9
Species of reed stands:		
Phragmites australis	186	79.5
Typha latifolia	168	71.8
Juncus effusus	136	58.1
Juncus conglomeratus	124	53.0
Carex rostrata	86	36.8

Table 9.1 (continued)

Plant species	Total occurrence in 234 waters	In %
Schoenoplectus lacustris	61	26.1
Eleocharis palustris	51	21.8
Typha angustifolia	42	17.9
Glyceria fluitans	28	11.9
Carex acutiformis	23	9.8
Alisma plantago-aquatica	16	6.8
Schoenoplectus tabernaemontani	12	5.1
Juncus articulatus	12	5.1
Sagittaria sagittifolia	9	3.8
Carex elata	8	3.4
Juncus acutiflorus	4	1.7

9.5.1 Ecological Characterization of Developmental Stages

Lignitization in the Lusatian coal mining district dates back to the Younger or sub-Sudetic formation of the Miocene period. In general, marine Oligocene overlays are absent. The Miocene layers contain large amounts of ferrous sulphidic material occurring as marcasite or pyrite. The sulphidic oxidation of marcasite and pyrite entails the formation of ferrous sulphate which, in turn, leads to the formation of bivalent iron and free mineral acid, especially sulphuric acid, where the latter is responsible for the reaction of groundwater, putting it into the extremely acid range. Therefore, the end of lignite mining is related to extremely growth-hostile site conditions in the residual waters. The water bodies are characterized by pH values between 1.9 and 2.3, free sulphuric acid (1.5 – 18.6 mmol/l H_2SO_4), very high iron quantities (45 – 620 mg/l $Fe^{2+/3+}$) and high sulphate contents (up to 1800 mg/l SO_4^{2-}). Therefore, the total hardness is in the range of "very hard" (up to 98 °GH).

Due to the oxidation of ferrous sulphidic compounds as marcasite and pyrite, larger amounts of free mineral acids are released, especially sulphuric acid which causes the extremely acid character of mining lake water bodies. The higher the pyrite content of the dumped virgin soil substrates, the higher the subsequent content of free mineral acids in the water bodies and the more extreme the developing acidity. Moreover, weathering entails the decomposition of sulphide and, hence, higher contents of sulphate, calcium and iron.

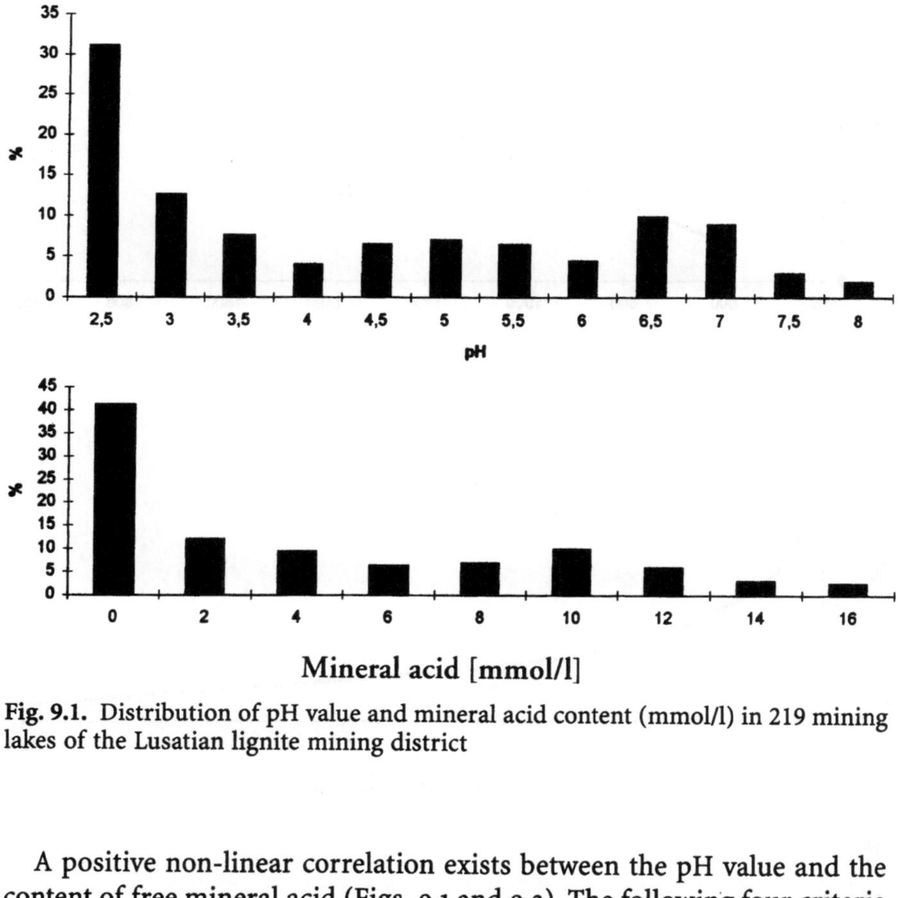

Fig. 9.1. Distribution of pH value and mineral acid content (mmol/l) in 219 mining lakes of the Lusatian lignite mining district

A positive non-linear correlation exists between the pH value and the content of free mineral acid (Figs. 9.1 and 9.2). The following four criteria specify the causes for the water body chemistry of mining lakes:

1. Weathering processes in the adjacent littoral area
2. Groundwater contact
3. Microbial processes – chemosynthesis of ferrous and sulphuric bacteria
4. Anthropogenic activities

In the initial stage, the ground and surface waters blended with pumped mine waters are extremely acid, with pH values between 1.9 and 2.9. In addition, they contain free mineral acid, particularly sulphuric acid. The water bodies contain large amounts of sulphate, iron and manganese; the total hardness is in the range between "very hard" and "extra hard". These are biologically sterile no-lime or low-lime calcium sulphate waters with negligible quantities of organic substances dissolved in water.

A positive linear correlation was found between the calcium and sulphate content on the one hand and the degree of total hardness on the

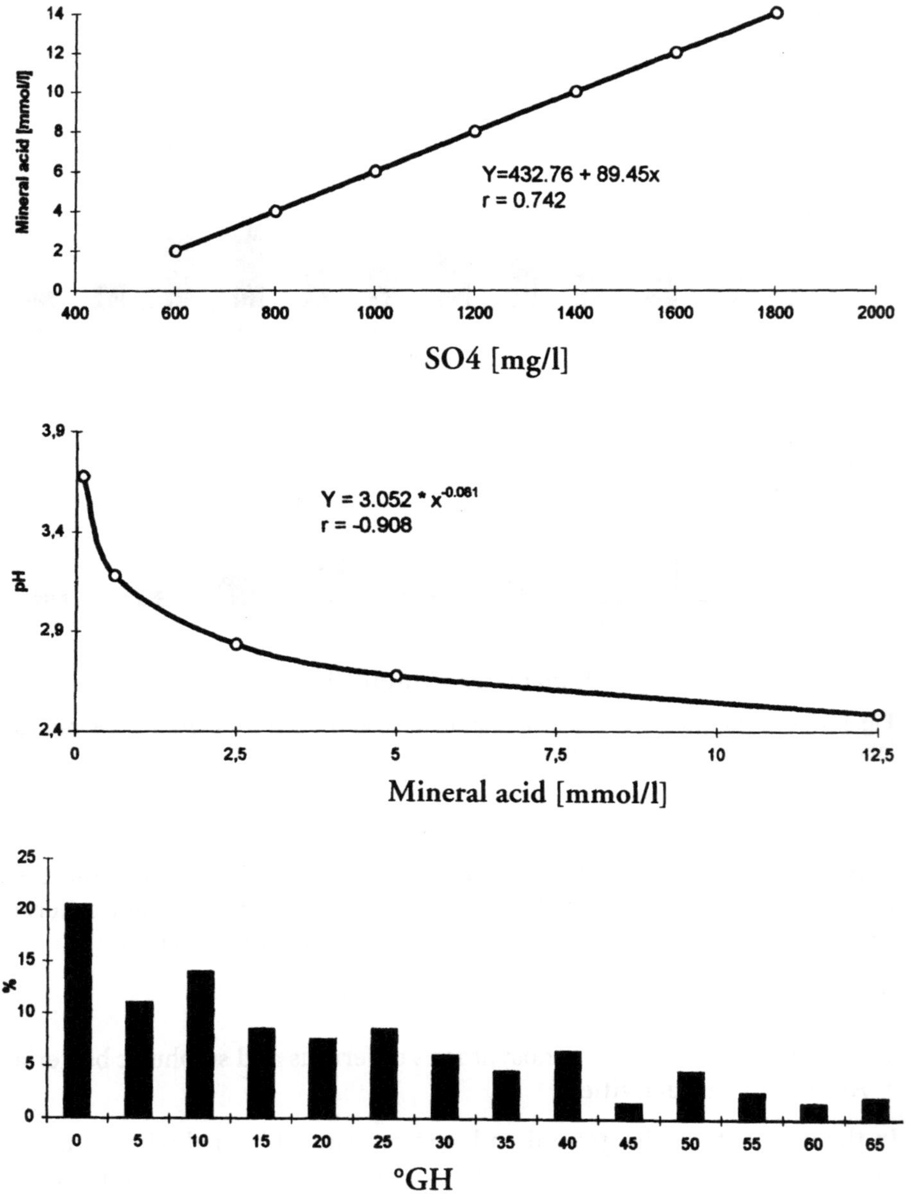

Fig. 9.2. Correlation between free mineral acid and sulphate, correlation between pH value and free mineral acid and distribution of the total hardness (°GH) in 219 mining lakes of the Lusatian lignite mining district

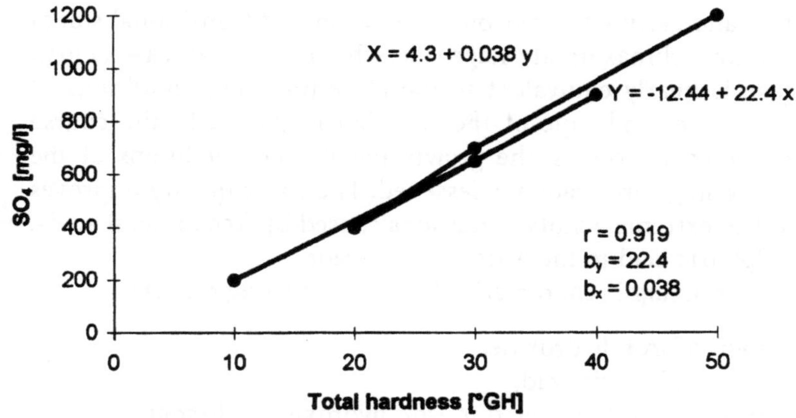

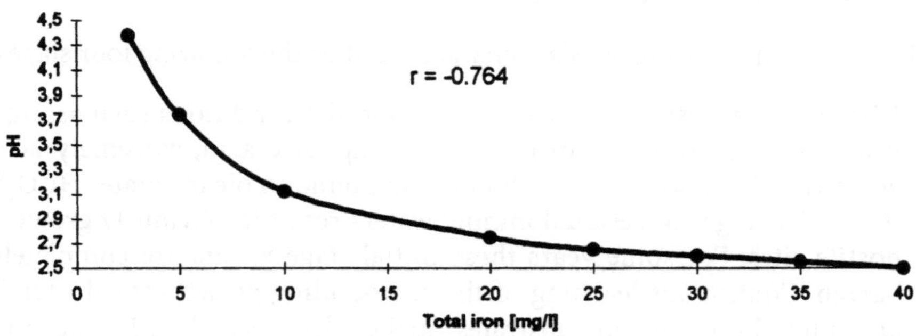

Fig. 9.3. Correlations between total hardness and sulphate content and between total iron and pH value in 219 mining lakes of the Lusatian mining district

other (Fig. 9.3). However, during the following decades a certain developmental stage is attained, where the residual waters are suddenly colonized by extensive mass stands of aquatic plant species. At the beginning, these species form monodominant stands of remarkable biomass production.

These species are obvious virgin soil pioneers colonizing the formerly growth-hostile sites of residual mining waters in the shallow littoral zone. *Juncus bulbosus* is the species which penetrates farthest into these waters and forms extensive submersed mats down to a water depth of 6 m. Southern and western parts of the mining lake Senftenberger See are not only covered by layers of dense, close mats, but also completely filled with these plants. In the course of decades, the physico-chemical conditions of mining lake water bodies and sediments in the Lusatian lignite mining district are subject to changes. In a developmental process, the waters pass an initial and early stage, various succession and transitional stages and finally attain a climax or late stage (Pietsch 1965, 1973, 1979 a–c, 1993). The late stage is nearly equivalent to the close-to-nature conditions of heath ponds in the landscape of the Lusatian lowlands. In the course of the developmental process, the growth-hostile site conditions of the initial and early stage are gradually lessened. The water quality improves remarkably. The extreme acidity conditions caused by free mineral acids, especially sulphuric acid in the water body, recede.

The process of aging is characterized by the following criteria:

- Sedimentation of iron hydroxide,
- Binding of free sulphuric acid,
- Absorption of free carbonic acid to iron hydroxide sediment,
- Sulphate elimination by binding of sulphate to calcium and sedimentation of $CaSO_4$ on the water bottom.

The process of metamorphosis is characterized by the following four stages:

1. *Waters in the initial stage.* Due to extreme site conditions such as high iron contents, large quantities of free sulphuric acid, extremely acid water reaction and lack of carbonic acid bound to bicarbonate (HCO_3), the initial stages of residual mining waters result in obviously growth-hostile sites. For some years these initial-stage waters are completely barren. Continuous leaching of the surrounding growth-hostile Tertiary sulphidic mine soil substrates further decreases the pH values to the extremely acid range (down to 1.9). The content of free mineral acids, mainly free sulphuric acid, dissolved in water is still steadily growing (up to 18.6 mmol/l). Iron content, total hardness and the offensive character of free carbonic acid (CO_2) are also increasing.
2. *Waters of the early stage.* These are waters of extremely acid water reactions with pH values between 1.9 and 4.0, large amounts of free sulphuric acid and high degrees of hardness (up to 98 °GH) which are caused by high sulphate contents. The total hardness is equivalent to the non-carbonate hardness since any carbonic acid bound to bicarbonate is missing. Instead, large amounts of free and offensive carbonic acid are present which, together with sulphuric acid, are the reason for high total acidities. The waters are very rich in iron, calcium, magnesium and sulphate; therefore they contain considerable amounts of dry residues. Moreover, due to geological reasons, very large quantities of NO_3^-- and NH_4^+-N are present. Since only small amounts of organic matter are dissolved in water, the permanganate value (PV; $KMnO_4$ mg/l) is very low, and the loss on ignition amounts to only a few percent of the dry residue. Waters of the early stage are mineral-

ogenic–acidotrophic waters following the initial stage or subject to negligible changes of chemistry over many years, where the latter has to be assigned to adverse geomorphological conditions.

3. *Waters of the transitional stage.* This stage includes waters under less extreme site conditions. They are characterized by acid water reactions at pH values between 4.1 and 6.0 and medium degrees of hardness. In the majority of waters the total hardness is still equivalent to the non-carbonate hardness; except for some single cases, carbonic acid bound to bicarbonate is still missing. Compared with the early stage, the content of free and offensive carbonic acid is reduced by 50%; however, the acidity is still high. Some waters still contain small quantities of free sulphuric acid. The contents of iron, sulphate, calcium and magnesium drop to about 50% of the early-stage values. On the other hand, the quantities of organic matter dissolved in water increase by one-third of the amounts present in early-stage waters. The age of these waters is between 10 and 30 years.
4. *Waters of the late stage.* These are mining lakes of subacid to neutral or subalkaline water reactions at pH values between 6.1 and 7.8, low degrees of hardness in the "very soft" range and small sulphate quantities which only amount to about 10% of the early-stage values. The total hardness is no longer equivalent to the non-carbonate hardness since up to 30% of the carbonate hardness may contribute to the total hardness. There are only minor amounts of free carbonic acid without any offensive properties. Free sulphuric acid is completely absent in these waters. Compared with waters of the early and transitional stage, the contents of iron, manganese, calcium, magnesium and silicon are very low. Due to minor sulphate contents, the residue of evaporation is only 10% of the early-stage value. The content of various parameters drops to the 30th or 50th part. NO_3^-- and NH_4^+-N are only detectable in low amounts. The waters of the late stage either came into being 50–80 years ago or are essentially younger and favourable geomorphological conditions resulted in a much earlier colonization by macrophytes.

9.5.2 Colonization by Plants and Development of Vegetation

In their initial stage, the residual mining waters are completely barren over many years. Only with the progressive process of development or maturation do the waters gradually lose their extreme properties and are colonized by first aquatic plants. These are not the same plant species as those generally known as aquatic plants of natural lakes and ponds, but rather it is a group of plants that is not widely known. These plants are specialists which can grow in acid waters of high iron contents, where combined carbonic acid (bicarbonate) is not present. These species are able to utilize free carbonic acid (CO_2) dissolved in water as carbon source for the process of assimilation.

Since particularly in the early and transitional stage the majority of mining lakes are especially rich in dissolved CO_2, this group of aquatic plants found optimal conditions. *Juncus bulbosus*, for instance, develops very dense, extensive submersed and floating dominance stands down to a water depth of 6 m. The vegetation of this pioneer colonization on extreme sites is characteristic of the early stage of all mining lakes in the Lusatian lignite mining district.

The development of vegetation in mining lakes is determined by the prevailing stage of development. Four stages can be distinguished in the colonization of waters by aquatic macrophytes:

1. *Waters of the initial and early stage free of macrophytes.* Because of adverse geomorphological conditions, extreme acidities and very high iron contents, the waters are completely barren. The water body colours are between reddish-brown and brown, and the littoral areas are heavily endangered by slides.
2. *Monodominant species stands of the early stage abounding with individuals.* The colonization by plants starts with initial stages and dominance stands of one species (monodominant species stands). After a short time, dense mats of this pioneer vegetation cover large areas of the whole littoral and shallow water zones. Floating and submersed aquatic mats of *Juncus bulbosus* and dense reed stands of the species *Phragmites australis, Schoenoplectus lacustris, Typha latifolia* and *T. angustifolia* grow on often thick iron hydroxide sludge layers and sandy-gravelly water bottoms of the littoral zones. In shallow residual waters – often with influxes of large coal sludge volumes – the whole water surface is covered with dense, thick cat-tail reeds.
3. *Vegetation mosaics of the transitional stage abounding with species.* The occurrence of at first small quantities of combined carbonic acid (bicarbonate) and receding extreme acidities in the water bodies are related to the appearance of submersed aquatic plants, e.g. *Potamogeton natans* and *P. polygonifolius*, the bladderwort species *Utricularia ochroleuca, U. intermedia, U. minor* and *U. australis*, and the white water lily (*Nymphaea alba*). The exclusive dominance of one species diminishes and is replaced by the formation of vegetation mosaics which, from the viewpoint of floristic-sociological structures, show first relationships to the vegetation of close-to-nature bog and heath

waters which are also characterized by one-sided ecological conditions. The predominance of monodominant early-stage pioneer vegetation is eliminated by the colonization of new species. These species penetrate into the older stands or grow around them. Various reed species provide the best example to demonstrate this development of vegetation; first reports refer to the mining lake Kabelbaggersee near Schwarzheide (Pietsch 1965).

4. *Formation of plant-sociological units – true plant associations in the late stage.* In small, shallow waters, e.g. in the fracture zone near Doebern and Weisswasser, and in some older residual mining waters in the region of Lauchhammer and Gruenewalde (Senftenberg district), a great number of true plant associations is already occurring (Heym 1971; Pietsch 1973, 1979 c). These are formations of the class of Utricularietea intermedio-minoris associations such as the association of Sparganietum minimi and Sphagno-Utricularietum minoris. In the shallow littoral zone of waters with subacid to neutral pH reactions, associations of the order Littorelletalia uniflorae, e.g. associations of Littorello-Eleocharitetum acicularis and Carici-Deschampsietum setaceae, colonize the fine sandy, partly gravelly water bottoms which are characterized by the absence of hydroxide sediments. The shallow littoral zones with partly thick iron hydroxide layers are colonized by stands of the order Juncetalia bulbosi. They include the following associations: Sphagno-Juncetum bulbosi, Junco bulbosi-Potametum polygonifolius, Junco bulbosi-Myriophylletum heterophyllae and the association Pilularetum globuliferae on sandy loamy soil. Due to the typical acidity conditions, true aquatic plant associations of water lilies and Potametea are rare in the residual waters of the Lusatian lignite mining district, i.e. associations of Hottonietum palustris, Potametum alpinae and Myriophyllo-Nupharetum.

Figure 9.4 shows the metamorphic process for eight selected waters by means of total salt content variations as an expression of the absolute ion content of main cations and anions. The first three mining lakes in the macrophyte-free initial stage (example 1) and in the early stage with *Juncus bulbosus* and *Typha* dominant stands (examples 2 and 3) show the highest total salt contents. These waters are represented by the largest circular diagrams in Fig. 9.4.

In the course of metamorphosis, the salt content steadily declines and the quantities of iron, calcium and sulphate decrease. Examples 4 and 5 characterize the waters of the transitional stage, with essentially smaller areas of the circular diagrams. These are *Juncus bulbosus* waters abounding with *Utricularia.* In *Juncus bulbosus* waters rich in *Sparganium mini-*

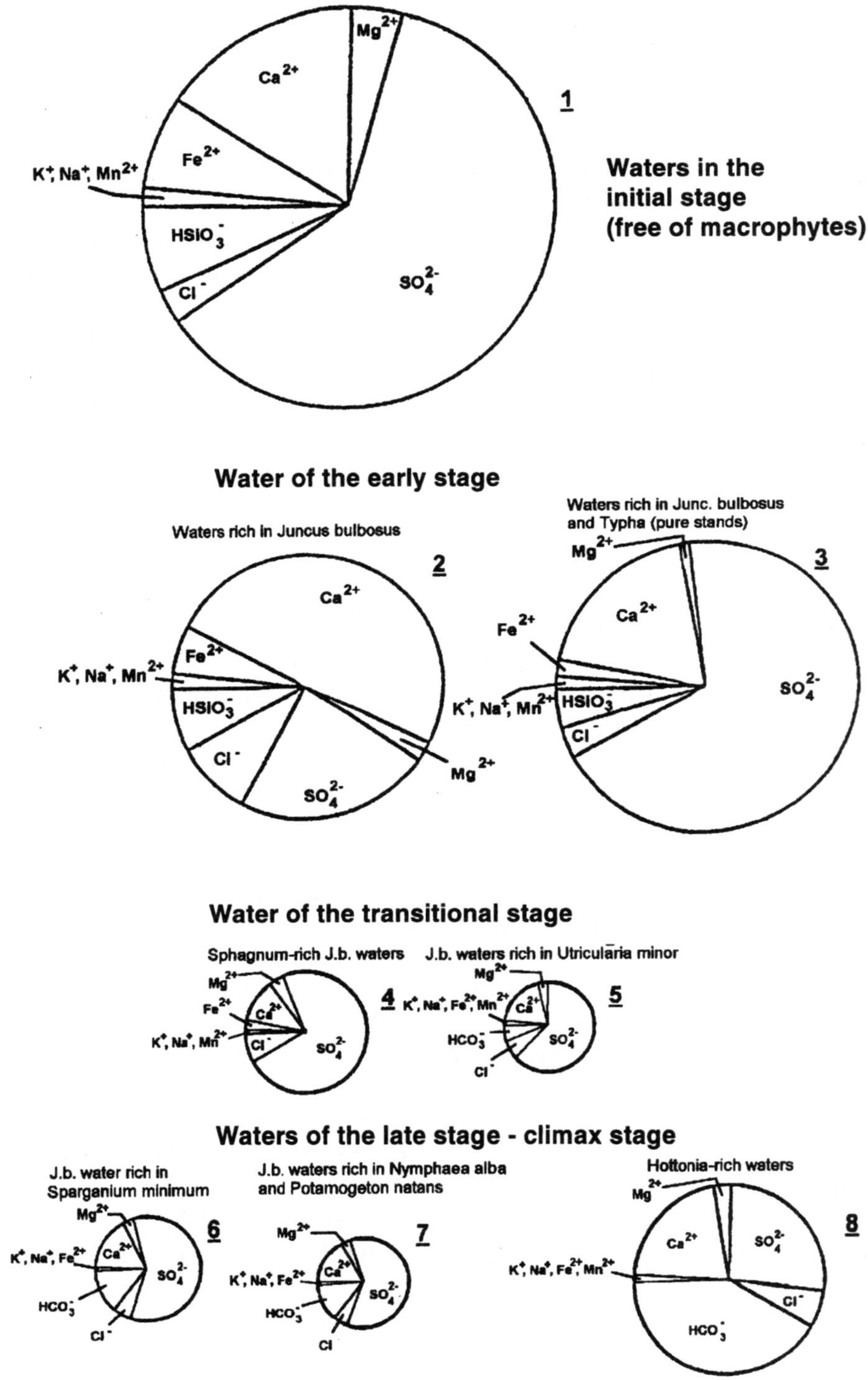
1
Waters in the initial stage (free of macrophytes)
Mg2+
Ca2+
Fe2+
K+, Na+, Mn2+
HSiO3-
Cl-
SO42-
Water of the early stage
Waters rich in Juncus bulbosus
2
Waters rich in Junc. bulbosus and Typha (pure stands)
3
Water of the transitional stage
Sphagnum-rich J.b. waters
4
J.b. waters rich in Utricularia minor
5
HCO3-
Waters of the late stage - climax stage
J.b. water rich in Sparganium minimum
6
J.b. waters rich in Nymphaea alba and Potamogeton natans
7
Hottonia-rich waters
8
K+, Na+, Fe2+, Mn2+

mum (example 6) or abounding with *Nymphaea alba* and *Potamogeton natans* (example 7), the conditions are even more balanced. These waters correspond to the late or climax stage. The lower total salt content, demonstrated by essentially smaller areas of the circular diagrams, is related to a considerably higher hydrogen carbonate content. The iron content of these waters is extremely low.

While the ion areas of the diagrams of examples 1–7 become continuously smaller, the water of example 8 shows an increasing absolute ion content. This is demonstrated by a larger area of the circle. Sulphate is replaced by hydrogen carbonate as the predominant anion. The mineralogically acidotrophic calcium-sulphate water developed into a calcium-hydrogen carbonate water. While acidophile representatives of the Juncetalia bulbosi association are missing, true aquatic plant species such as *Hottonia palustris, Potamogeton crispus* and *Myriophyllum spicatum* developed and form a characteristic Potametea vegetation.

Distribution and composition of the aquatic macrophyte vegetation depending on the water genesis are shown in Fig. 9.5. The colonization capacity amplitude includes bicarbonate-free mining lakes of extreme acidity and high iron contents of the initial and early stage, as shown in the upper part of Fig. 9.5, and the late stage of waters with higher bicarbonate concentrations and subacidic to neutral or subalkaline water qualities, as shown in the lower part of Fig. 9.5. In the course of the metamorphic process, the mineralogic-acidotrophic calcium sulphate waters change from carbonate-free conditions to stages of high hydrogen carbonate contents, and finally calcium-hydrogen carbonate waters. In

Fig. 9.4. Total salt content and absolute ion content of some mining lakes in different stages of aging, as at 8 September 1972. *Example 1* Residual water E Schlabendorf of the open-cast mine Schlabendorf Nord, district of Luckau: pH 2.7; age 2 years. *Example 2* Mining lake Viktoria near Freienhufen, district of Senftenberg: pH 2.5; age 7 years; dominant stands of *Juncus bulbosus. Example 3* Mining lake Kabelbaggersee near Schwarzheide, district of Senftenberg: pH 3; age 15 years; dense *Juncus bulbosus* mats and monodominant *Typha* stands. *Example 4* Mining lake Skaska, district of Hoyerswerda: pH 4.4; age about 50 years; floating *Sphagnum* mats. *Example 5* Mining lake near Welzow: pH 6; age about 50 years; *Juncus bulbosus* stands rich in *Utricularia* and *Potamogeton natans. Example 6* Residual water of the fracture zone near Gross-Koelzig, district of Forst: pH 7; age about 60 years; *Juncus bulbosus* water abounding with *Sparganium minimum. Example 7* Residual water of the open-cast mine Alte Conradsgrube near Gross-Koelzig, district of Forst: pH 7; age about 80 years; *Juncus bulbosus* water abounding with *Nymphaea alba* and *Potamogeton. Example 8* Lake Wehlenteich near Lauchhammer, district of Senftenberg: pH 7.5; age about 65 years; very dense vegetation of *Potamogeton crispus, P. lucens, Myriophyllum spicatum* and *Hottonia palustris*

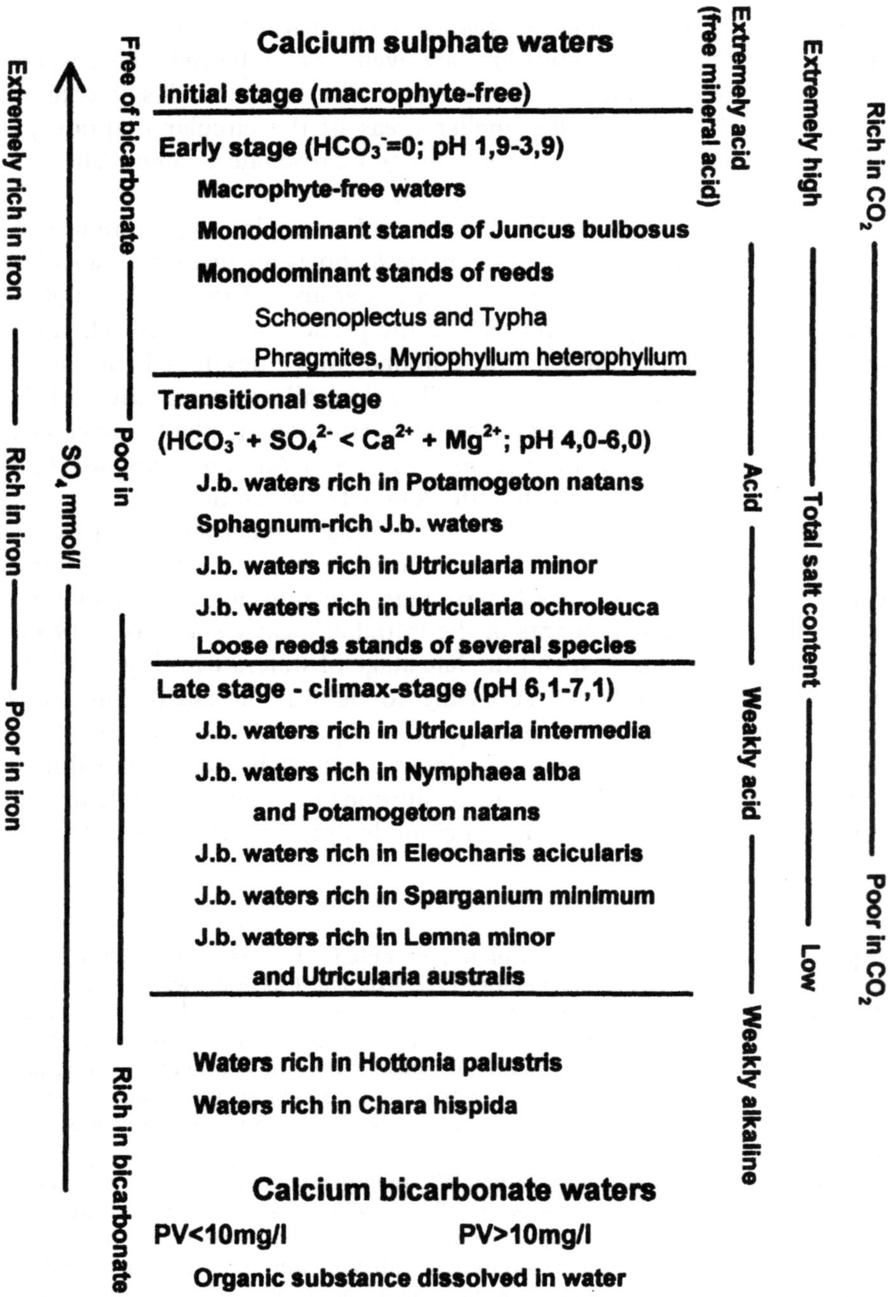

Fig. 9.5. Distribution of macrophyte vegetation in mining lakes of the Lusatian lignite mining district dependent on the hydrochemical conditions of waters. *PV* Permanganate value

the residual waters of the Lusatian lignite mining district, changes in the vegetation structure with the process of metamorphosis can also be observed, i.e. a certain sequence of aquatic and bog plants. The vegetation stages allow us to draw conclusions on the developed or present stage of waters.

The upper part of the survey (Fig. 9.5) includes the macrophyte-free waters of the initial and early stage. This is followed by monodominant stands of *Juncus bulbosus* and various reed species. The middle part of the survey is characterized by aquatic plant species and the vegetation of the transitional stage. The lower part shows various plant associations of the late stage. Floristic–sociological structures and the ecological quality of waters colonized by these plants are similar to the close-to-nature bog and heath waters.

9.6 Indicator Values of Aquatic Macrophyte Vegetation

The quality of water bodies and bottom substrates in water areas down to a depth of 6 m can be easily characterized by structures of the prevailing aquatic macrophyte vegetation. Seven major parameters, subdivided into frequency classes, are used to characterize the indicator value of the aquatic plant vegetation. They include frequency classes of the pH value, sulphate and hydrogen carbonate content, total residue of evaporation, chloride and total iron content and total hardness. Tables 9.2 and 9.3 show the concentration range of the frequency classes. The figures used in Table 9.2 classify the developmental stages of residual mining waters on the basis of bioindicators of aquatic macrophyte stands. The following is a brief presentation and explanation of indicator properties of 14 major water stages.

1. Macrophyte-free type of waters (water type 0). This includes mining lakes of the initial or early stage with extremely acid water reactions (pH 1.9–3.5), high contents of free mineral acid – particularly sulphuric acid –, iron, manganese, aluminium and sulphate. In connection with adverse geomorphological conditions, a shallow littoral zone is missing, and forms of deep basins are exposed to steady mass influxes.
2. Monodominant *bulbosus* rush dominant stands (water type 1)
   - *Juncus bulbosus* dominant stands – these stands constitute the pioneer vegetation of residual mining waters in the initial and early stages with extreme hydrochemical but favourable geomorphological conditions. They are obvious virgin soil pioneers and form the most common type of vegetation in the Lusatian lignite mining district.

Table 9.2. Vegetation type indicator values of aquatic macrophyte species in residual waters of the Lusatian lignite mining district

Water types	Types of vegetation	Hydrochemical quality parameters according to frequency classes						
		pH	SO_4	HCO_3	Residue of evaporation	Cl	Fe	Total hardness
0	Macrophytefree waters	2–3	8–10	0	10–11	7–9	10–11	10–12
1	Monodominant *Juncus bulbosus* stands	2–3	6–10	0	7–11	3–8	6–11	5–11
2	Waters abounding with reeds							
2a	Monodominant reed stands	2–3	5–10	0	6–11	3–9	6–11	4–11
2b	Reed stands of several species	4–7	0–5	0–1	0–6	0–4	0–5	0–5
3	*Juncus bulbosus* waters abounding with *Myriophyllum heterophyllum*	3–6	3–6	9	4–8	3–7	5–9	4–9
4	*J. bulbosus* waters abounding with *Potamogeton natans*	4–6	3–5	0–1	3–5	1–3	3–5	1–3
5	*J. bulbosus* waters abounding with *Utricularia minor*	5–6	4–5	1–2	4–5	1–3	1–3	1–2
6	*J. bulbosus* waters abounding with *Sphagnum*	4–5	2–3	0–1	2–3	1–2	3	1–2
7	*J. bulbosus* waters abounding with *Pilularia globulifera*	4–6	0–2	1–3	1–2	1–2	0	2–3

Table 9.2 (continued)

Water types	Types of vegetation	Hydrochemical quality parameters according to frequency classes						
		pH	SO_4	HCO_3	Residue of evaporation	Cl	Fe	Total hardness
8	*J. bulbosus* waters abounding with *Utricularia intermedia* and *Sparganium minimum*	6–7	0–1	3–4	0–1	0–1	0–1	0–1
9	*J. bulbosus* waters abounding with *Nymphaea alba* and *Potamogeton natans*	6–7	0	1–4	0	0–1	0-(1)	0–1
10	Waters abounding with *Eleocharis acicularis* and *Littorella uniflora*	7–8	0	4–5	0	0	0	0
11	Waters abounding with *Potamogeton crispus* and *Myriophyllum spicatum*	7–9	0–1	4–11	0–1	0–1	0	0–1
12	Waters abounding with *Alisma plantago–aquatica*	7–9	0	6–8	0	0	0	0
13	Waters abounding with *Chara hispida*	7–9	0–1	6–11	1–4	0–1	0	2–3

Table 9.3. Survey of frequency classes for seven water quality classes

Frequency class	pH	Water quality parameters					
		Residue of evaporation) (mg/l	Sulphate (mg/l)	Bicarbonate (mg/l)	Total iron (mg/l)	Chloride (mg/l)	Total hardness (°GH)
0	0–0.95	0–199	0–199	0	0–2.5	0–9.9	0–5.0
1	1.0–1.95	200–399	200–399	0.1–10	2.6–5.0	10–19.9	5.1–10
2	2.0–2.95	400–599	400–599	10.1–20	5.1–7.5	20–29.9	10.1–15
3	3.0–3.95	600–799	600–799	20.1–30	7.6–10	30–39.9	15.1–20
4	4.0–4.95	800–999	800–999	30.1–40	10.1–15	40–49.9	20.1–25
5	5.0–5.95	1000–1199	1000–1199	40.1–50	15.1–20	50–59.9	25.1–30
6	6.0–6.95	1200–1399	1200–1399	50.1–60	20.1–30	60–69.9	30.1–35
7	7.0–7.95	1400–1599	1400–1599	60.1–70	30.1–40	70–79.9	35.1–40
8	8.0–8.95	1600–1799	1600–1799	70.1–80	40.1–50	80–89.9	40.1–45
9	9.0–9.95	1800–1999	1800–1999	80.1–90	50.1–75	>90	45.1–50
10	10.0–10.95	2000–2200	>2000	90.1–100	75.1–100		50.1–55
11	>11	>2200		>100	>100		55.1–60
12							>60

3. Waters abounding with reeds (water types 2a and 2b)
   - Monodominant stands abounding with individuals (*Phragmites australis, Schoenoplectus lacustris, Typha angustifolia, T. latifolia*) – these grow on thick iron hydroxide sludge layers or in the sandy littoral zone of early-stage waters, mostly interspersed with dense *Juncus bulbosus* stands.
   - Loose reed stands consisting of several species – these grow in littoral zones of transitional-stage mining lakes.
4. *Bulbosus* rush stands abounding with water milfoil (water type 3)
   - *Juncus bulbosus* stands abounding with *Myriophyllum heterophyllum* – the whole water body is filled with dense floating stands of *Juncus bulbosus* and *Myriophyllum heterophyllum* (Pietsch and Jentsch 1984). In some cases the floating mats are interspersed with reed stands (*Phragmites australis, Schoenoplectus lacustris, Typha latifolia* and *Sparganium emersum*). Occasionally, stands of *Pilularia globulifera* developed on sandy-gravelly soil substrates with low humus contents. These are oligotrophic waters deficient in lime with pH values in the acid range and high contents of bivalent iron, calcium, sulphate and dissolved free carbonic acid (CO_2). These waters are characterized by negligible contents of ammonium and organic matter dissolved in water. The water bottom mainly consists of deposits with high iron hydroxide concentrations covering the originally sandy virgin soil substrate of the waters.
5. *Bulbosus* rush waters abounding with pond weed (water type 4)
   - *Juncus bulbosus* stands abounding with *Potamogeton natans* – stands of *Potamogeton natans* and *Juncus bulbosus* are typical in mining lakes of the transitional stage, i.e. less extreme chemical conditions of water bodies and bottoms: pH 4.2; 7–25 mg Fe/l; total hardness 5–25°/GH; first occurrence of low combined carbonic acid quantities (hydrogen carbonate). Free sulphuric acid is not present in these water bodies.
6. *Bulbosus* rush waters abounding with bladderwort (water type 5)
   - *Juncus bulbosus* stands abounding with *Utricularia minor* – in mining lakes of the transitional stage, with acid water reactions but higher concentrations of organic matter dissolved in water (pH 4.5–6.2; organic matter < 15 mg $KMnO_4$/l; < 400 mg sulphate/l), *Utricularia minor* and the previously occurring primary stages of *Juncus bulbosus* form a characteristic floating vegetation which is superimposed by floating *Utricularia* layers. Compared with type 4 waters, stands abounding with *Utricularia* are characterized by higher iron contents.

7. *Bulbosus* rush waters abounding with bog moss (water type 6)
   - *Juncus bulbosus* stands abounding with *Sphagnum* – floating bog mosses (*Sphagnum cuspidatum, Sph. obesum* and *Sph. inundatum*) and *bulbosus* rush form a dense floating vegetation which is characteristic of waters of the transitional stage with higher concentrations of ammonium, iron and organic matter dissolved in the water. The penetration of bog mosses from the littoral zones results in pH changes from the acid into the superacid range, with simultaneous reductions of the total salt content (Pietsch 1970): pH 3.6–5.2; 15–40 mg total iron/l; 2–4.5 mg NH_4^+/l; organic substance >25 mg $KMnO_4$/l.
8. Waters abounding with pillwort (water type 7)
   - Stands abounding with *Pilularia globulifera* – *Pilularia globulifera* forms an extensive, dense, floating vegetation and prefers sites with residues of loamy mine spoil material on water bottoms and in the littoral zones. This water type includes residual waters with high contents of electrolyte, calcium and sulphate characterized by acid-to-subacid water reactions. There are no free mineral acids in the water body; the latter is rather characterized by low concentrations of total iron, ammonium and organic matter dissolved in water. On these sites thick iron hydroxide layers are absent. Special features of the water sediments are: increasing quantities of aluminium, manganese and molybdenum, pH 4.4–7.1; 0.01–1.1 mg total iron/l; organic matter <30 mg $KMnO_4$/l; free carbonic acid 9–12 mg CO_2/l; small amounts of hydrogen carbonate 0–15.3 mg HCO_3^-/l and sulphate 108–292 mg SO_4^{2-}/l.
9. *Bulbosus* rush waters abounding with bladderwort and bur reed (water type 8)
   - *Juncus bulbosus* stands abounding with *Utricularia intermedia* and *Sparganium minimum* – *Utricularia intermedia* and *Sparganium minimum* are typical in oligotrophic late-stage residual mining waters of minor iron contents and in small, shallow, residual waters of fracture zones. Water type 8 represents the most common water type in the folded arch region of Muskau and always refers to acid water reactions with low amounts of organic matter in the water body (pH 6.1–7.2; sulphate 78–200 mg SO_4^{2-}/l; total iron <2 mg Fe_t/l; organic substance <20 mg $KMnO_4$/l). The waters are characterized by clear water colours and high transparencies. This type of vegetation also corresponds with the vegetation conditions of close-to-nature bog and heath ponds and the hollow associations in transition moor areas (Sparganietum minimae, Junco bulbosi-Utricularietum intermediae).

10. *Bulbosus* rush waters abounding with water lilies and pond weed (water type 9)
    - *Juncus bulbosus* stands abounding with *Nymphaea alba* and *Potamogeton natans* – *Potamogeton natans, P. alpinus, P. polygonifolius* and *Nymphaea alba*, together with *Juncus bulbosus*, form a dense, floating vegetation completely filling the basins of small residual waters. In larger mining lakes, loose stands penetrate to a water depth of 3 m. These stands are mainly common in waters of the late stage. The iron hydroxide sludge layers at the water bottoms are superimposed with thick covers of organic matter. The water quality is between subacid and neutral, and the water body contains essentially lower quantities of calcium, iron, manganese, sulphate and chloride than the water types mentioned above.
11. Waters abounding with needle rush and *Littorella* (water type 10)
    - Stands abounding with *Eleocharis acicularis* and *Littorella uniflora* – mining lakes of the late stage with subacid to subalkaline water qualities and very low contents of dissolved matter are characterized by *Eleocharis acicularis, Littorella uniflora, Deschampsia setacea* and *Elatine hydropiper*. The water substrates of the shallow littoral zones are fine sandy to gravelly with subdeposits of argillitic–clayey masses. Since iron hydroxide sludge layers are absent, this rare water type is especially suited for swimming. The vegetation conditions are similar to those in fish ponds and heath waters of the Lusatian lowlands.
12. Waters abounding with pond weed and cereal water milfoil (water type 11)
    - Stands abounding with *Potamogeton crispus* and *Myriophyllum spicatum* – the waters of the late stage with subalkaline water reactions (pH 7.1 – 7.4) and remarkable bicarbonate contents (>30 mg HCO_3^-/l) are characterized by *Myriophyllum spicatum, Potamogeton crispus, Hottonia palustris* and *Elodea canadensis*. The iron concentration is <0.1 mg Fe /l. These are calcium bicarbonate waters, where the bicarbonate content exceeds the sulphate concentration. Due to the all-year alkaline water reaction, *bulbosus* rush is not present in these waters. This plant is restricted to waters of the calcium sulphate type, i.e. the mineralogenic–acidotrophic waters of types 1 – 10.
13. Waters abounding with water plantain and floating *Glyceria* (water type 12)
    - Stands abounding with *Alisma plantago-aquatica* and *Glyceria fluitans* – *Alisma plantago-aquatica, Sagittaria sagittifolia* and *Glyceria fluitans* form small reed stands which are characteristic

of shallow bays and littoral zones of late-stage residual waters. The water condition is between neutral and subalkaline with sandy-gravelly bottom substrates, where iron hydroxide sludge layers and thick organic deposits are absent. The vegetation cor-responds with the small reed stands of ditches and ponds which are typical of the Lusatian region.

14. Waters abounding with water horsetail (water type 13)
    - Dominant stands of *Chara hispida* – as dense submersed mats in a depth of 1.6–3.2 m, *Chara hispida* colonizes large areas of water bottoms and forms extensive submersed mats. The bottom substrates consist of oligotrophic calciferous material with low humus contents and negligible iron, manganese and aluminium concentrations. Occasionally, various *Potamogeton* species occur in the shallow water zones, while *Juncus bulbosus* is completely absent in these all-year residual waters. Only some rare examples of this water type are found in the Lusatian lignite mining area, since Miocene overburden material usually does not contain any calciferous marl layers. This water type is predominant in the majority of mining lakes of the central German and Cologne lignite mining districts (Herbst 1966; Bauer 1970; Pietsch 1991).

9.7 Conclusions

Intensive mining activities in the Lusatian region destroyed most of the waters colonized by aquatic plants. After coal extraction, however, the residual or developing waters and mining lakes were recolonized by acidophile littoral associations of the order Juncetalia bulbosi within the class of Littorelletea. Without anthropogenic acceleration of the water genesis this colonization was a process of several decades. It is doubtful whether before the beginning of mining activities the occurrence of Atlantic aquatic plant species was so widespread in the Lusatian region as at the present time.

The residual mining waters of the Lusatian lignite mining district with their naturally occurring virgin soil substrates are ideal initial stages for the formation and, simultaneously, refuge for the maintenance of rare plant species of the Atlantic floral element which are so typical of the aquatic plant vegetation of the Lusatian lowlands.

However, the mineralogenic–acidotrophic residual waters do not provide suitable conditions for aquatic plant species of the classes Potametea and Lemnetea which, in the past, used to be characteristic of the old branches of the River Schwarze Elster. The establishment of suitable ecological priority areas for waters of reduced acidity conditions and higher amounts of combined carbonic acid and nutrients was a prerequisite to provide habitats for water nut (*Trapa natans*) and water soldier (*Stratiotes aloides*).

References

Barber E (1893) Beiträge zur Flora des Elstergebietes in der preußischen Oberlausitz. Abh Naturf Ges Görlitz 20:147–166

Bauer HJ (1970) Untersuchungen zur biozönologischen Sukzession im ausgekohlten Kölner Braunkohlenrevier. Nat Landschaft 45:210–215

Herbst HV (1966) Limnologische Untersuchungen von Tagebaugewässern in den Rekultivierungsgebieten der Braunkohlenindustrie im Kölner Raum. Min Ernährung, Landwirtschaft/Forsten, Nordrhein-Westfalen, pp 1–20

Heym W-D (1971) Die Vegetationsverhältnisse älterer Bergbau-Restgewässer im westlichen Muskauer Faltenbogen. Abh Ber Naturkundemus Görlitz 46:1–40

Müller HJ (1961) Zur Limnologie der Restgewässer des Braunkohlenbergbaus. Verh Int Ver Theor Angew Limnol 14:850–854

Pietsch W (1965) Die Erstbesiedlungsvegetation eines Tagebaugewässers. Limnologica (Berl) 3:177–222

Pietsch W (1966) Wasserchemie und Vegetationsentwicklung in den Tagebauseen des Lausitzer Braunkohlenrevieres. Niederlaus Flor Mitt 2:34–41

Pietsch W (1970) Ökophysiologische Untersuchungen an Tagebaugewässern der Lausitz. Habilschrift, Fakultät für Bau-, Wasser- und Forstwesen, Technische Universität Dresden

Pietsch W (1973) Vegetationsentwicklung und Gewässergenese in den Tagebauseen des Lausitzer Braunkohlen-Revieres. Arch Naturschutz Landschaftsforsch 13: 187–217

Pietsch W (1979a) Zur hydrochemischen Situation der Tagebaurestgewässer des Lausitzer Braunkohlen-Revieres. Arch Naturschutz Landschaftsforsch 19:97–115

Pietsch W (1979b) Klassifizierung und Nutzungsmöglichkeiten der Tagebaugewässer des Lausitzer Braunkohlen-Revieres. Arch Naturschutz Landschaftsforsch 19:187–215

Pietsch W (1979c) Zur Vegetationsentwicklung in den Tagebaugewässern des Lausitzer Braunkohlen-Reviers. Natur Landschaft Bez Cottbus NLBC 2:71–83

Pietsch W (1988) Vegetationskundliche Untersuchungen im NSG "Welkteich". Brandenburgische Naturschutzarbeit in Berlin und Brandenburg 24:82–95

Pietsch W (1990) Erfahrungen über die Wiederbesiedlung von Bergbaufolgelandschaften durch Arten des atlantischen Florenelementes. Abh Ber Naturkundemus Görlitz 64:65–68

Pietsch W (1991) Landschaftsgestaltung im Bezirk Cottbus, dargestellt am Beispiel des Senftenberger Sees. Abh Sächs Akad Wiss Leipzig Math Nat Klasse 57:29–38

Pietsch W (1993) Restoration of green environments and harmonious landscapes in the Lusatian lignite area in Germany, based on the ecological situation. XVth international botanical congress, Yokohama, Japan, Pacifico Yokohama, 28 Aug–3 Sept 1993, 2.9.1–14:1–16

Pietsch W, Jentsch H (1984) Zur Soziologie und Ökologie von Myriophyllum heterophyllum Mich. in Mitteleuropa. Gleditschia 12:303–335

Rothmaler W (1994) Exkursionsflora von Deutschland, vol 4: Gefäßpflanzen: Kritischer Band, 15th edn. Fischer, Jena

Part 3
Inorganic Processes of Acidification

10 Pyrite Chemistry: The Key for Abatement of Acid Mine Drainage

V.P. Evangelou

Department of Agronomy, N-122 Agricultural Science Center North, University of Kentucky, Lexington, Kentucky 40546-0091, USA

10.1 Introduction

Pyrite oxidation produces extremely acidic drainages (as low as pH 2) enriched with Fe, Mn, Al, SO_4, and often heavy metals such as Pb, Hg, Cd, etc. (Table 10.1). Pyrite oxidation takes place when the mineral is exposed to air and water. The process is complex because it involves chemical, biological, and electrochemical reactions, and varies with environmental conditions. Factors such as pH, pO_2, specific surface and morphology of pyrite, presence or absence of bacteria and/or clay minerals, as well as hydrological factors determine the rate of oxidation. There is, therefore, no single rate law to describe kinetics of pyrite oxidation for all cases (Evangelou 1995b; Evangelou and Zhang 1995).

According to Singer and Stumm (1970), Fe^{3+} is the major pyrite oxidant in the acidic pH region, while O_2 is expected to be the direct pyrite oxidant at neutral-to-alkaline pH. The essential feature of the above reaction mechanisms (low or high pH) is initiation of pyrite oxidation by adsorption of O_2 onto partially protonated pyrite surface. However, investigations on pyrite oxidation carried out by Moses et al. (1987), Moses and Herman (1991), and Brown and Jurinak (1989) demonstrated that Fe^{3+} may be a very effective oxidant at circumneutral pH. The conclusion that Fe^{3+} could be the major direct pyrite oxidant at circumneutral pH was supported by theoretical considerations based on magnetic properties and/or on molecular orbital theory of the reactants involved (Moses et al. 1987; Luther et al. 1992). Experimental results by Moses et al. (1987) indicated that pyrite oxidation over the pH range 2–9 was favored in the presence of Fe^{3+} as opposed to dissolved O_2, and a low concentration of Fe^{3+} was very effective in oxidizing pyrite.

Pyrite can vary significantly in grain size and morphology, depending on the environment of formation (Arora et al. 1978; Evangelou 1995b and references therein). Ainsworth et al. (1982) studied the morphology

Table 10.1. Concentrations of environmentally important constituents (mg l^{-1}) in acid mine drainages in the USA and Canada

Substance	Coal mine drainage throughout the United States	Acid mine drainage from Vancouver, Canada	Waste rock seepage from Saskatoon, Canada	Metal mine drainage from Colorado, USA	Drinking water standard, USA
			mg l^{-1}		
As	0.002–0.20	12	5.4–9.7	0.020	0.05
Cd	0.01–0.10	2.0	–	0.03	0.01
Cu	0.01–0.17	190	–	1.6	1.0
Fe	0.6–200	2300	0.1–9.6	50	0.3
Mn	0.3–12	313	7–92	32	0.05
Pb	0.01–0.40	–	–	0.01	0.05
pH	3.2–7.9	–	3.94–5.20	2.6	6.5–8.5
SO_4^{2-}	–	20,000	86–1060	2100	250
Zn	0.03–2.2	273	–	10	5.0
Reference[a]	1	2	3	4	5

[a] 1, USEPA (1982); 2, Fyson et al. (1994); 3, Rowley et al. (1994); 4, Wildeman (1991); 5, compiled from US Goverment Printing Office (1988).

of pyrite isolated from Pennsylvanian-age shale in Missouri and reported their findings as follows: (1) Pyrite with smooth crystal surface which includes octahedral, cubic, and pyritohedral structures; (2) conglomerates with irregular surfaces composed of many cemented particles; and (3) framboids in which the cemented crystals formed a smooth sphere. They also reported that the conglomerates were the predominant form. Zhang (1993) characterized the chemical and morphological properties of pyrite separated from scrubber sludge, gob, and slurry of coal residue. Their results indicated that scrubber sludge contained a trace amount of pyrite in the form of smooth-rounded conglomerates and framboidal forms. Both gob and slurry contained 2–5% pyrite, with the predominant forms as conglomerates and framboids or polyframboids. Numerous octahedral, cubic, pyritohedral, and twinned crystals were also found.

Framboid and polyframboid pyrites are of interest because they are more reactive than conglomeritic pyrite due to high specific surface and high porosity (Caruccio et al. 1977; Evangelou 1995b and references therein). It has been well documented that pyrite oxidation is a surface-controlled reaction (Singer and Stumm 1970; Hoffmann et al. 1981; Moses et al. 1987; Moses and Herman 1991).

10.2 Mechanisms of Pyrite Oxidation

It is believed that two electron transfer mechanisms account for the pyrite-surface oxidation process: (1) formation of a pyrite/surface free radicle; and (2) transfer of an oxygen atom from the pyrite/surface-adsorbed water to the sulfur on the pyrite surface (S_B) (Taylor et al. 1984a, b). Each one of the mechanisms is responsible for the transfer of three electrons up to a total of six. The reactions are summarized below:

$$FeS_2 + 6Fe(H_2O)_6^{3+} + 3H_2O \rightarrow Fe^{2+} + S_2O_3^{2-} + 6Fe(H_2O)_6^{2+} + 6H^+ . \quad (1)$$

The transfer of six electrons from pyritic sulfur (S) to six Fe^{3+} produces one $S_2O_3^{2-}$ and six H^+. In the presence of excess Fe^{3+}, $S_2O_3^{2-}$ is rapidly transformed to SO_4^{2-} through the transfer of eight more electrons as shown below:

$$5H_2O + S_2O_3^{2-} + 8Fe^{3+} \rightarrow 8Fe^{2+} + 10H^+ + 2SO_4^{2-} . \quad (2)$$

Summation of Eqs. (1) and (2) gives:

$$8H_2O + FeS_2 + 14Fe^{3+} \rightarrow 15Fe^{2+} + 16H^+ + 2SO_4^{2-} . \quad (3)$$

The limiting step in the oxidation of pyrite is oxidation of Fe^{2+} by dissolved O_2:

$$Fe(H_2O)_6^{2+} + 1/4O_2 + H^+ \rightarrow Fe(H_2O)_6^{3+} + 1/2H_2O. \qquad (4)$$

According to Moses and Herman (1991), Fe^{3+} is an effective and direct pyrite oxidant at low pH as well as at circumneutral pH and the role played by dissolved O_2 is to sustain the reaction by regenerating Fe^{3+}. It is well known that the rate of Fe^{2+} oxidation increases as pH increases (Singer and Stumm 1970; Millero and Izaguirre 1989). The role of OH^- in oxidizing Fe^{2+} has been postulated by Luther et al. (1992) to be due to the potential increase in frontier molecular orbital electron density of Fe^{2+} upon binding to oxygen by coordinating OH^-. Such coordination increases Fe^{2+} basicity and stabilizes the Fe^{3+} formed. Luther et al. (1992) demonstrated that an increase in electron density also increases the potential of Fe^{2+} to oxidize rapidly to Fe^{3+} when the former (Fe^{2+}) is in the form of a complex with a ligand containing oxygen as the ligating atom. The above, according to Luther et al. (1992), also explains the observed increase in pyrite oxidation as soluble Fe^{3+}-organic complexes increase.

Hood (1991) studied kinetics of pyrite oxidation in alkaline suspensions at a constant pH by maintaining a constant HCO_3^-/CO_3^{2-} ratio and increasing concentration of the two species. Hood (1991) observed that the rate of pyrite oxidation increased with increasing carbonate concentration. Hood (1991) concluded that the cause of the observed increase in pyrite oxidation was formation of a pyrite/surface $Fe(II)–CO_3$ complex which facilitates electron transfer to O_2 and consequently rapid oxidation of ferrous iron. Hood (1991), however, did not provide direct molecular evidence for the ferrous–carbonate complex formed on the pyrite surface. Evangelou and Huang (1994a), using FTIR (Fourier transform infrared) spectroscopy, showed that pyrite exposure to atmospheric air leads to formation of pyrite surface $Fe(II)–CO_3$ complexes and/or a pyrite/surface carboxylic groups which may promote pyrite oxidation (Evangelou 1995b and references therein; Evangelou and Zhang 1995).

The spectra in Fig. 10.1 represent mineral pyrite exposed to atmospheric air at 100% relative humidity in a chamber for 14 days. The infrared absorption band near 1621 cm^{-1} shown in Fig. 10.1 is assigned to the H_2O deformation band. The band at 1650 cm^{-1} suggested the presence of an additional pyrite/surface chemical species. When the oxidized mineral pyrite was exposed to desiccation at room temperature or heated to 160 °C for 3 and 8 h under a nitrogen gas atmosphere, the absorption band at 1650 cm^{-1} decreased in intensity, while the 1621 cm^{-1} band and the

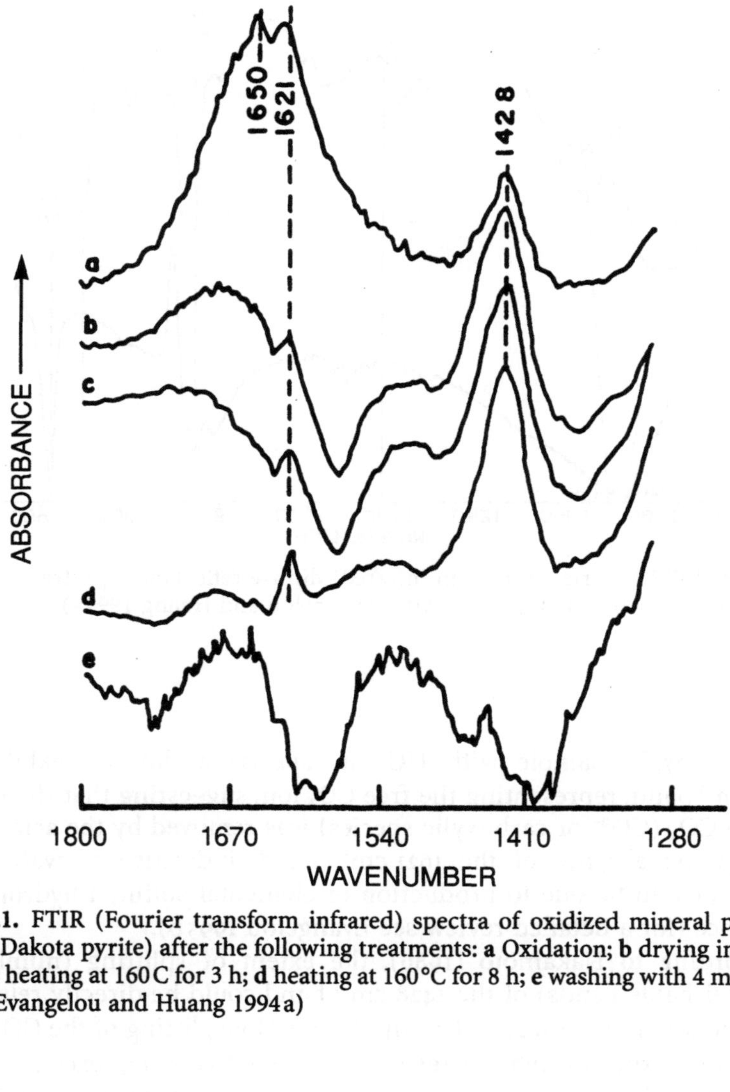

Fig. 10.1. FTIR (Fourier transform infrared) spectra of oxidized mineral pyrite 2 (South Dakota pyrite) after the following treatments: **a** Oxidation; **b** drying in desiccator; **c** heating at 160C for 3 h; **d** heating at 160 °C for 8 h; **e** washing with 4 mol HCl. (After Evangelou and Huang 1994a)

1428 cm^{-1} band remained intact. This suggested that the 1650 cm^{-1} absorption band was most likely due to carboxylate (COO^-). When the pyrite sample was heated, adsorbed water decreased and, therefore, surface acidity increased, leading to carboxylate protonation (Harter and Ahlrichs 1967; Mortland 1968; Mortland and Raman 1968); consequently, the 1650-cm^{-1} band was eliminated. The continuing presence of the 1621 cm^{-1} band may be due to the rapid rehydration of the pyrite surface prior to scanning or due to strongly adsorbed water. After washing the

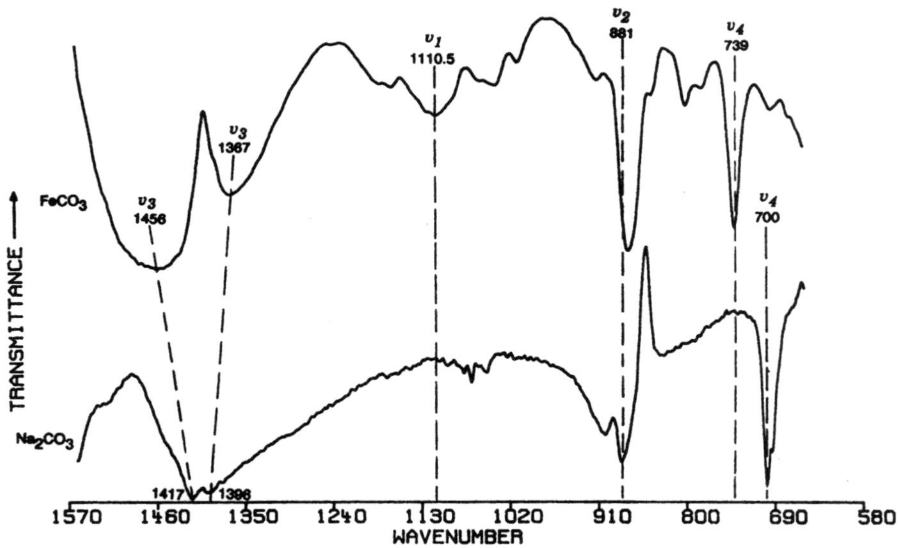

Fig. 10.2. FTIR (Fourier transform infrared) diffuse reflectance spectra of natural siderite ($FeCO_3$) and Na_2CO_3 salt. (After Evangelou and Huang 1994a)

oxidized pyrite sample with HCl, its spectrum did not exhibit the 1428 cm^{-1} band, representing the free CO_3 ion, suggesting that the pyrite/surface CO_x (CO_3^{2-} or carboxylic species) was removed by the acid. Additionally, the absence of the 1621 cm^{-1} band, indicative of water loss/absence, could be due to production of elemental sulfur, a hydrophobic substance (for a detailed review see Evangelou 1995b).

According to Nakamoto (1986), the extent of splitting (number of distinguishable bands) of the 1428 cm^{-1} band could be directly related to the bond strength between CO_3^{2-} and Fe(II). Nonsplitting of the CO_3 band in Fig. 10.1 indicates that pyrite/surface Fe(II)–CO_3 complexes if present are most likely weak electrostatic complexes or that the 1428 cm^{-1} band represents symmetric C-O vibrations of a carboxylic group. When the spectrum of bulk $FeCO_3$ is compared with that of Na_2CO_3 (Fig. 10.2) or oxidized pyrite (Fig. 10.3), it is shown that in the case of bulk $FeCO_3$ the 1428 cm^{-1} band splits into two clearly distinguishable bands, signifying a unidentate complex (Nakamoto 1986), while in the case of Na_2CO_3 or coal pyrite the 1428 cm^{-1} absorption region appears to form as expected more of a single band, signifying electrostatic interaction. The above infrared spectroscopic evaluation clearly supports the potential adsorption of CO_2 by the surface of pyrite either as CO_3^{2-} [weakly adsorbed by pyrite/surface

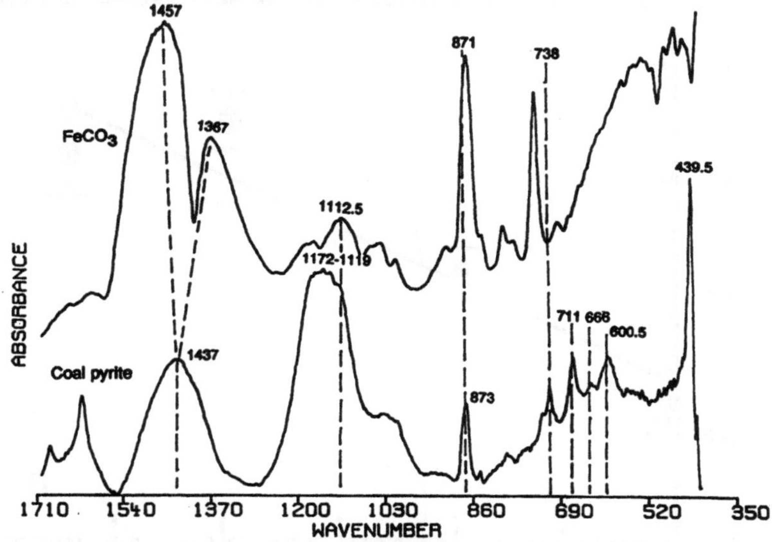

Fig. 10.3. FTIR (Fourier transform infrared) diffuse reflectance spectra of natural siderite ($FeCO_3$) and coal pyrite after a several-month exposure to atmospheric air at room temperature. (After Evangelou and Huang 1994a)

Fe(II)] and/or as a pyrite surface carboxylic acid group (Evangelou and Zhang 1995).

Based on (1) experimental observations that at constant pH but at increasing solution CO_3^{2-} concentration abiotic pyrite oxidation rate increases (Fig. 10.4) (Brown and Jurinak 1989; Hood 1991), and (2) FTIR spectra supporting the presence of pyrite/surface-CO_3 complexes and/or pyrite/surface carboxylic groups (Evangelou and Huang 1994a), the following model is proposed to account for abiotic pyrite oxidation enhancement at pH near or above neutral (Fig. 10.5) in environments enriched with CO_2/CO_3 and O_2:

$$-Fe(H_2O)_5\,OH^+ + CO_2 \rightarrow -Fe(H_2O)_5HCO_3^+ \quad (5)$$

and

$$-Fe(H_2O)_5HCO_3^+ + 1/2\,H_2O + 1/4\,O_2 \rightarrow -Fe(H_2O)_4(OH)_2^+ + CO_2 . \quad (6)$$

According to Millero and Izaguirre (1989), Fe^{2+} oxidation is promoted in the presence of HCO_3^-/CO_3^{2-} due to Fe^{2+}-carbonate complex formation. Based on the above, the surface-adsorbed $-Fe(H_2O)_4(OH)_2^+$ forms a persulfido bridge and pyrite oxidation proceeds according to Eq. (3). An alternate CO_2 involvement in pyrite oxidation proposed by Evangelou

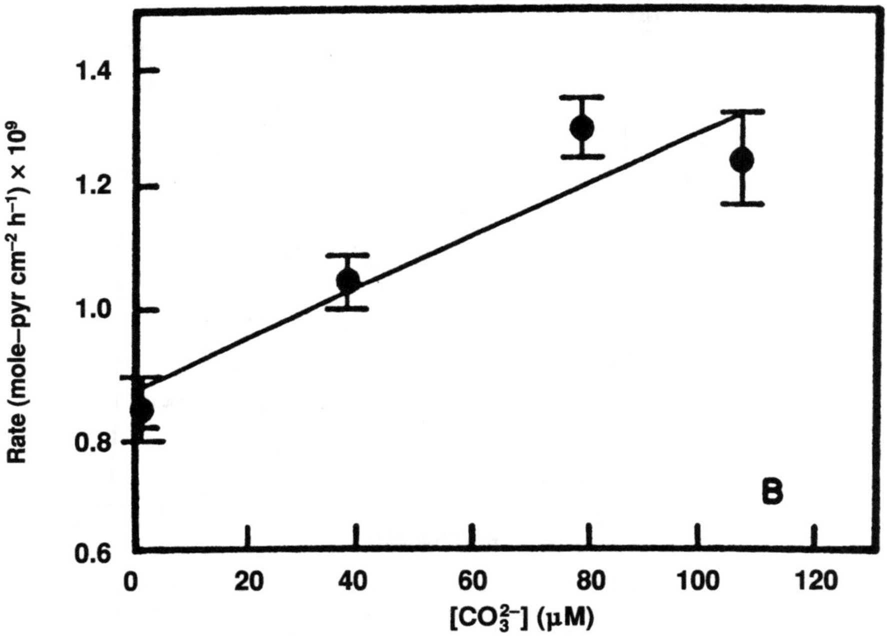

Fig. 10.4. Kinetics of pyrite oxidation as influenced by carbonate. (After Hood 1991)

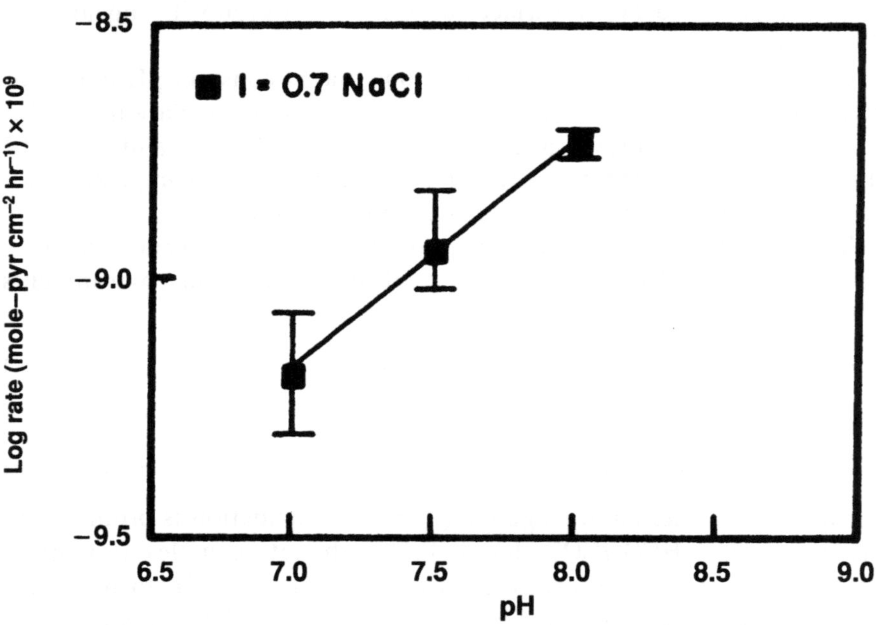

Fig. 10.5. Kinetics of pyrite oxidation as influenced by pH. (After Hood 1991)

and Zhang (1995) is through formation of a pyrite/surface carboxylic acid group:

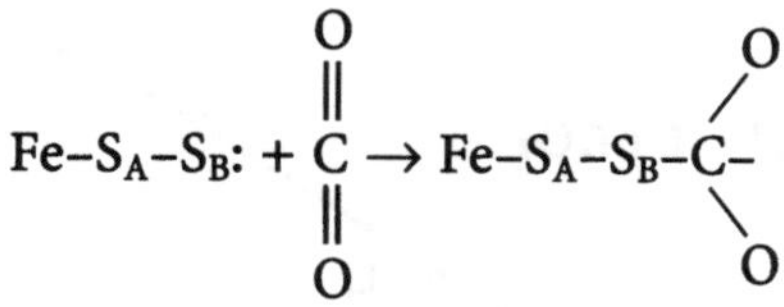
(7)

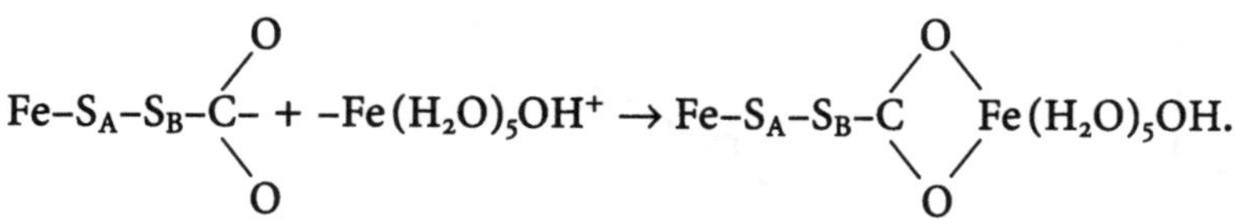
(8)

The Fe–S_A–S_B–COO-Fe(H_2O)OH complex would promote Fe^{2+} oxidation due to the fact that the carboxylate ligand contains oxygen as the ligating atom (Luther et al. 1992). Oxidation of Fe^{2+} to Fe^{3+}, however, is expected to produce a metastable pyrite/surface carboxylate–Fe^{3+} complex, leading to pyrite/surface decarboxylation and formation of a persulfido bridge. The latter facilitates electron transfer from S_B to Fe^{3+}. The overall reaction (unbalanced) is shown below:

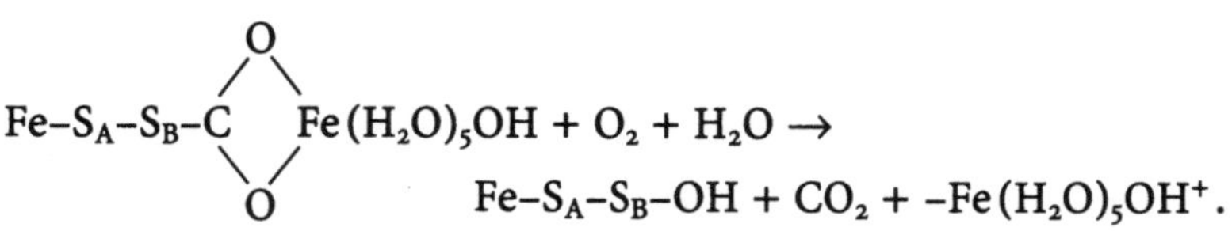
(9)

The introduction of three OH^- groups to the S_B leads to the formation of thiosulfate (S_2O_3), which in the presence of Fe^{3+} rapidly oxidizes to SO_4 as presented in Eq. (2) (Evangelou 1995b).

Based on the above, at neutral-to-alkaline pH the abiotic rate of Fe^{2+} oxidation rises rapidly (Fig. 10.6), but Fe^{3+} concentration also decreases greatly due to precipitation of $Fe(OH)_3$. Because there is probably very little bacterial participation in pyrite oxidation at neutral to alkaline pH, some researchers suggested that in such environments O_2 is a more important pyrite oxidant than Fe^{3+} (Goldhaber 1983; Hood 1991). However, recent findings showed that Fe^{3+} was the preferred pyrite oxidant at circumneutral pH and the major role played by O_2 was to oxidize Fe^{2+} and thereby sustain the pyrite oxidation cycle (Luther 1987, 1990; Moses et al. 1987; Moses and Herman 1991).

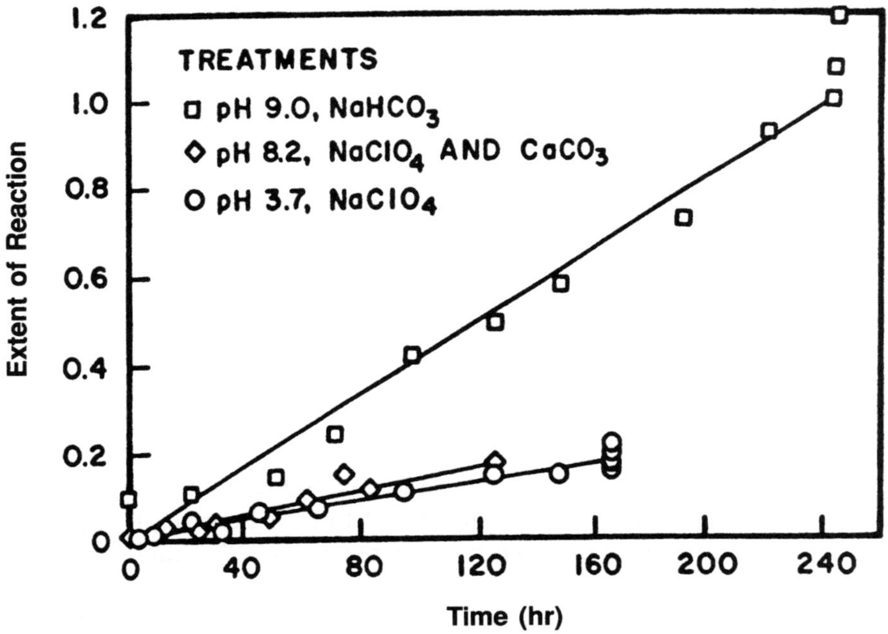

Fig. 10.6. Kinetics of pyrite oxidation as influenced by pH. (After Brown and Jurinak 1989)

10.3 Acid Mine Drainage Treatments

Methods to treat the products of acid mine drainage (AMD) include addition of alkaline material, anoxic limestone drains, phosphatic materials, and wetlands. Methods of preventing AMD include application of organic wastes, altering mine site hydrology, inundation, and bactericides. One new prevention technology is a pyrite particle coating to prevent oxidation. No single technology is appropriate for all situations, and in many cases a combination of technologies offers the best chance of success (Evangelou 1995b).

10.3.1 Alkaline Materials

Alkaline chemicals are commonly applied or pumped into active mines where AMD is a problem. Alkalinity derived from limestone acts as pH buffer; it also precipitates heavy metals as hydroxides. Materials, such as alkaline fly ash and topsoil or their mixtures with lime have been found to significantly reduce drainage iron concentration as well as manganese and sulfate concentrations (Jackson et al. 1993).

A more recent approach in controlling AMD production is through the use of alkaline recharge, e.g. NaOH, $CaCO_3$ (Caruccio and Geidel 1985; Ladwig et al. 1985). Neutralizers such as sodium hydroxide, because of their high solubility, can be easily moved with percolating water deep in the strata to sites where acid drainage is produced. However, effectiveness may last only as long as someone supplies neutralizers.

Limestone has considerable cost advantage over other alkaline materials in treating AMD. However, limestone, because of its limited solubility at near-neutral pH and its tendency to armor with ferric hydroxide precipitates, is not as effective in controlling AMD as one might expect (US EPA 1982; Evangelou 1995b). When mine waters enriched with iron contact limestone in an oxidizing environment, the limestone is coated rapidly with ferric hydroxide precipitates (Table 10.2).

10.3.2 Anoxic Limestone Drains

Anoxic limestone drains (ALD) are given attention owing to their effectiveness in treating AMD (Turner and McCoy 1990; Brodie et al. 1991; Nairn et al. 1991, 1992; Watzlaf and Hedin 1993). An ALD is an excavation filled with limestone and then covered with plastic and clay to inhibit oxygen penetration and loss of carbon dioxide gas. Under these conditions, limestone dissolves at a higher rate due to lowering of pH caused by the higher partial pressure of carbon dioxide, and iron armoring of limestone is limited due to inhibition of iron oxidation. The water discharged from ALD contains a significant concentration of HCO_3 – and this strongly buffered alkaline water is then aerated and metal oxidation, hydrolysis, and precipitation occur in a settling pond or constructed wetland (Evangelou and Zhang 1995).

A disadvantage of ALD is the incongruent dissolution of limestone through formation of $FeCO_3$ and/or $MnCO_3$ gels. This process occurs because the solubility product constant of $FeCO_3$ and/or $MnCO_3$ is approximately 100-fold smaller than that of $CaCO_3$. Accumulation of such heavy-metal carbonate gels diminishes the potential of ALD to function properly (Evangelou 1995b).

Table 10.2. Average chemical composition of spoil leachate after an 8-month reaction period. (Lekhakul 1981)

Type of Spoil	Lime treatment[a] (metric t/m)	pH		Ca^{2+}	Mg^{2+}	Na^{+}	K^{+}	Al^{3+}	Mn^{2+}	Fe^{2+}	SO_4^{2}
		After a 4-month period	After an 8-month period	$mmol_c\ l^{-1}$							
Black Shale	0	1.8	1.8	29.66	337.76	5.41	0.21	301.91	42.66	1166.18	2304.02
	125	4.0	2.2	26.05	508.88	6.28	0.98	41.54	16.40	80.39	734.31
	250	7.0	4.2	24.19	350.50	7.18	1.77	0.47	4.30	0.09	398.85
Siltstone	0	3.8	3.0	23.12	148.86	6.72	0.57	10.40	6.88	1.42	201.54
	67.5	6.8	4.2	23.45	116.07	7.27	0.74	0.72	0.72	0.06	140.27
	135	7.1	4.5	23.31	106.07	7.26	0.76	0.45	0.45	0.05	128.14
Black shale:	0	1.8	1.9	28.68	382.25	4.84	0.22	168.36	41.49	589.81	1332.97
siltstone (1:1)	10	4.3	2.2	24.78	354.95	6.13	0.80	20.13	9.20	24.96	458.07
	20	7.7	5.3	23.35	222.79	6.83	1.21	0.16	0.88	0.0	260.64

[a] Maximum lime rate represents the total measured acidity in the sample.

10.3.3
Phosphate

The potential of Fe^{3+} to act as a pyrite oxidant can be reduced by the addition of phosphate through precipitation of Fe^{3+} as a relatively insoluble $FePO_4$ or $FePO_4 \cdot 2H_2O$ (strengite) (Baker 1983; Hood 1991; Huang and Evangelou 1994). Spotts and Dollhopf (1992) evaluated two sources of apatite and two phosphate by-products applied at a concentration of 30 g kg^{-1} by weight to pyritic mine overburden. Their results indicated that the materials are effective for the control of pyrite oxidation. However, Huang and Evangelou (1992) carried out a series of experiments trying to demonstrate that indeed phosphate applications limit pyrite oxidation. The results of these studies indicated that this inhibition is only temporary because of iron armoring (Evangelou et al. 1992; Huang and Evangelou 1992).

10.3.4
Wetlands

The wetland processes (Hammer 1989) identified as having the potential for removing metals from AMD include: adsorption of metals by ferric oxyhydroxides; plant and algae uptake of metals; complexation of metals by organic materials; and precipitation of metals into oxides, oxyhydroxides, and sulfides. From the above processes, only precipitation as either oxides or sulfides has long-term metal removal potential. The rest of the listed processes are considered to be insignificant or to have finite capacity. Table 10.3 (modified from Brodie et al. 1988) summarizes the chemical and physical characteristics of influent and effluent from several constructed wetlands. Water quality generally improved in all cases, and most meet the State effluent guidelines for total Fe < 3 mg l^{-1}, total Mn < 2 mg l^{-1}, pH 6–9, and nonfilterable residues (NFR) < 35 mg l^{-1}.

Studies are now focused on optimizing the activity of sulfate-reducing bacteria in the anaerobic zone (Kleinman 1989). These sulfate-reducing bacteria consume acidity, and most of the hydrogen sulfide they produce reacts with heavy metals to yield insoluble precipitates. Results from column and reactor studies simulating reducing conditions indicated that trace metals such as Co, Cu, Cd, Ni, Pb, and Zn can be removed as sulfides (Hammack and Edenborn 1991; Staub and Cohen 1992).

Table 10.3. Summary of the treatment of the Tennessee Valley Authority (TVA) acid drainage wetlands. (Modified from Brodie et al. 1988)

Wetlands system	Date of initiated operation	Area (m^2)	Number of cells	Influent water parameters ($mg\ l^{-1}$)				Effluent water parameters ($mg\ l^{-1}$)						Treatment area ($m^2\ mg^{-1}\ min^{-1}$)	
				pH	Fe	Mn	NFR*	pH	Fe	Mn	NFR*	Flow ($l\ min^{-1}$)		Fe	Mn
												Av.	Max.		
Imp 1	5–85	5700	4	6.3	30.0	9.1	57.0[a]	6.5	0.9	2.1	2.8	53	227	3.6	11.8
Imp 2	6–86	11,000	5	3.1	40.0	13.0	9.0	3.1	3.4	14.0	0.8[b]	400	2200	0.7	2.1
Imp 3	10–86	1200	3	6.3	13.0	5.0	28.0	6.8	0.8	1.9	4.7	87	379	1.1	2.8
Imp 4	11–85	2000	3	4.9	135.0	24.0	42.0	4.6	3.0	4.0	6.0	42	49	0.4	2.0
King 006	10–87	9300	3	4.2	153.0	4.9	40.0					379	2271	0.2	5.0
RT-2	9–87	7300	3	5.7	45.2	13.4		6.7	0.8	0.2	2.0	238	681	0.7	2.3
WC 018	6–86	4800	3	5.6	150.0	6.8		3.9	6.4	6.2		70	1495	0.2	4.2
WC 019	6–86	25,000	3	5.6	17.9	6.9		4.3	3.3	5.9		492	6360	2.8	7.4
Col 013	10–87	9200	5	5.7	0.7	5.3		6.7	0.7	13.5		288	408	45.6	6.0

* Nonfilterable Residues.

[a] From preconstruction in-stream sample.

[b] One sample, July 1987.

10.4
Acid Mine Drainage Prevention

10.4.1
Organic Waste

Organic waste may inhibit pyrite oxidation via various mechanisms: (1) A number of organic compounds, from simple aliphatic acids to amino acids, sugars, and alcohols, are inhibitory to *Thiobacillus ferrooxidans* and *T. thiooxidans*; (2) formation of Fe–organic complexes which are not readily oxidized or reduced; (3) specific adsorption of organic materials on pyrite surface, preventing the approach or release of oxidants (either Fe^{3+} or dissolved oxygen) or oxidizing microbes (Pichtel and Dick 1991); (4) organics may combine with Fe oxide and result in the formation of stable precipitate of micellar colloids (Hiltunen et al. 1981).

Organic waste, however, may also promote pyrite oxidation by solubilization of $Fe(OH)_3$ through formation of Fe^{3+}–carboxylate complexes. Such complexes, especially if positively charged, could adsorb onto the pyrite surface and act as electron acceptors in an outer sphere mode (Luther et al. 1992).

Pichtel and Dick (1991) determined the inhibiting effect of various organic amendments on pyrite oxidation. The amendments include composted sewage sludge, composted papermill sludge, water-soluble extract from composted sewage sludge, and pyruvic acid (Table 10.4). After 28 days of incubation of pyritic spoil and amendments, they found that pH, sulfate-S, and total soluble Fe of the amended spoil were affected less than the control. They concluded that the organic material reduced acid production from pyrite by preventing Fe oxidation and by removing soluble Fe from solution.

10.4.2
Macropore Flow

Solutes in spoils, assuming that macropores are not involved, tend to move as a pulse and, therefore, the solute concentration in this pulse tends to increase with depth of the spoil profile (Evangelou et al. 1982; Evangelou and Phillips 1984). On the other hand, macropore flow produces highly diluted groundwaters and when such waters become surface waters their quality is relatively high (Evangelou 1995b).

Table 10.4. pH value and concentrations of sulfate-S and total soluble Fe in spoil suspensions incubated with various organic amendments. (Pichtel and Dick 1991)

Amendment	pH Time of incubation (days)					Sulfate-S Time of incubation (days)					Total soluble Fe Time of incubation (days)				
	0	7	14	21	28	0	7	14	21	28	0	7	14	21	28
						(mmol kg^{-1} spoil)					(mmol kg^{-1} spoil)				
Non-amended	5.90	4.96	4.35	4.06	3.80	5.2	16.7	22.9	67.7	81.3	5.5	3.4	21.5	87.7	91.3
Composted sewage sludge	5.90	5.75	5.22	5.15	5.00	5.2	29.1	31.2	26.6	37.5	1.2	0.4	0.5	0.5	0.5
Composted papermill sludge	5.90	8.05	8.07	7.90	7.50	10.4	33.9	49.5	44.3	78.7	1.2	0.7	0.0	0.8	0.5
Water-soluble extract (composted sewage sludge)	5.90	6.30	4.90	4.80	4.70	5.2	16.1	27.1	31.8	66.2	6.1	6.3	9.8	2.7	8.9
Pyruvic acid	5.90	5.75	5.60	5.75	5.70	5.2	10.9	18.2	27.1	45.3	6.4	7.0	11.5	13.6	13.4
$LSD_{0.05}$	–	0.17	0.10	0.27	0.23	0.5	13.1	10.9	38.9	25.1	0.7	1.3	2.5	49.6	29.4

10.4.3 Inundation

Underwater disposal of pyritic materials has been used by the metal mining industries with some success (Ritcey 1991). Kleinmann and Crerar (1979) reported that there was no significant growth of *T. ferrooxidans* in water-saturated environments. A water cover on acid-generating uranium tailings in a 65-ha field test by Dave and Vivyurka (1994) indicated that a shallow water cover (0.5–1 m in depth) was effective in controlling acid generation. After 3 months of water cover, acidity (mg $CaCO_3$ l^{-1}) and SO_4^{2-} concentrations were maintained at less than 100 mg l^{-1} and effluent pH was maintained between 6 and 7. However, complete inhibition of pyrite oxidation by flooding may never be possible because of the availability of Fe^{3+} as an alternate oxidant.

Additional concerns with underwater disposal include the potential to maintain complete and continuous water saturation. It is known that biotic oxidation of pyrite is not limited until pore gas oxygen is reduced to less than 1% (Carpenter 1977; Hammack and Watzlaf 1990).

10.4.4 Bactericides

Anionic surfactants (common cleaning detergents), organic acids, and food preservatives have been used as pyrite oxidation inhibitors by controlling bacterial growth (Kleinmann 1981; Erickson and Ladwig 1985; Dugan 1987). In the presence of such compounds, hydrogen ions in the acidic environment move freely into or through bacteria cell membranes, causing their deterioration. Acid production has been reduced by 60–95% in coal refuse piles and isolated zones of fresh pyrite material (Kleinman 1989). However, wide use of anionic surfactants is limited for three reasons: (1) anionic surfactants are very soluble and move with water; thus, repeated treatments are required to prevent bacteria repopulation; (2) anionic surfactants may be adsorbed on the surfaces of other minerals and may not reach the pyrite–bacteria interface (Erickson and Ladwig 1985; Shellhorn and Rastogi 1985); and (3) bactericides do not have much of an effect on acid metallic drainages produced prior to treatment and the strata continue to discharge "acid effluents" during water flows (see also Evangelou 1995b).

10.4.5 Coating Technologies

Recently, Huang and Evangelou (1994) and Evangelou (1995a), using small leaching columns, have developed three new microencapsulation (coating) methodologies for preventing pyrite oxidation and acid production in coal pyritic waste. The *first coating methodology* involves leaching coal waste with a solution composed of low but critical concentrations of H_2O_2, KH_2PO_4, and a pH buffer. During the leaching process, H_2O_2 oxidizes pyrite and produces Fe^{3+} so that iron phosphate precipitates as a coating on pyrite surfaces. The purpose of pH buffer in the coating solution is to eliminate the inhibitory effect of the protons, produced during pyrite oxidation, on the precipitation of iron phosphate. The coating process is shown schematically in Fig. 10.7.

The dotted lines in Fig. 10.7B signify physical bonding between pyrite and $FePO_4$. When iron (Fe^{3+}) reacts with PO_4^{3-} it forms an acid-resistant ferric-phosphate ($FePO_4$) coating, which inhibits oxidation of pyrite as shown in Fig. 10.8A. It is important to note that the technology of coating pyrite as described above is not to be confused with field application of rock phosphate (Flynn 1969). Rock phosphate complexes dissolve iron [Fe(II)], thus reducing the potential of Fe(III) production, and reducing the potential for pyrite oxidation as well (Stumm and Morgan 1970). Rock phosphate *does not* coat pyrite; rather, it complexes released Fe(II) from the oxidizing pyrite (Evangelou et al. 1992). Instead, a rock-phosphate surface-coating with Fe(II) forms, reducing rock phosphate dissolution. Therefore, the effectiveness of rock phosphate in controlling pyrite oxidation is short-lived.

A *second coating methodology* involves leaching pyritic waste with a solution composed of low but critical concentrations of H_2O_2 and a pH buffer. During the leaching process, H_2O_2 oxidizes pyrite and produces an iron-oxide coating on the surface of pyrite (Fig. 10.8C). The purpose of a

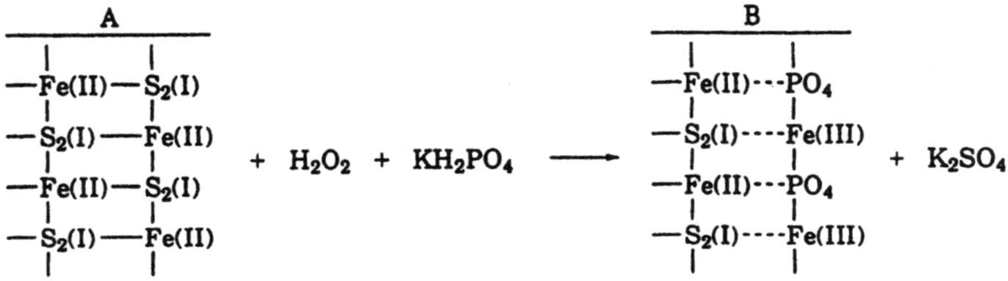

Fig. 10.7. Schematic of H_2O_2-induced oxidation proof surface coating on iron sulfides (Evangelou and Huang 1994b; US Patent No. 5,286,522)

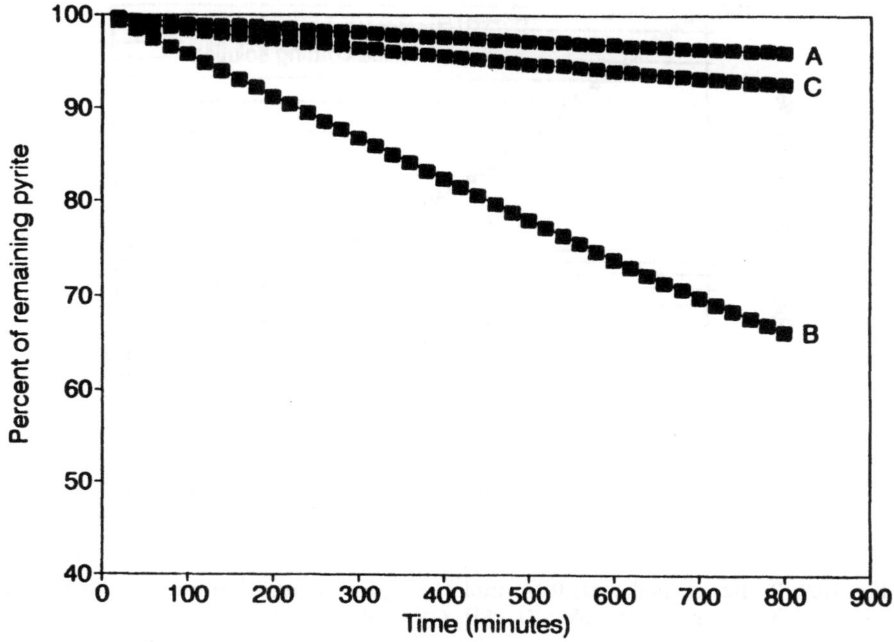

Fig. 10.8. Oxidation kinetics of pyrite leached with the following three pH-6 solutions: *A* 0.01 mol sodium acetate (NaAC) plus 0.106 mol H_2O_2 and 0.001 mol KH_2PO_4; *B* 0.01 mol NaAC plus 0.106 mol H_2O_2 and 0.013 mol EDTA [ethylene diamine tetra-acetic (ethanoic) acid]; *C* 0.01 mol NaAC plus 0.106 mol H_2O_2

pH buffer in this case is to buffer the solution during coating formation at a pH between 5 and 7 where iron oxide formation is promoted. When EDTA [ethylene diamine tetra-acetic (ethanoic) acid] was added to the acetate/H_2O_2 solution, pyrite oxidation was not inhibited, suggesting that EDTA did not permit formation of ferric-hydroxide coating on the surface of pyrite (Fig. 10.8 B).

A *third coating methodology* is that of an iron-oxide/silica coating. The data in Fig. 10.9 show the oxidation potential of framboidal pyrite by 0.145 mol hydrogen peroxide (H_2O_2) l^{-1} in the presence and absence of 50 mg dissolved silica l^{-1} at pH 5 adjusted with 0.01 mol sodium acetate l^{-1}. As can been seen, silica significantly suppressed the potential of H_2O_2 to oxidize pyrite. The explanation for this behavior is that oxidation of pyrite by H_2O_2 in the presence of Si and sodium acetate lead to the formation of an iron-oxide/silica coating, as shown schematically in Fig. 10.10.

In order to demonstrate that this coating is resistant to low pH values or even strong acid attack, the following experiment was carried out.

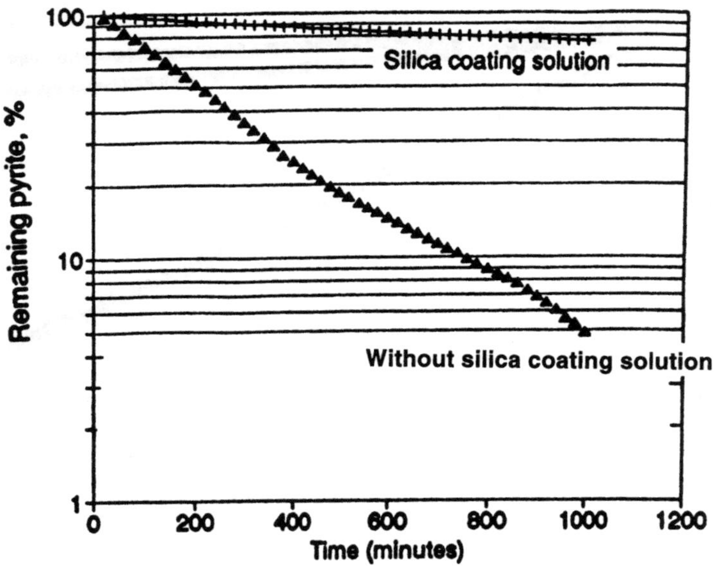

Fig. 10.9. Pyrite leached with 0.145 mol H_2O_2 plus 50 mg l^{-1} silica (Si) [having as source sodium metasilicate ($Na_2SiO_3 \cdot 5H_2O$)] at pH 5 buffered with 0.01 mol sodium acetate (coating solution) and with 0.145 mol H_2O_2 alone at room temperature (Zhang and Evangelou, unpubl. data, 1994)

First, framboidal pyrite was coated with iron-oxide/silica coating. This coated pyrite was then oxidized with 0.145 mol H_2O_2 l^{-1} at room temperature. This is demonstrated in Fig. 10.11. The data from 0 to 900 min represents the silica coating process of pyrite. After 900 min, the data labeled A represent leaching of iron-oxide/silica-coated pyrite with oxygenated water alone. No pyrite oxidation was apparent. The data labeled B represent coated pyrite oxidation with 0.145 mol H_2O_2 l^{-1}, a strong pyrite oxidizer. These data show that the iron-oxide/silica coating protected pyrite from oxidizing by inhibiting H_2O_2 diffusion to the pyrite surface.

In order to demonstrate that the coating on the surface of pyrite was acid-resistant because it was composed of two distinct layers, an iron-oxide layer (acid-sensitive) and a silicon-oxide (silica) layer (acid-resistant), an iron-oxide/silica-coated pyrite sample was leached with 50 ml 4 mol hydrochloric acid (HCl) l^{-1}. The purpose of this treatment was to dissolve the iron-oxyhydroxide or remove the iron from the pyrite silica coating, leaving behind the silica coating by itself. After removal of the iron-oxyhydroxide, the pyrite sample was oxidized with 0.145 mol H_2O_2 l^{-1}.

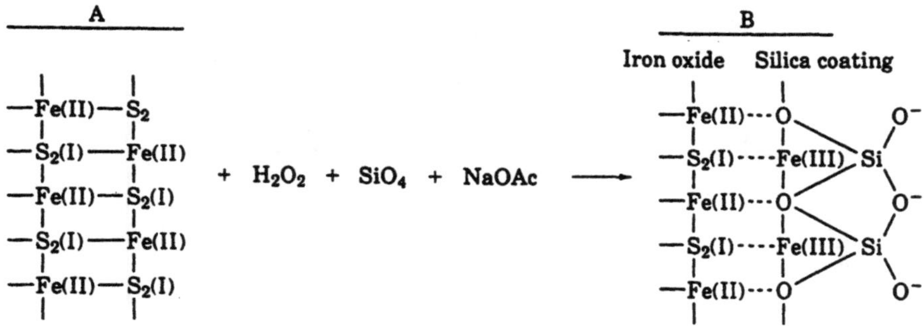

Fig. 10.10. Schematic of H_2O_2-induced, oxidation-proof, iron-oxide/silica surface coating (Evangelou 1996a, US Patent No. 5,494,703)

Another pyrite-coated sample was treated with 50 ml 4 mol hydrofluoric acid (HF) l^{-1}. The purpose of this treatment was to remove the iron-oxyhydroxide coating as well as the silica coating (HF is a strong acid and specifically decomposes silica; HF is not present in acid drainages emanating from pyritic waste). After removal of the iron-oxyhydroxide–silica coating, the pyrite sample was oxidized with 0.145 mol H_2O_2 l^{-1}. The purpose of these two treatments (HCl versus HF) was to demonstrate that silica coating on the surface of pyrite was produced and that this coating offered substantial resistance to acid attack. The results are shown in Fig. 10.11 C, D. The data in Fig. 10.11 C (representing 4 mol HCl treatment l^{-1}) show that oxidation of pyrite by 0.145 mol H_2O_2 $l^{-1}$was greatly suppressed relative to that treated with 4 mol HF l^{-1} (Fig. 10.11 D). This strongly suggests that the silica part of the coating alone offers substantial protection to pyrite from H_2O_2 (a very strong oxidizer) attack due to the fact that silica is not completely soluble in acid.

Successful application of such coating methodologies in the field could mean long-term solution (perhaps even permanent solution) to certain types of acid mine drainage problems. These coating methodologies are expected to be cost-effective since they involve readily available materials and only cover the surface of pyrite particles. Furthermore, the coating solution could be applied to any permeable coal mine waste; thus, little or no physical disturbance of coal mine waste during treatment would be necessary.

Disclaimer. The views and conclusions contained in this chapter are those of the author and should not be interpreted as necessarily representing the official policies or recommendations of the Interior Department's US Bureau of Mines, US Geological Survey (USGS) or of the US Government.

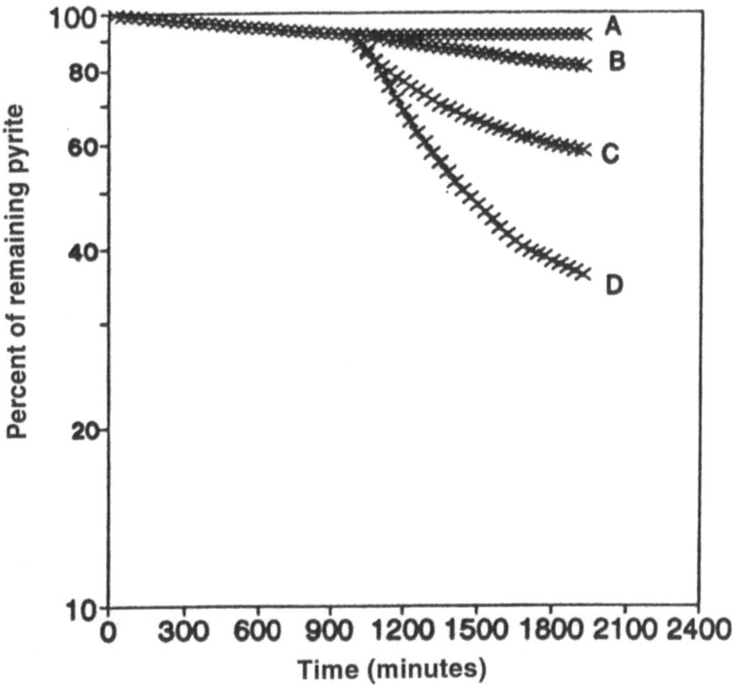

Fig. 10.11. Various pyrite samples first (up to 900 min) leached with 0.145 mol H_2O_2, 50 mg silica (Si) l^{-1} at pH 5 buffered with 0.01 mol sodium acetate at room temperature. Each of the pyrite-coated samples was treated as follows: Sample *A* was leached with water; sample *B* was leached with 0.145 mol H_2O_2 l^{-1}; sample *C* was first leached with 50 ml 4 mol hydrochloric acid (HCl) l^{-1} and then leached with 0.145 mol H_2O_2 l^{-1}; sample *D* was first leached with 50 ml 4 mol hydrofluoric acid (HF) l^{-1} and then leached with 0.145 mol H_2O_2 l^{-1} (Zhang and Evangelou, unpubl. data 1994)

Acknowledgements. The work for this chapter was funded by the US Bureau of Mines, US Geological Survey and the Kentucky Agricultural Experimental Station and is published with the approval of the Director.

References

Ainsworth CC, Blanchar RW, King EJ (1982) Morphology of pyrite from Pennsylvanian-age shales in Missouri. Soil Sci 134:244–251

Arora HS, Dixon JB, Hossner LR (1978) Pyrite morphology in lignitic coal and associated strata of east Texas. Soil Sci 125:151–159

Baker BK (1983) The evaluation of unique acid mine drainage abatement techniques, Master's Thesis, West Virginia University, Morgantown

Brodie GA, Hammer DA, Tomljanovich DA (1988) Constructed wetlands for acid drainage control in the Tennessee Valley. In: Mine drainage and surface mine reclamation, vol 1: mine water and mine waste. US Bureau of Mines Information Circular IC 9 (183), Pittsburgh, p 325

Brodie GA, Britt CR, Tomaszewski TM, Taylor HN (1991) Use of the passive anoxic limestone drains to enhance performance of acid drainage treatment wetlands. In: Oaks W, Bowden J (eds) Proceedings reclamation 2000: technologies for success, Durango, Colorada, pp 211–222

Brown AD, Jurinak JJ (1989) Mechanism of pyrite oxidation in aqueous mixtures. J Environ Qual 18:545–550

Carpenter PL (1977) Microbiology. Saunders, Philadelphia, pp 218–220

Caruccio FT, Geidel G (1980) The assessment of a stratum's capacity to produce acidic drainage. Proc Natl Symp on Surface mine hydrology, sedimentology and reclamation, University of Kentucky, Lexington, pp 437–444

Caruccio FT, Geidel G (1985) The prediction of acid mine drainage from coal strip mines. Proc Reclam Aband Acid Spoils Missouri, Dept of Nat Resour Land Reclam Commission, Jefferson City, Missouri, 21 pp

Caruccio FT, Ferm J C, Horne J, Geidel G, Baganz B (1977) Paleoenvironment of coal and its relation to drainage quality. EPA-600/7-77-067, Environmental Protection Agency, Washington, DC

Dave NK, Vivyurka AJ (1994) Water cover on acid generating uranium tailings – laboratory and field studies. In: Proc Int Land reclamation and mine drainage Conf and 3rd Int Conf on the Abatement of acidic drainage, Pittsburgh, 24–29 April, vol 1, p 297

Dugan PR (1987) Prevention of formation of acid drainage from high-sulfur coal refuse by inhibition of iron- and sulfur-oxidizing microorganisms. II. Inhibition in "Run of Mine" refuse under simulated field conditions. Biotech Bioeng 29: 49–54

Erickson PM, Ladwig KJ (1985) Control of acid formation by inhibition of bacteria and by coating pyritic surfaces. Final report to the West Virginia Department of Energy Division of Reclamation Charleston, West Virginia

Evangelou VP (1995a) Potential microencapsulation of pyrite by artificial inducement of $FePO_4$ coatings. J Environ Qual 24:535–542

Evangelou VP (1995b) Pyrite oxidation and its control. CRC Press, Boca Raton

Evangelou VP (1996a) Schematic of H_2O_2 induced, oxidation proof iron-oxide/silica surface coating. US Patent no 5,494,703

Evangelou VP (1996b) Pyrite surface silica coating for inhibiting oxidation. US Patent no 5, 494, 703

Evangelou VP, Huang X (1994a) Infrared spectroscopic evidence of an iron(II)-carbonate complex on the surface of pyrite. Spectrochim Acta 50A:1333–1340

Evangelou VP, Huang H (1994b) Peroxide induced oxidation proof phosphate surface coating on iron sulfides. US Patent no 528–522

Evangelou VP, Phillips RE (1984) Ionic composition of pyritic coal spoil leachate interactions and effect on saturated hydraulic conductivity. Reclam Reveg Res 3:65–76

Evangelou VP, Zhang YL (1995) A review: pyrite oxidation mechanisms and acid mine drainage prevention. Critical reviews. Environmental science and technology, vol 25/2. CRC Press, Boca Raton, pp 141–199

Evangelou VP, Phillips RE, Shepard JS (1982) Salt generation pyritic coal spoils and its effect on saturated hydraulic conductivity. Soil Sci Am J 46:456–460

Evangelou VP, Sainju UM, Huang X (1992) Evaluation and quantification of armoring mechanisms of calcite, dolomite and rock phosphate by manganese. In: Younos T, Diplas P, Mostaghimi S (eds) Land reclamation: advances in research and technology. American Society of Ag Engineers, Nashville, pp 304–316

Flynn JP (1969) Treatment of earth surface and subsurface for prevention of acidic drainage from the soil. US Patent 3,443,882, 13 May

Fyson A, Kalin M, Adrian LW (1994) Arsenic and nickel removal by wetland sediments. In: Proc Int Land reclamation and mine drainage Conf and the 3rd Int Conf on the Abatement of acidic drainage, Pittsburgh, vol 1, p 109

Goldhaber MB (1983) Experimental study of metastable sulfur oxyanion formation during pyrite oxidation at pH 6–9 and 30 °C. Am J Sci 283:913–217

Hammack RW, Edenborn HM (1991) The removal of nickel from mine waters using bacteria sulfate reduction. In: Proc 1991 Natl Meet American Society of Surface Mining and Reclamation, West Virginia, vol 1, p 97

Hammack RW, Watzlaf GR (1990) The effect of oxygen on pyrite oxidation. In: Proc Mining and reclamation Conf, Charleston, West Virginia, 23–26 April 1990, pp 257–264

Hammer DA (ed) (1989) Constructed wetlands for wastewater treatment: municipal, industrial and agricultural. Lewis, Chelsea, Michigan

Harter RD, Ahlrichs JL (1967) Determination of clay surface acidity by infrared spectroscopy. Soil Sci Soc Am Proc 31:30–33

Hiltunen P, Vuorinen A, Rehtijarvi P, Tuovinen OH (1981) Release of iron and scanning electron microscopic observations. Hydrometallurgy 7:147–157

Hoffmann MR, Faust BC, Panda FA, Koo HH, Tsuchiya HM (1981) Kinetics of the removal of iron pyrite from coal by microbial catalysis. Appl Environ Microbial 42:259–271

Hood TA (1991) The kinetics of pyrite oxidation in marine systems. PhD Diss, University of Miami, Coral Gables, Florida

Howarth RW, Teal JM (1979) Sulfur reduction in New England salt marsh. Limnol Oceanogr 24:999–1013

Huang X, Evangelou VP (1992) Abatement of acid mine drainage by encapsulation of acid producing geologic materials. US Department of the Interior, Bureau of Mines, Pittsburgh, 60 pp

Huang X, Evangelou VP (1994) Kinetics of pyrite oxidation and surface chemistry influences. In: Alpers CN, Blowes DW (eds) The environmental geochemistry of sulfide oxidation. American Chemical Society, Washington, DC, pp 562–573

Jackson ML, Stewart BR, Daniels WL (1993) Influence of fly ash, topsoil, lime and rock-P on acid mine drainage from coal refuse. In: Proc 1993 Natl Meet American Society for Surface Mining and Reclamation, Spokane, Washington, 16–19 May 1993, pp 266–276

Kleinmann RLP (1981) The US Bureau of Mines acid mine drainage research program. In: Proc 2nd West Virginia Surface Mine Drainage Task Force Symp, Clarksburg, West Virginia

Kleinmann RLP, Crerar DA (1979) *Thiobacillus ferrooxidans* and the formation of acidity in simulated coal mine environments. Geomicrobiol J 1:373–388

Ladwig KJ, Erickson PM, Kleinmann RLP (1985) Alkaline injection: an overview of recent work. In: Control of acid mine drainage. Bureau of Mines IC 9027 USDA, Bureau of Mines, Pittsburgh, pp 35–40

Lekhakul S (1981) The effect of lime on the chemical composition of surface mined coal spoils, and the leachate from spoil. PhD Diss, Agronomy Dept, Univ of Kentucky, Lexington, Kentucky, p 134
Luther GW III (1987) Pyrite oxidation and reduction: molecular orbital theory consideration. Geochem Cosmochem Acta 51:3193–3199
Luther GW III (1990) The frontier-molecular-orbital theory approach in geotechnical processes. In: Stumm W (ed) Aquatic chemical kinetics. Wiley, New York, pp 173–198
Luther III GW, Kostka JE, Church TM, Sulzberger B, Stumm W (1992) Seasonal iron cycling in the salt-marsh sedimentary environment: the importance of ligand complexes with Fe (II) and Fe (III) in the dissolution of Fe (III) minerals and pyrite, respectively. Mar Chem 40:81–103
Millero FJ, Izaguirre M (1989) Effect of ionic strength and ionic interactions on the oxidation of Fe^{2+}. J Solut Chem 18:585–599
Mortland MM (1968) Protonation of compounds on clay mineral surfaces. 9th Int Congr Soil Sci I:691–699
Mortland MM, Raman KV (1968) Surface acidity of smectites in relation to hydration, exchangeable cation and structure. Clays Clay Min 16:393
Moses CO, Herman JS (1991) Pyrite oxidation at circumneutral pH. Geochim Cosmochim Acta 55:471–482
Moses CO, Nordstrom DK, Herman JS, Mills AL (1987) Aqueous pyrite oxidation by dissolved oxygen and by ferric iron. Geochim Cosmochim Acta 51:1561–1571
Nairn RW, Hedin RS, Watzlaf GR (1991) A preliminary review of the use of anoxic limestone drains in the passive treatment of acid mine drainage. In: Proc 12th Annu West Virginia Surface Mine Drainage Task Force Symp, Morgantown, West Virginia, pp 23–38
Nairn RW, Hedin RS, Watzlaf GR (1992) Generation of alkalinity in an anoxic limestone drain. In: Proc 9th Annu Meet American Society for Surface Mining and Reclamation. Duluth, Minnesota, 14–18 June 1992
Nakamoto K (1986) Infrared and Raman spectra of inorganic and coordination compounds. Wiley, New York
Pichtel JR, Dick WA (1991) Influence of biological inhibitors on the oxidation of pyritic mine spoil. Soil Biol Biochem 23:109–116
Ritcey GM (1991) Deep water disposal of pyritic tailings. In: Proc 2nd Int Conf on the Abatement of acidic drainage, 16–18 Sept, Montreal, Quebec, pp 421–442
Rowley MV, Warkentin DD, Yan VT, Piroshco BM (1994) The biosulfide process: integrated biological/chemical acid mine drainage treatment – result of laboratory piloting. In: Proc Int Land reclamation and mine drainage Conf and 3rd Int Conf on the Abatement of acidic drainage, vol 1: mine drainage, Pittsburgh, p 205
Shellhorn M, Rastogi V (1985) Practical control of acid mine drainage using bactericides. In: Proc 6th West Virginia Surface Mine Drainage Task Force Symp, Morgantown, West Virginia
Singer PC, Stumm W (1970) Acid mine drainage: rate-determining step. Science 167:1121–1123
Spotts E, Dollhopf DJ (1992) Evaluation of phosphate materials for control of acid production in pyritic mine overburden. J Environ Qual 21:627–634
Staub MW, Cohen RRH (1992) A passive mine drainage treatment system as a bioreactor: treatment efficiency, pH increase, and sulfate reduction in two parallel reactors. In: Achieving land use potential through the reclamation. Proc 9th

Natl Meet American Society of Surface Mining and Reclamation, 14–18 June 1992, Duluth, Minnesota, pp 550–562

Stumm W, Morgan JJ (1979) Aquatic chemistry. Wiley, New York

Taylor BE, Wheeler MC, Nordstrom DK (1984a) Oxygen and sulfur compositions of sulfate in acid mine drainage: evidence for oxidation mechanism. Nature 308:538–541

Taylor BE, Wheeler MC, Nordstrom DK (1984b) Stable isotope geochemistry of acid mine drainage: experimental oxidation of pyrite. Geochim Cosmochim Acta 48:2669–2678

Turner D, McCoy D (1990) Anoxic alkaline drain treatment system, a low cost acid mine drainage treatment alternative. In: Graves DH, De Vore RW (eds) Proc 1990 Natl Symp on Mining, Lexington, Kentucky, pp 73–75

US Environmental Protection Agency (1982) Development document for effluent limitations guidelines and standards for coal mining, Washington, EPA 440/1–82/057, 660 pp

US Government Printing Office (1988) Code of federal regulations, title 40, protection of the environment, Washington, DC, parts 100–149, pp 530–531, p 608

Watzlaf GR, Hedin RS (1993) A method for predicting alkalinity generated by anoxic limestone drains. In: Proc 1993 West Virginia Surface Mine Drainage Task Force Symp, Morgantown, West Virginia, 27–28 April

Wildeman TR (1991) Drainage from coal mine. In: Peters DC (ed) Geology in coal resource utilization. Techbooks, Fairfax, Virginia, p 499

Zhang YL (1993) Chemistry of fly ash scrubber sludge components in plant–soil–water systems. PhD Diss, University of Missouri, Columbia

11 Chemical Reactions in Aquifers Influenced by Sulfide Oxidation and in Sulfide Oxidation Zones

F. Wisotzky

Ruhr-Universität Bochum, Fakultät für Geowissenschaften, Lehrstuhl Geologie III, Universitätsstr. 150, 44801 Bochum, Germany

11.1 Introduction

Pyrite oxidation is widely recognized as one of the most serious inorganic pollution sources for ground and surface waters. Because of the environmental significance, research on sulfide oxidation has been undertaken for decades (Lowsen 1982; Nordstrom 1982; Bierns De Haan 1991). Sulfides occur in metallic ores, black shales and in overburden sediments of lignite and hard coal deposits. Sulfide oxidation results in the release of large amount of sulfuric acid. This release of acidity leads to the formation of acid mine drainage (AMD), or acid rock drainage (ARD; Krothe et al. 1980; Dubrovsky et al. 1985; Van Berk 1987; Morrison et al. 1990; Blowes et al. 1991; Hedin et al. 1994; Wisotzky 1994; Wisotzky and Obermann 1995).

The area of investigation of this study is the Rhineland lignite mining area in Germany. With a stock of 55×10^9 t lignite it is the largest lignite deposit in Western Europe. Besides its lignite, the Rhineland mining area has also important aquifers of Quaternary and Tertiary age. The open pit mines in the Rhineland mining area produce 110×10^6 t lignite a year. With an average overburden (m^3) to lignite (t) ratio of 6 to 1 (Thole 1992), approximately 40×10^9 m^3 sandy overburden will have to be moved and dumped by the end of mining in the year 2045. Overburden is predominantly unlithified sandy sediments (Fig. 11.1) with an average pyritic-sulfur content of about 0.1–0.2 wt% (Wisotzky 1994). Due to the insufficient supply of oxygen and pyrite oxidation kinetics, only 10–20% of the whole pyritic-sulfur is oxidized (Obermann et al. 1993). For the opencast Garzweiler I/II, 0.036 wt% oxidized pyritic-sulfur content has been predicted (Wisotzky 1994).

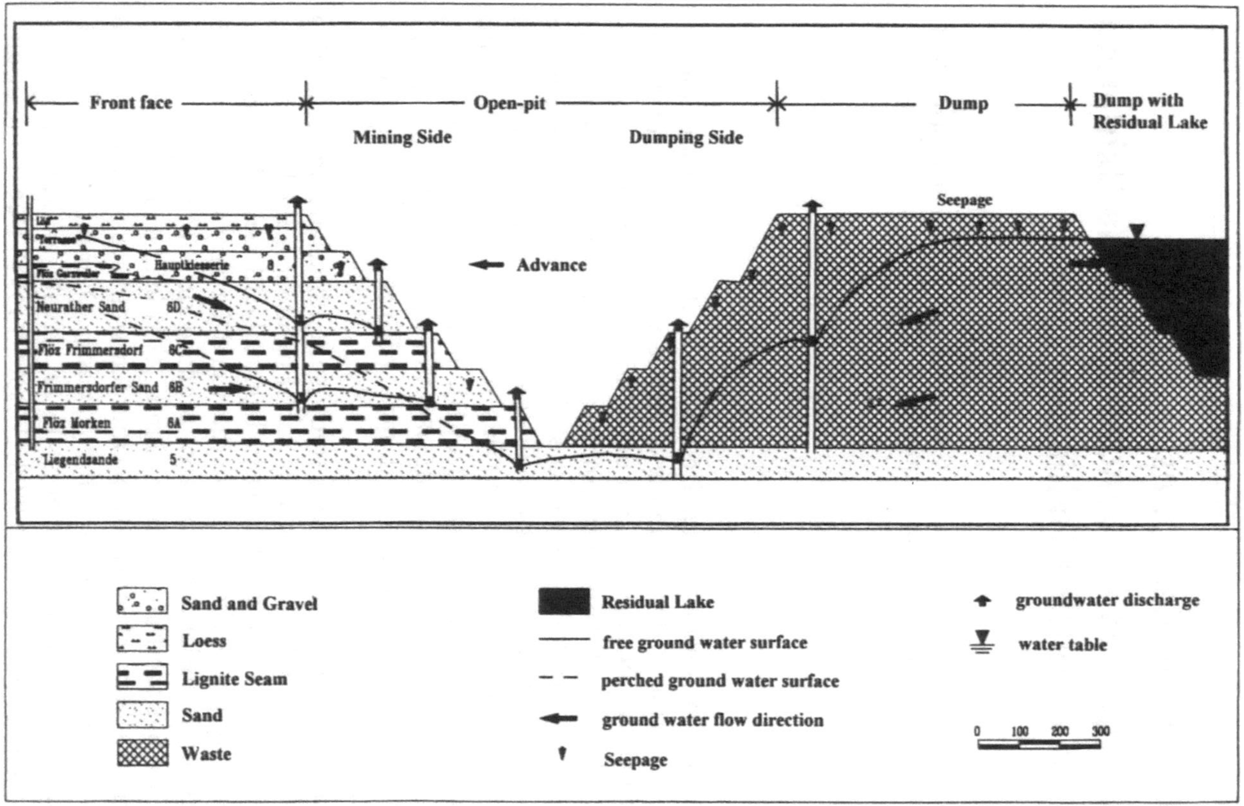

Fig. 11.1. Schematic profile across the active mining sector with location of the lowered groundwater table(s) and residual lake

11.2
Methods

Water samples were collected in experimental dumps and in overburden dump aquifers. The pH, Eh, temperature, oxygen content and electrical conductivity of the water samples were measured on site in a flow-through cell. The water was sampled by filtering a 200-ml aliquot through a Millipore filter (0.45 m) and acidified with ultrapure sulfuric acid before analyses of iron, aluminium and cation trace elements by ICP-AES (inductively coupled plasma–atomic emission spectroscopy). Another 200-ml aliquot was sent to the laboratory for determination of chloride, sulfate, nitrate, sodium, potassium, calcium and magnesium by ion chromatography. Alkalinity and dissolved carbon dioxide determination were made by titration or calculated after measuring total inorganic carbon (TIC). The used computer code PHREEQE (Parkhurst et al. 1980) is designed to perform a wide variety of aqueous geochemical calculations. Concentration of elements, molalities and activities of aqueous species, pH, pE, saturation indices and mole transfers of phases to achieve equilibrium can be calculated as a function of specified reversible and irreversible geochemical reactions.

11.3
Pyrite Oxidation and Its Modelling

Pyrite is thermodynamically stable under reducing conditions in which it is formed. In contact with an oxidizing agent, such as the oxidizing atmosphere, sulfides weather. Pyrite oxidation can be subdivided into the oxidation of sulfide-sulfur (first oxidation step) and the oxidation of ferrous iron (second oxidation step).

$$FeS_2 + 14Fe^{3+} + 8H_2O \Rightarrow 15Fe^{2+} + 2SO_4^{2-} + 16H^+ . \quad (1)$$

$$14Fe^{2+} + 7/2\ O_2 + 14H^+ \Rightarrow 14Fe^{3+} + 7H_2O. \quad (2)$$

$$FeS_2 + 7/2\ O_2 + H_2O \Rightarrow Fe^{2+} + 2SO_4^{2-} + 2H^+ . \quad (3)$$

Sulfide-sulfur oxidation requires mainly oxygen, yet the actual oxidation takes place via ferric iron [Eq. (1)]. Since large amounts of ferrous iron are released by the weathering process, there has to be a reoxidation of ferrous to ferric iron in order to maintain the process [Eq. (2)]. Equation (3) represents the overall reaction of sulfide-sulfur oxidation by oxygen acting as the oxidizing agent. The acidity produced during pyrite oxidation can be subdivided into sulfide acidity [Eq. (3)] and iron acidity

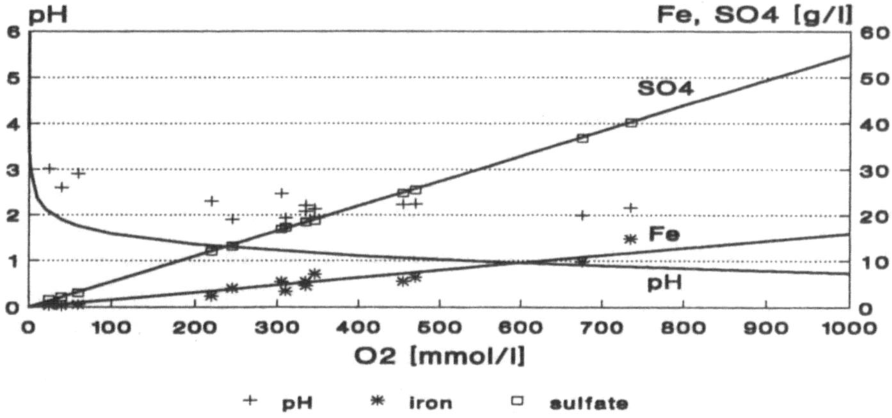

Fig. 11.2. Calculated chemistry of water (*solid lines*) by unlimited pyrite content and increasing oxygen and measured values (*dots*) of seepage water in experimental dumps at pH <4

[Eq. (10)]. The following conditions for an intensive pyrite oxidation must be fulfilled:

- Pyrite or other reactive sulfide minerals must be present.
- Contact and supply of an oxidizing agent (O_2, Fe^{3+}, NO_3^-).
- A sufficient reaction rate or reaction time respectively with a large surface area of sulfides.
- The overburden must contain only small amounts of pH-buffering minerals such as carbonates.

Using the computer code PHREEQE (Parkhurst et al. 1980), the pyrite oxidation was modelled in a stepwise manner. An unlimited quantity of pyrite was brought in contact with an increasing amount of oxygen. Without considering pH-buffering reactions with the solid phase and kinetics, the development of the water composition was modelled (Fig. 11.2). Starting with water at a neutral pH value, the sulfate and iron concentration increase linearly and the pH value decreases. The addition of 1000 mmol O_2 oxidizes 285.7 mmol disulfide-sulfur. The associated pH value is 0.74. Comparable, low pH values were found in column experiments by Kölling (1990) and Bergmann (1993). The released iron is predominantly ferrous iron (pE 0–6). The modelled water chemistry can be compared with measured values of seepage water in a pyrite oxidation zone. This comparison shows the effects of kinetics and reactions of the pyrite oxidation products with the solid phase. The measurements were carried out in experimental dumps with pyritic-sulfur content up to 0.6 wt%. The experimental dumps had a length of 8 m, a width of 6 m and a maximum depth of 3 m. Natural precipitation led to the formation of seepage water which was collected at the bottom of the dumps (for details see Wisotzky 1994). The downward-moving water collected the products of pyrite oxidation and thus reflected the water chemistry in pyrite oxidation zones. Because of the measured and modelled sulfate concentration it is possible to correlate the results (Fig. 11.2). Figure 11.2 compares measured and calculated iron concentrations. Pyrite oxidation in experimental dumps led to a stoichiometric ratio of 1 (iron) to 2 (sulfate) in seepage water at pH values < 4, corresponding to Eq. (3). The coupled pH values of the seepage water decreased, but the low values calculated were not reached. Besides pH-buffering in the aquatic phase [Eqs. (4) and (5)], this is an indication of pH-buffering reactions with the solid phase [e.g. Eq. (6)].

$$SO_4^{2-} + H^+ \Leftrightarrow HSO_4^- \qquad pK = -1.987. \tag{4}$$

$$HCO_3^- + H^+ \Leftrightarrow H_2CO_3^* \qquad pK = 6.352. \tag{5}$$

This led to an increase in pH values. Because of the minor TIC (total inorganic carbon; 0.007 wt%), it is assumed that only silicate minerals were responsible for pH-buffering in the experimental dumps. The Tertiary sediments contain quartz (> 90 wt%), clay minerals (kaolinite, illite) and only a small amount of feldspar. Because of low concentrations of sodium and potassium (< 100 mg/l), it is postulated that pH-buffering by Na–K–feldspar is not an important process. An equilibration with Na–K–feldspars would lead to sodium and potassium concentrations of about 1000 mg/l (PHREEQE calculations). Basset et al. (1992) demonstrated the involvement of clay minerals (kaolinite and montmorillonite) in pH-buffering reactions coupled with geogenic pyrite oxidation in the Rocky Mountains. Because of this, an equilibrium with kaolinite was assumed by further computations [Eq. (6); Fig. 11.3).

$$Al_2Si_2O_5(OH)_4 + 6H^+ \Leftrightarrow 2Al^{3+} + 2H_4SiO_4 + H_2O \qquad pK = -5.708. \tag{6}$$

The pH values increased and aluminium was released in the aquatic phase. This aluminium would precipitate as $Al(OH)_3$ with rising pH values [Eq. (7)].

$$Al^{3+} + 3H_2O \Leftrightarrow Al(OH)_3(am) + 3H^+ \qquad pK = 8.049. \tag{7}$$

Figure 11.3 shows the simulated concentrations of pH and aluminium in equilibrium with kaolinite, and the measured values in the seepage water of the experimental dumps. At equilibrium with kaolinite, pH values and Al-concentrations would increase and reach the simulated values. Due to the incomplete equilibrium with kaolinite, the measured values (pH, Al) did not reach the calculated values. However, the measured values indi-

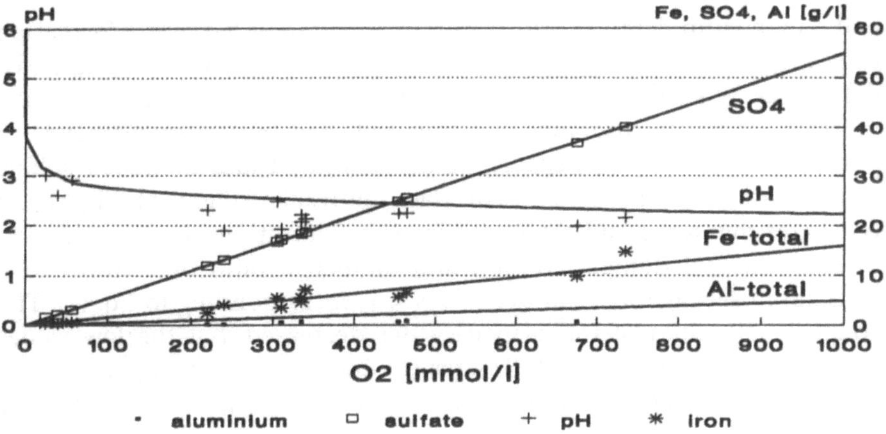

Fig. 11.3 Calculated chemistry of water (*solid lines*) with unlimited pyrite content, increasing oxygen and equilibrium with kaolinite and measured values (*dots*) of seepage water in experimental dumps at pH < 4

cated a development in the direction of calculated pH values and Al concentrations. This indicates that the dissolution of silicate minerals takes place in AMD environments. Due to the slow reaction rate of silicate dissolution (Kölling 1990), the measured values (pH, Al) were between calculated values with (Fig. 11.3) and without (Fig. 11.2) equilibrium with kaolinite. Above pH 4 only small concentrations of iron were observed in the seepage water of the dumps. In spite of sulfate concentrations of up to 1000 mg/l (corresponding to iron in pyrite: 290 mg/l) only about 10 mg/l iron was measured. Saturation index (SI) calculations [log (IAP; ion activity product)/(K; equilibrium constant)] indicated a saturation or supersaturation with respect to ferric hydroxide above pH 4 in the water samples. Therefore, the iron concentrations were limited by a precipitation of $Fe(OH)_3$ under aerobic conditions at pH values > 4 [Eqs. (8)–(10)].

$$Fe^{2+} + 1/4O_2 + H^+ \Leftrightarrow Fe^{3+} + 1/2H_2O. \quad (8)$$

$$Fe^{3+} + 3H_2O \Leftrightarrow Fe(OH)_3(am) + 3H^+. \quad (9)$$

$$Fe^{2+} + 1/4O_2 + 2.5H_2O \Leftrightarrow Fe(OH)_3(am) + 2H^+. \quad (10)$$

The reaction mechanism of ferrous iron oxidation and hydrolysis of ferric iron leads to a further release of protons [Eq. (10)]. The system develops to low pH values. Below pH 4, the precipitated ferric iron dissolves (Fig. 11.4). This is important, because the release of ferric iron is responsible for an intensification of pyrite oxidation [Eq. (1)]. By an addition of pH-buffering substances such as calcite, the pH value and the coup-

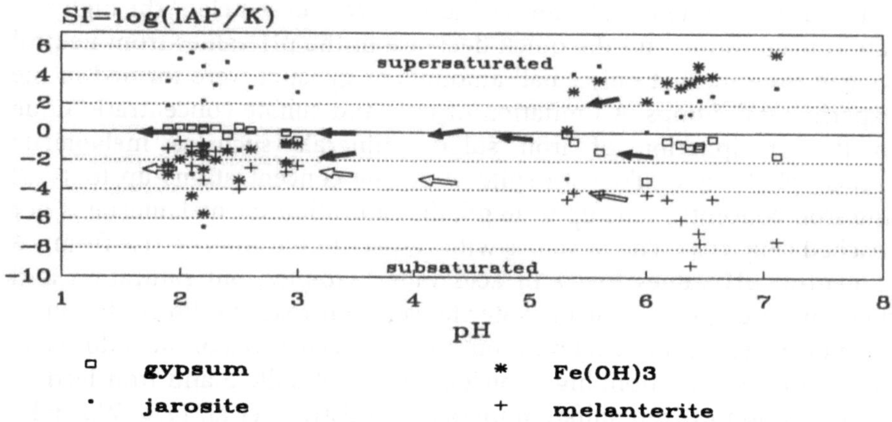

Fig. 11.4. Calculated saturation index (SI) values [log (IAP; ion activity product/K; equilibrium constant)] of secondary mineral phases versus pH values as master variable of seepage water samples from experimental dumps (*arrows* show development by increasing pyrite oxidation)

led precipitation of ferric iron decrease the oxidation rate of pyrite. The addition of pH-buffering substances has therefore two important effects on pyrite oxidation:

1. Neutralization of released acidity.
2. Decrease in pyrite oxidation rate.

The complete oxidation of sulfide-sulfur and ferrous iron leads to the release of 4 mol protons per mole pyrite [Eq. (11)].

$$FeS_2 + 15/4\ O_2 + 3.5\ H_2O \Rightarrow Fe(OH)_3(am) + 2SO_4^{2-} + 4H^+. \qquad (11)$$

The second oxidation step of pyrite (oxidation of ferrous iron) takes place in pyrite oxidation zones after oxidation of sulfide-sulfur or by contact of an anaerobic iron-rich groundwater with atmospheric oxygen. In residual lakes this leads to pH values between 2 and 4 depending on the ferrous iron concentration of the groundwater (Pietsch 1979; Obermann et al. 1993; Klapper and Schultze 1994).

Besides saturation calculations for $Fe(OH)_3$, calculations were done for the secondary minerals melanterite ($FeSO_4 \cdot 7H_2O$), hydronium-jarosite [$HFe_3(OH)_6(SO_4)_2$] and gypsum ($CaSO_4 \cdot 2H_2O$). These minerals were observed in the Garzweiler open-cast mine (Reichel 1991; Wisotzky 1994) and were found in AMD environments (Nordstrom et al. 1979; Blowes et al. 1991; Cravotta 1994). Above pH 5.5, the seepage water is subsaturated (undersaturated) with respect to gypsum. However, water samples with pH values below 3 have SI values of about zero. Nevertheless, the high sulfate concentration and the quick decrease in the pH values from neutral to acid indicate that only small amounts of gypsum were formed in the experimental dumps. A limitation of iron and sulfate concentration due to the precipitation of iron sulfate minerals such as melanterite ($FeSO_4 \cdot 7H_2O$) is unlikely. In spite of sulfate concentrations up to 55 g/l and iron concentrations up to 16 g/l, the saturation for melanterite is not reached (Fig. 11.4). The SI values with respect to melanterite rise from −6 at neutral pH values to −2 in acid water samples, but saturation and therefore precipitation of melanterite were not reached. Precipitation of melanterite seems to be relevant only at the surface area of the sediments. Because of evaporation, high concentrations of sulfate and iron lead to the formation of iron sulfate minerals (Nordstrom et al. 1979; Wisotzky 1994). The strong supersaturation with respect to hydronium-jarosite was also found by other researchers in AMD environments (Nordstrom et al. 1979; Blowes et al. 1991). The oxidation products (SO_4^{2-}, H^+, $Fe^{2+/3+}$) are stored in acidic overburden material in the connate water or in the form of soluble mineral phases. Besides the main oxidation products, pyrite oxidation leads to the release of Ca, Al, Co, Ni, Zn and As in the Rhineland lignite mining area (Wisotzky 1994). After the overburden dump is filled with groundwater, stored products of pyrite oxidation are leached into the aquatic phase.

11.4 Chemical Reactions in Lignite Overburden Dumps

Without seepage and groundwater flow the stored pyrite oxidation products react only with the solid phase in the pyrite oxidation zones. By filling the overburden dump with groundwater the soluble pyrite oxidation products (mainly SO_4^{2-}, H^+, $Fe^{2+/3+}$) are leached into the water (Fig. 11.1). Acidity, sulfate and ferrous iron are transported by the migrating water to the non-acidic overburden material in the dump aquifer. The non-acidic overburden material is predominant (about 80 wt%) in the dumps. Carbonate phases (immobile phase) preserved in non-acidic overburden react with the migrating products of pyrite oxidation (mobile phase). Contrary to carbonate in the pyrite oxidation zones, e.g. of the experimental dumps, the carbonate mineral content cannot be neglected in dump aquifers.

Up to now there are only two overburden dumps with partially oxidized pyritic overburden which have been filled with groundwater in the Rhineland lignite mining area. One of these dumps is the Berrenrath overburden dump south-east of the mining area. The area of the dump is approximately 100 ha with an average thickness of 40 m (20 m unsaturated). The dump contains unlithified sediments with varying sediment compositions, especially with respect to different contents of oxidized pyritic-sulfur. Thus, it is impossible to investigate the hydrochemical interactions between solid and aquatic phases on a groundwater flow path. However, the increasing content of pyrite oxidation products in the sediments is coupled with an increasing release of protons into the groundwater. Therefore it is possible to use the pH value as a master variable in dump aquifers. As indirect evidence of the reactions, saturation index (SI) calculations are used in this chapter.

The groundwater of the Berrenrath dump has pH values between 7.5 and 3.5. Therefore the pH values in the groundwater are higher than the pH values in pyrite oxidation zones (Figs. 11.2–11.4). Besides dilution effects, this is caused by an interaction of acid and pH-buffering sediment contents. For example, the released sulfuric acid reacts with calcite and leads to pH values between 4 and 8 [Eqs. (12)–(14)].

$$H_2SO_4 + 2CaCO_3 \Rightarrow 2Ca^{2+} + SO_4^{2-} + 2HCO_3^- \qquad pH \approx 8.2. \quad (12)$$

$$3/2H_2SO_4 + 2CaCO_3 \Rightarrow 2Ca^{2+} + 3/2SO_4^{2-} + HCO_3^- + H_2O + CO_2 \qquad pH \approx 6.3. \quad (13)$$

$$2H_2SO_4 + 2CaCO_3 \Rightarrow 2Ca^{2+} + 2SO_4^{2-} + 2H_2O + 2CO_2 \qquad pH \approx 4.3. \quad (14)$$

The neutralization of the sulfuric acid "consumes" calcite. To neutralize 1 mol acid, 1.3–2 mol calcite is necessary, using the simplified reaction of Eqs. (12) and (13). However, the release of the "cation acid" ferrous iron by pyrite oxidation requires also calcite to neutralize [Eq. (17)] the overall acidity (sulfide and iron acidity). The iron acidity is not neutralized by using the stoichiometry of Eqs. (12) and (13). Because of the resulting pH value of Eq. (14) (pH ≈ 4.3), we cannot call this reaction "neutralization".

In the groundwater of the Berrenrath dump, bicarbonate concentrations up to 1000 mg/l were found at a pH value of about 6.3. The dissolution of calcite also enhances the calcium concentration of the groundwater. This leads to the precipitation of gypsum. Figure 11.5 shows a saturation of the groundwater with respect to calcite between pH values of 7.5 and 6.3. Below pH 6.3, the water samples are calculated and measured unsaturated with respect to calcite, but saturated with respect to gypsum (Fig. 11.5). Because of the release of sulfuric acid, the unstable calcium-phase calcite dissolves and the stable calcium-phase gypsum is formed (Fig. 11.5) in dump aquifers. If the measured pH value is below 6, a complete dissolution of calcite is found in the dump aquifer (Fig. 11.5). In spite of saturation of the groundwater with respect to gypsum below pH 6, the sulfate concentrations are not constant and reach values up to 7000 mg/l.

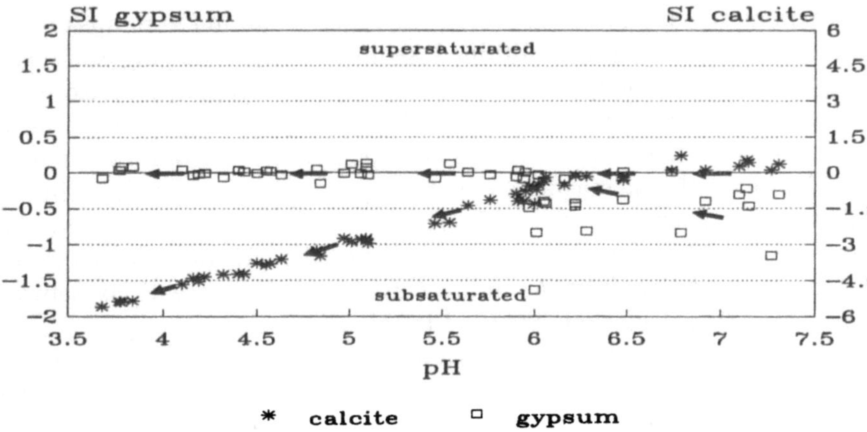

Fig. 11.5. Calculated saturation index (SI) values with respect to calcite and gypsum versus pH values of groundwater samples from the Berrenrath dump aquifer (*arrows* show development with increasing pyrite oxidation products)

This is possible due to the formation of complexes ($FeSO_4$, $CaSO_4$, etc.) and because of the decrease in activity coefficients with increasing ionic strength.

The dissolution of calcite leads to a release of calcium and CO_2-species (CO_3^{2-}, HCO_3^-, $H_2CO_3^*$). Due to decreasing pH values, high concentrations of CO_2 were found in the groundwater and in the soil gas above the groundwater. The CO_2 concentration in the unsaturated zone reached values up to 22 vol% (Wisotzky 1994). Highest CO_2 concentrations in the soil gas were found directly above the groundwater table, indicating the outgassing of CO_2. Figure 11.6 shows the calculated CO_2 partial pressures of groundwater samples versus pH values. Starting with pH 7.3, the CO_2 partial pressure increases with decreasing pH values. Maximum values are found at pH 6.3, if calcite dissolves completely. Besides dissolution of calcite, the transformation of bicarbonate to carbon dioxide takes place:

$$HCO_3^- + H^+ \Rightarrow CO_2 + H_2O. \tag{15}$$

Below pH 6 a decrease in CO_2 partial pressure to values of about 0.25 atm was found in the Berrenrath dump. The development of the SI values (Fig. 11.5) and the CO_2 partial pressure (Fig. 11.6) indicate the complete dissolution of calcite below pH 6.3. The addition of crushed limestone would prevent the acidification and lead to an increasing precipitation of gypsum in dump aquifers. Above pH 6.3 the iron concentrations are small (< 100 mg/l) despite sulfate concentrations of up to 1600 mg/l. The released iron of the pyrite oxidation process can decrease

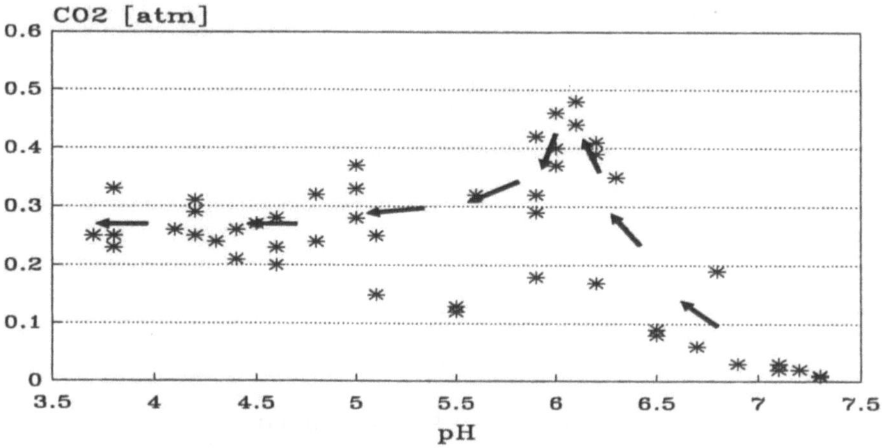

Fig. 11.6. Calculated CO_2 partial pressure (PHREEQE calculations) of water samples of the Berrenrath dump versus pH values (*arrows* show development with increasing pyrite oxidation products)

under aerobic [Eq. (10)] and anaerobic [Eqs. (16) and (17)] conditions. In the dump aquifer the concentration of ferrous iron decreases because of cation exchange reactions [Eq. (16)] and/or Ca-siderite precipitation [Eq. (17)] at pH > 6.3.

$$Ca{-}X + Fe^{2+} \Rightarrow Fe{-}X + Ca^{2+}. \quad (16)$$

$$CaCO_3 + Fe^{2+} \Rightarrow Fe_{0.9}Ca_{0.1}CO_3 + 0.9\ Ca^{2+} + 0.1\ Fe^{2+}. \quad (17)$$

Investigations of contaminant migration from uranium-tailing impoundments in Canada indicate the replacement of calcite by siderite (Morin and Cherry 1986; Ptacek and Blowes 1994). The formation of Ca-siderite leads to a decrease of dissolved ferrous iron but also to a decrease in the calcite content of the dump sediments [Eq. (17)]. Water samples from the Berrenrath dump with pH values > 6.3 contain a maximum iron concentration of 100 mg/l, because Ca-siderite is formed (Fig. 11.7). Between pH values of 6.3 and 5 the iron concentration in the groundwater increases with decreasing pH up to 500 mg/l. In this pH range the secondarily produced Ca-siderite buffers the pH value after calcite is completely removed (Fig. 11.7). Iron concentrations up to 3500 mg/l were observed in the groundwater at pH values below 5. Generally, the iron concentrations increase with decreasing pH values in the groundwater. In spite of the high ferrous iron concentration in the water, a saturation with respect to Ca-siderite is not reached at pH values < 5 (Fig. 11.7). Hydrogeochemical calculations of Van Berk and Wisotzky (1995) show that

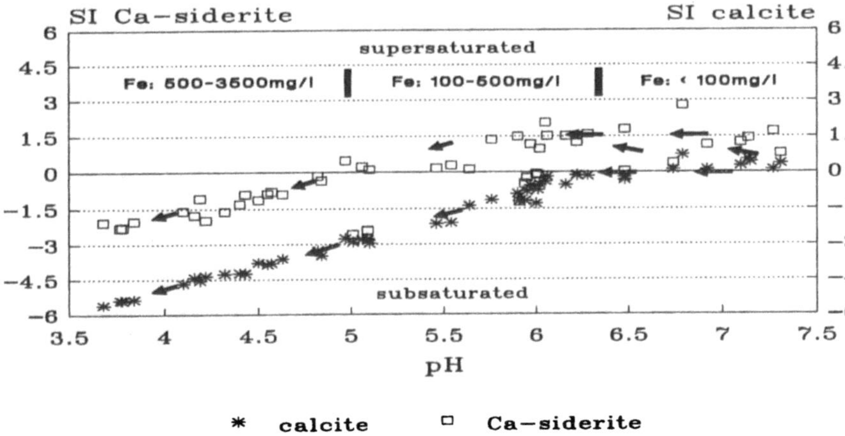

Fig. 11.7. Calculated saturation index (SI) values with respect to calcite and Ca-siderite versus pH values of groundwater samples from the dump Berrenrath (*arrows* show development with increasing pyrite oxidation products)

the groundwater composition in dump aquifers can be explained in principle by a simultaneous equilibrium of Ca-siderite, gypsum and iron hydroxide.

It has been shown that the computations and measurements indicate the importance of carbonate phases in pyrite oxidation zones and in dump aquifers. So, an addition of crushed limestone to overburden sediments leads, in contact with the pyrite oxidation products, to increasing pH values and a decreasing dissolved iron concentration. This decrease in dissolved iron is associated with a decrease in trace metals such as Co, Ni and Zn (Wisotzky 1994). The replacement of calcite by Ca-siderite removes a substantial portion of the dump aquifer acid neutralization capacity (ANC). This must be taken into account by calculating the necessary calcite addition to dumps to prevent the formation of AMD.

11.5 Conclusions

The mobility of iron in acid mine drainage (AMD) environments is limited by the precipitation of different minerals under aerobic and anaerobic conditions. In pyrite oxidation zones the formation of ferric hydroxide [$Fe(OH)_3$] at $pH > 4$ is important. This precipitation has an effect on the pyrite oxidation rate, because ferric iron is involved in sulfide oxidation reactions. At pH values < 4, iron and sulfate as pyrite oxidation products are observed in a stoichiometric ratio of 1 to 2 in the aqueous phase. In dump aquifers under anaerobic conditions, the replacement of calcite by Ca-siderite limits the ferrous iron and trace metal concentrations (Co, Ni, Zn) in groundwaters at pH values > 6.3. After depletion of calcite, iron carbonate acts as a pH-buffer and high concentrations of ferrous iron are observed in the groundwater. The released sulfuric acid by pyrite oxidation reacts with calcite predominantly in the groundwater, and gypsum is formed. Both the replacement of calcite by Ca-siderite and the neutralization of sulfuric acid "consume" the calcite or acid neutralization capacity (ANC) of dump sediments. This must be taken into account by calculating the necessary calcite addition to dumps to prevent the formation of acid mine drainage.

References

Basset RL, Miller WR, McHugh J (1992) Simulation of natural acid sulfate weathering in an alpine watershed. Water Resour Res 28:2197–2209

Bergmann A (1993) Chemische und mikrobiologische Ansätze zur Verminderung der Pyritoxidation in sandigem Abraummaterial des Braunkohlentagebaus Garzweiler I in Laborversuchen. Ruhr-Universität Bochum, Bochum, p 151 (unpublished)

Bierns De Haan S (1991) A review of the rate of pyrite oxidation in aqueous systems at low temperature. Earth Sci Rev 31:1–10

Blowes DW, Reardon EJ, Jambor JL, Cherry JA (1991) The formation and potential importance of cemented layers in inactive sulfide mine tailings. Geochim Cosmochim Acta 55:965–978

Cravotta CA (1994) Secondary iron sulfate minerals as sources of stored acidity and ferric ions in acidic groundwater at a reclaimed coal mine in Pennsylvania. In: Alpers CN, Blowes DW (eds) Environmental geochemistry of sulfide oxidation. Am Chem Soc, Washington, DC, pp 345–364

Dubrovsky NM, Cherry JA, Reardon EJ (1985) Geochemical evolution of inactive pyrite tailings in the Elliot Lake Uranium District. Can Geotech J 22:110–128

Hedin RS, Narin RW, Kleinmann LP (1994) Passive treatment of coal mine drainage. In: United States Department of the Interior, Bureau of Mines (ed) Information Circular 1994, Pittsburgh, p 35

Klapper H, Schultze M (1994) Erarbeitung der wissenschaftlichen Grundlagen für die Gestaltung, Flutung und Wassergütebewirtschaftung von Bergbaurestseen. GKSS – Institut für Gewässerforschung, Magdeburg, unpublished report, p 30

Kölling M (1990) Modellierung geochemischer Prozesse im Sickerwasser und Grundwasser. Berichte aus dem Fachbereich Geowissenschaften der Universität Bremen 8, p 135

Krothe NC, Edkins JE, Schubert JP (1980) Leaching of metals and trace elements from sulfide-bearing coal waste in southwestern Illinois. National symposium of surface mining, hydrology, sedimentology and reclamation. University of Kentucky, Lexington, pp 455–463

Lowsen RT (1982) Aqueous oxidation of pyrite by molecular oxygen. Chem Rev 82:461–497

Morin KA, Cherry JA (1986) Trace amounts of siderite near a uranium-tailing impoundment, Elliot Lake, Ontario, Canada, and its implication in controlling contaminant migration in a sandy aquifer. Chem Geol 56:117–134

Morrison JL, Scheetz BE, Strickler DW, Willams EG, Rose AW, Davis A, Parizk R (1990) Predicting the occurrence of acid mine drainage in the Alleghenian coal-bearing strata of western Pennsylvania – an assessment by simulated weathering (leaching) experiments and overburden characterisation. In: Chyi LL, Chou C-L (eds) Recent advances in coal geochemistry. Geol Soc Am Spec Pap 248:87–99

Nordstrom DK, Jenne EA, Ball JW (1979) Redox equilibria of iron in acid mine waters. In: Jenne EA (ed) Chemical modeling in aqueous systems. Am Chem Soc Symp Ser 93:51–79

Nordstrom DK (1982) Aqueous pyrite oxidation and the consequent formation of secondary iron minerals. In: Kittrick JA (ed) Acid sulfate weathering. Soil Sci Soc Am Spec Publ 10:37–57

Obermann P, Van Berk W, Wisotzky F (1993) Endbericht über die Untersuchungen zu den Auswirkungen der Abraumkippen im Rheinischen Braunkohlenrevier auf die Grundwasserbeschaffenheit. Ruhr-Universität Bochum, Bochum, p 181 (unpublished report)

Parkhurst DL, Thorstenson DC, Plummer LN (1980) PHREEQE – a computer program for geochemical calculations. In: US Geol Survey (ed) US Geol Survey Water Resour Invest Rep 80–96, Washington, DC, p 210

Pietsch W (1979) Zur hydrochemischen Situation der Tagebauseen des Lausitzer Braunkohlen-Reviers. Arch Naturschutz Landschaftsforsch 19:97–115

Ptacek CJ, Blowes DW (1994) Influence of siderite on the pore-water chemistry of inactive mine-tailings impoundments. In: Alpers CN, Blowes DW (eds) Environmental geochemistry of sulfide oxidation. Am Chem Soc, Washington, DC, pp 172–189

Reichel F (1991) Geologische Verhältnisse sowie geochemisch/mineralogische Veränderungen bei Braunkohlennebengesteinen im Bereich der Tagebaue Garzweiler I und II der Rheinbraun AG. Ruhr-Universität Bochum, Bochum, p 100 (unpublished)

Rheinbraun AG (1992) Braunkohle – Beitrag zur Energieversorgung der Bundesrepublik Deutschland. Cologne, p 36

Thole B (1992) Das Rheinische Braunkohlenrevier. Braunkohle 7:47–56

Schwertmann P, Süsser P, Nätscher L (1987) Protonenpuffer – Substanzen in Böden. Z Pflanzenernähr Bodenk 150:174–178

Van Berk W (1987) Hydrochemische Stoffumsetzungen in einem Grundwasserleiter – beeinflußt durch eine Steinkohlenbergehalde. Besondere Mitt Dtsch Gewässerkdl Jahrb 49:175

Van Berk W, Wisotzky F (1995) Sulfide oxidation in brown coal overburden and chemical modeling of reactions in aquifers influenced by sulfide oxidation. Environ Geol 26:192–196

Wisotzky F (1994) Untersuchungen zur Pyritoxidation in Sedimenten des Rheinischen Braunkohlenreviers und deren Auswirkungen auf die Chemie des Grundwassers. Besondere Mitt Dtsch Gewässerkl Jahrb 58:153

Wisotzky F, Obermann P (1995) Hydrogeochemie der Pyritoxidation am Beispiel des Rheinischen Braunkohlenreviers. In: Niedersächsisches Landesamt für Bodenforschung (ed) Grundwassergüteentwicklung in den Braunkohlegebieten der neuen Länder 1. Schweizerbart'sche Verlagsbuchhandlung, Stuttgart, pp 167–183

12 Oxygen as a Limiting Factor for Pyrite Weathering in the Overburden of Open Pit Lignite Areas*

A. Prein[1] *and R. Mull*[2]

[1] Professor Mull & Partner GmbH, Osteriede 5, 30827 Garbsen, Germany
[2] Institute of Water Resources, Hydrology and Agricultural Engineering, University of Hannover, Appelstr. 9 a, 30167 Hannover, Germany

12.1 Introduction

The overburden of lignite areas has to be moved in order to obtain coal. This overburden very often contains pyrite. During hauling, pyrite weathering begins due to drastic changes in environmental conditions in the overburden. Oxidation processes create sulphate, iron and hydrogen ions. Iron bacteria catalyse the reaction of pyrite weathering (Singer and Stumm 1970; Brierley 1978; Stumm and Morgan 1984; Tuovinen and Kelly 1984; Karavaiko 1988; Ahonen and Tuovinen 1989; Kölling 1990; Beveridge and Southam 1992; Curutchet 1992; Prein 1993). The components resulting from the weathering reaction are washed out into the groundwater, where they cause groundwater pollution.

The amount of weathering products depends on the availability of oxygen. A so-called mixed-cell model was applied to simulate the transport of oxygen in the soil air and the seepage water, the biochemical processes and the transport of the components resulting from the weathering of pyrite. As a result of these calculations, the expected groundwater pollution in lignite areas can be estimated.

Figure 12.1 shows a technological cross-section of a mining pit. The overburden is picked up on the right-hand side in Fig. 12.1 and deposited on the left-hand side. This process creates temporary surfaces. These surfaces are exposed to atmospheric influence for a relatively short time (exposure time), which depends on the progress of winning coal. The overburden investigated here is assumed to be situated in the vadose zone.

---

* This contribution was made within the scope of a graduate study by A. Prein, "substance fluxes in water and soil" at the University of Hannover, sponsored by the Deutsche Forschungsgemeinschaft.

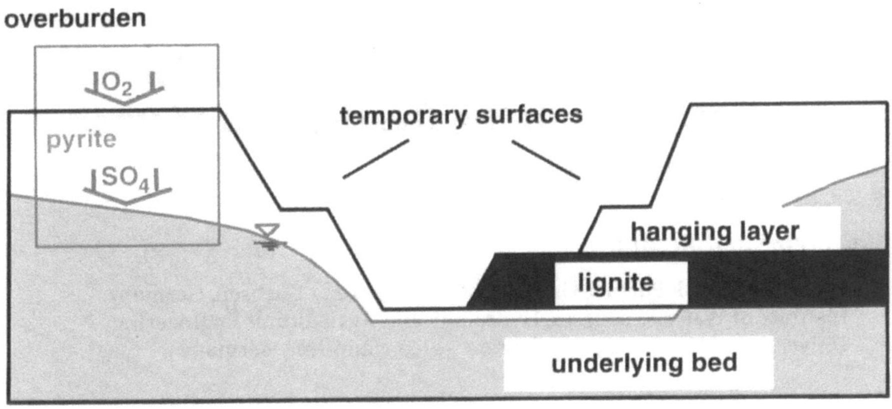

Fig. 12.1. Technological cross section of a mining pit

12.2 Chemical Model and Microbiology of Pyrite Weathering

12.2.1 Chemical Model

The direct oxidation of pyrite follows Eq. (1) in Fig. 12.2. However, this initial reaction is of subordinate importance for further pyrite weathering because of the slow reaction rate in comparison with Eq. (2). The pyrite oxidation by Fe^{3+} with increased acid production [Eq. (3)] is essential. It depends on the quantity of Fe^{3+}, which is produced following Eq. (2). Here, the Fe^{3+}-production is determined by the availability of oxygen at the reaction spot. Therefore the reaction following Eq. (2) is the rate-limiting step of the pyrite weathering. Figure 12.3 shows that iron bacteria catalyse the oxidation of Fe^{2+} to Fe^{3+} to obtain energy for their metabolism. By a system of reactions of hydrolysis and solution the pH drops to values of about 2–3, which are optimal for the bacteria.

12.2.2 Environmental Factors

In the literature, many differing statements concerning the optimal living conditions of *Thiobacillus* bacteria can be found. The essential explanation for this is the great variety of species, but also the fact that in nature

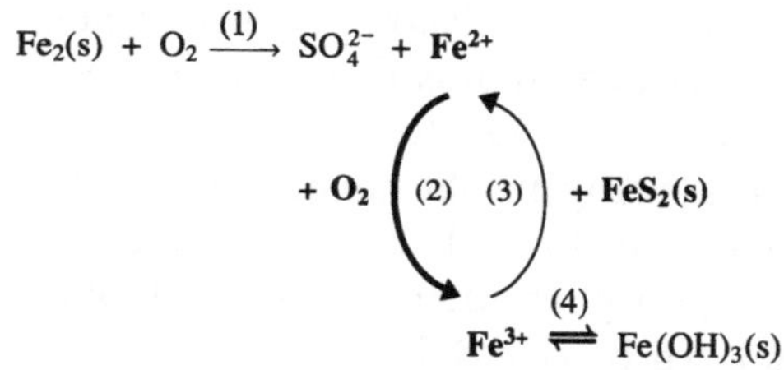

Fig. 12.2. Chemical model of pyrite oxidation. (Modified after Stumm and Morgan 1984)

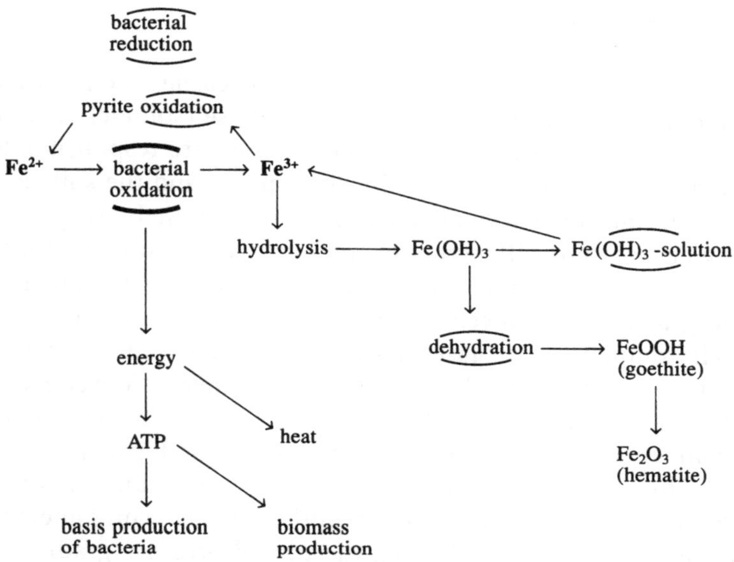

Fig. 12.3. Pyrite oxidation with *Thiobacillus ferrooxidans*. (Modified after Karavaiko 1988)

only mixed populations exist. Acidophile iron bacteria grow in a pH range from 0.5 to 6. The optimum lies between pH 1.6 and 4.5. Concerning the temperature, Näveke (1991) supposed and Prein (1993) showed that the temperature rise caused by the pyrite weathering in lignite areas is low.

Pyrite weathering is catalysed by *Thiobacillus ferrooxidans* and mixed populations dominated by them. The activity of these bacteria depends decisively on the availability of oxygen. If there is a lack of oxygen the weathering reaction breaks down. The pyrite-oxidizing bacteria are carbon-autotrophic, i.e. they supply the demand on carbon from carbon dioxide, reducing it to organic cell components. Even in a low-carbonated environment, sufficient carbon dioxide is released by the reaction of bacteria-originated acid with carbonate.

Now the nutrients nitrogen, phosphate, sulphur and iron have to be considered. The pyrite-oxidizing bacteria are able to assimilate nitrogen from the air. Their phosphate demand is low (approx. 3% of bacteria dry substance) and can be covered from the surrounding rock. Sulphur can be assimilated from amply available sulphate released during pyrite weathering. Also, iron is available sufficiently.

12.3 Transport Mechanisms

Oxygen necessary for the weathering reaction reaches the reaction zone by convection and diffusion in the gaseous phase (soil air) as well as in the liquid phase (seepage water). Several factors are responsible for the convective gas exchange between the atmosphere and the soil air: oscillations of groundwater level, displacement of soil air by seepage water, changing air pressure, differences in air density as a result of temperature difference and because of the pressure and suction effects of wind. However, the gradients of concentration created by consumption or release of soil air components or gases dissolved in water lead to the diffusive transport of gases in the direction of the gradient. Convective movement of soil air can be caused by the consumption of one of the soil air components (e.g. oxygen) and the so formed gradient of total pressure. The maintenance of this slight pressure gradient is only possible by simultaneous diffusion of the other soil air components (here nitrogen and carbon dioxide) along the partial pressure gradient in the opposite direction.

With reference to the reaction zone, the following concentration gradients for components of the soil air result (Fig. 12.4). The concentration of oxygen decreases from the surface of overburden to the reaction zone due to the oxygen-consuming weathering reaction, besides other oxygen-consuming processes, and cannot increase below the reaction zone because of the absence of sources. The concentration of nitrogen increases from the surface of overburden to the reaction zone because of the consumption of oxygen in the reaction zone and the formation of a total pressure gradient in the direction of the reaction zone. Due to this, a moving of soil air is possible. It brings oxygen, nitrogen and carbon dioxide to the reaction zone (inert gases are neglected). The so formed concentration gradient in the direction of the overburden surface causes a diffusive transport in this direction. As a result of buffer reaction (carbo-

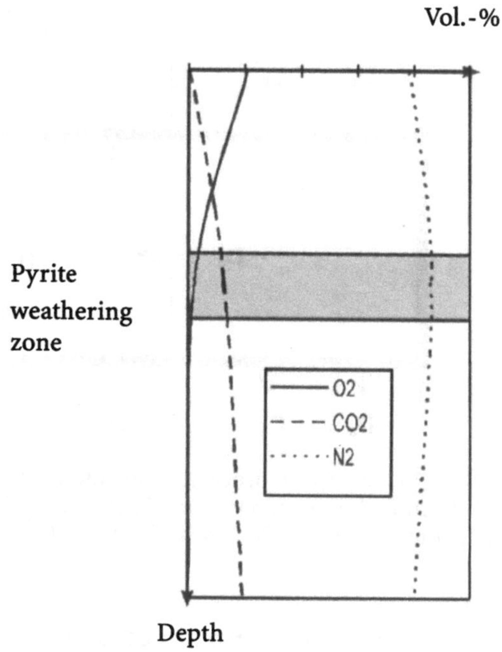

Fig. 12.4. Concentration gradients for oxygen, nitrogen and carbon dioxide in the gaseous phase

nate buffer), a source of carbon dioxide from the groundwater below the reaction zone is possible. Thereby a total pressure gradient in the upward direction can be created, which would induce a convective and diffusive transport in this direction.

12.4 Model Assumptions

The investigated part of the overburden is bordered by the atmosphere at the top. The lower boundary lies in the water-unsaturated soil. Oxygen needed for pyrite weathering is transported to the reaction zone with the soil air and the seepage water. For one element of the mixed-cell model the various fluxes considered in the numerical model are shown in Fig. 12.5. The following boundary conditions are imposed: The air pressure on the top of the column oscillates in time within a determined amplitude depending on local conditions (Farrier and Massmann 1992) and is assumed constant. The air pressure below the column is equal to the air pressure in the lowest element of the column. Below the column, the water-saturated zone begins. This is why the concentration gradients of all components in the gaseous phase are zero.

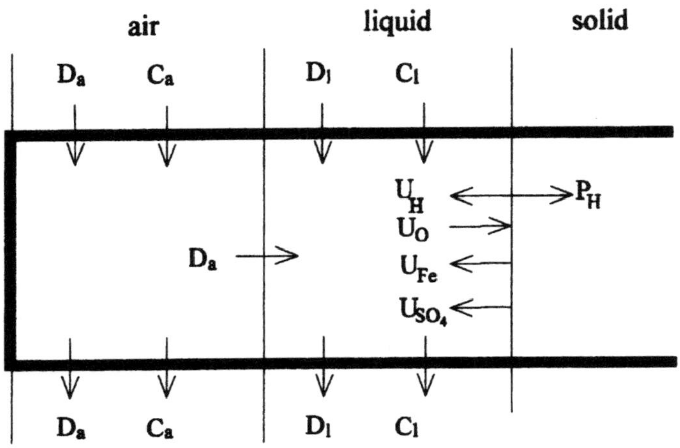

Fig. 12.5. The transport processes considered for an element of the mixed-cell model. *D* Diffusion; *C* convection; *U, P* chemical source reaction. Indices: *a* air; *l* liquid; *H* hydrogen ion; *O* oxygen; *Fe* iron; SO_4 sulphate ion

For $z = 0$: $p_{air} = f(t)$ or $p_{air} = \text{constant}$.

For $z = z_{\text{lower boundary}}$: $dp/dz = 0$ and $dc/dz = 0$.

For the simplification of the liquid movement an average of the recharge is considered. The rate of pyrite weathering is according to Singer and Stumm (1970):

For pH > 4.5: $-d[Fe^{2+}]/dt = k[Fe^{2+}][OH^-]^2 \, pO_2$,
with $k = 1.33 \cdot 10^{12} \, l^2 \, mol^{-2} \, (1013.25 \, hPa)^{-1} \, s^{-1}$.

For lower pH: $-d[Fe^{2+}]/dt = k[Fe^{2+}] \, pO_2$,
with $k = 1.67 \cdot 10^{-9} \, (1013.25 \, hPa)^{-1} \, s^{-1}$.

12.5 Extension and Dynamic Displacement of the Pyrite Weathering Zone in the Overburden

Considering all described processes, it is possible to show differences in the distribution of pyrite for sufficient and limited oxygen supplies. If the oxygen supply is not limited, pyrite is weathered over the whole depth. If the oxygen supply is limited though, a pyrite-weathering zone is formed. Such zones are observed in existing overburden soils in the Lausitz region, about 100 km south of Berlin (Fig. 12.6). Their extension ranges between a few decimeters and a few meters. In the course of the years,

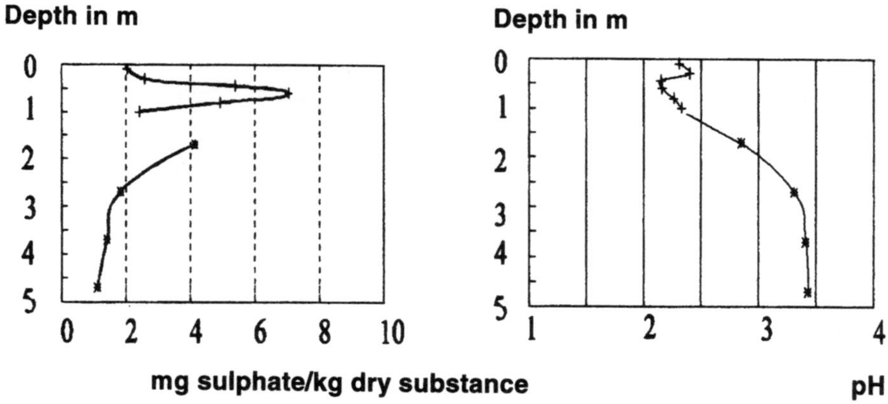

Fig. 12.6. Contents of sulphate in and pH values of the water of soil samples originating from overburden

they are displaced to greater depth, whereby the reaction rate decreases because of the greater diffusion distance.

12.6 Examples – Sulphate Input into Groundwater Below Idealised Overburden in the Lausitz Region

In order to estimate the sulphate output from the overburden over time, the formation of the overburden has to be considered. In the lignite mining area of the Lausitz region, the thickness of several layers of overburden is determined by the applied mine bridge technology, a special technology to move the overburden (see also Prein 1993). In the top and bottom layers, Quaternary material is situated. This material has a low content of pyrite, a low proton buffering capacity and a small air-filled porosity. The material of the two layers located in the middle of the overburden is Tertiary with a high content of pyrite, a high proton buffering capacity and a high air-filled porosity (Fig. 12.7).

Since it is difficult to obtain precise data concerning the composition of the overburden, an idealised structure of an overburden is assumed. The distribution of the pyrite content and the buffer capacity are shown in Fig. 12.8. If the overburden is built up in four layers as shown in Fig. 12.7, there are four stages of initial distribution of oxygen and sulphate in the liquid phase, as shown in Fig. 12.9. During this time, the pH does not drop below 3. As initial conditions, saturation of the seepage water and of the soil air in the overburden with oxygen is assumed. This

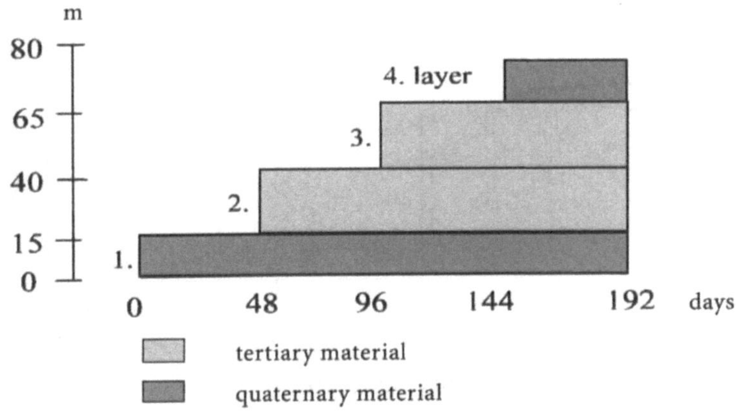

Fig. 12.7. Schematic development of overburden with time

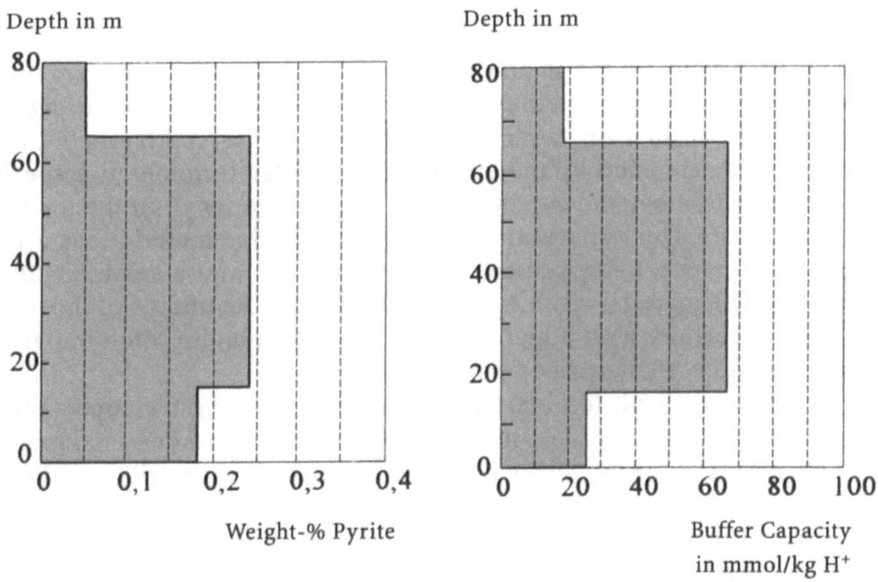

Fig. 12.8. Distribution of pyrite content and buffer capacity of the examined idealised overburden

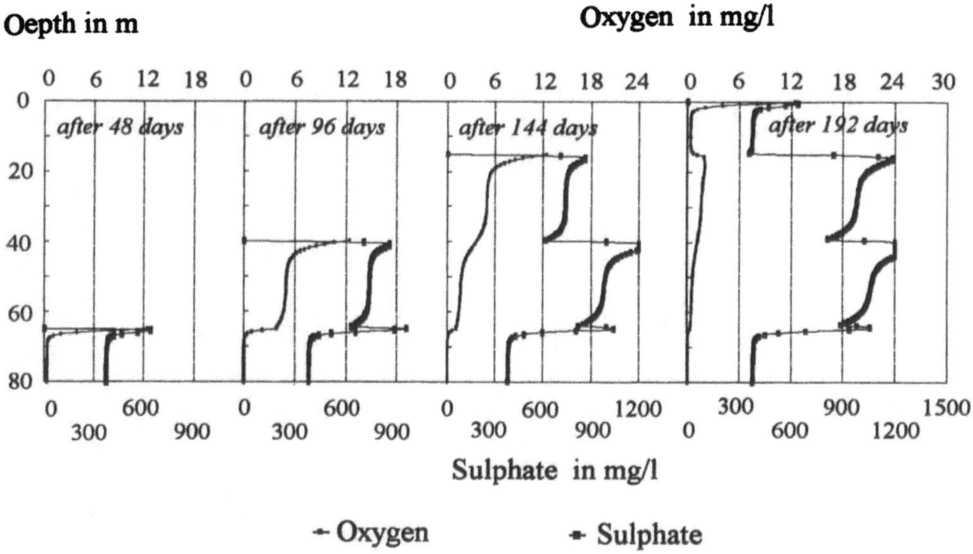

Fig. 12.9. Stages of initial distribution of oxygen and sulphate in the overburden layers during the stepwise buildup of overburden in case of available buffer capacity

is a justified assumption considering the mixing during the transport of overburden material. The initial sulphate content resulting from the weathering process before the dumping of the overburden is given as well.

Thus, from the formation of the overburden the initial distribution for a long-time simulation is obtained. The results of this simulation are shown in Fig. 12.10. In the range of the pyrite weathering zone, which establishes near the surface of the overburden, concentrations of up to 1200 mg SO_4^{2-}/l are computed. The transport of the initially high sulphate content in the overburden and the displacement of the pyrite weathering zone from the top of the overburden to the bottom are recognisable. Simultaneous with this displacement, the release of sulphate decreases in this range. The reason is the smaller amount of oxygen reaching the reaction zone because of the growing diffusion distance. Due to the rise of the groundwater level, the lower model boundary shifts upwards. This accelerates the input of sulphate into the groundwater. Figure 12.11 shows the calculated sulphate input at the rising groundwater surface for a period of 68 years.

Regarding the sulphate input into groundwater, two time periods have to be distinguished. In the first period, in this example the first 30 years, the sulphate input into the groundwater depends decisively on the formation of the overburden. The characteristics of the overburden material and

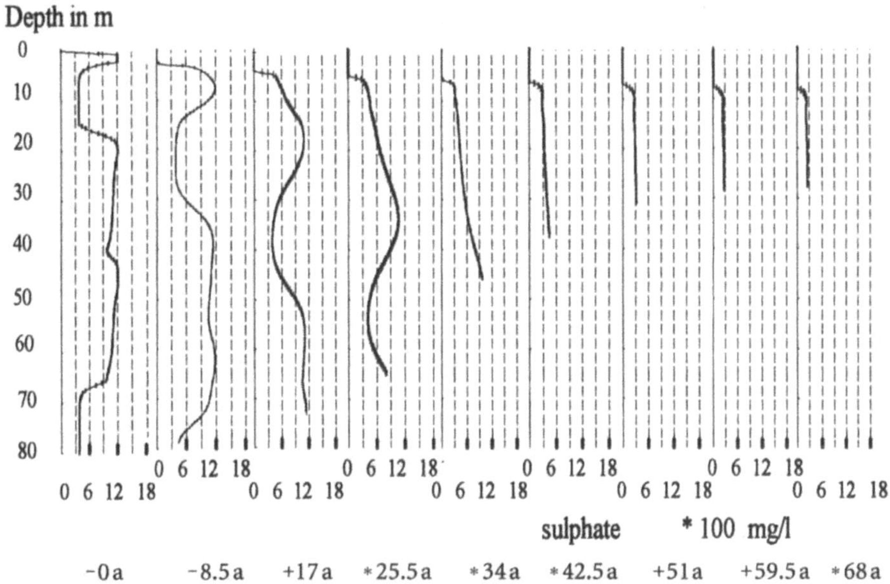

Fig. 12.10. Sulphate distribution with depth after completed buildup of overburden at different times in case of available buffer capacity

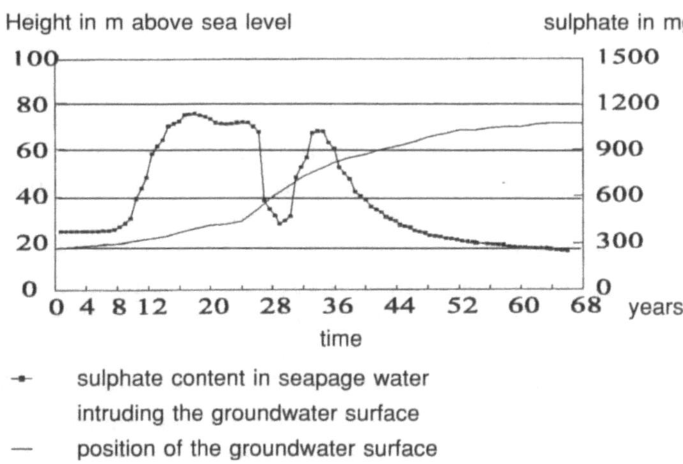

Fig. 12.11. Rise of groundwater level and sulphate input at the groundwater surface in case of available buffer capacity

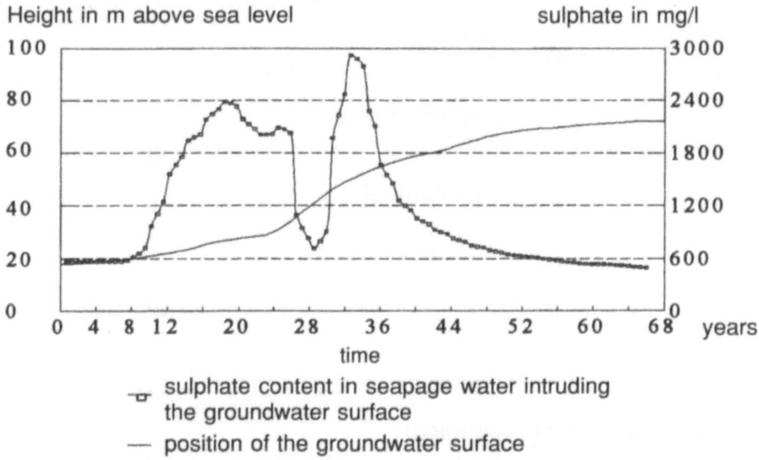

Fig. 12.12. Rise of groundwater level and sulphate input at the groundwater surface in case of lacking buffer capacity

the exposition times of several layers are essential. In the second period, the sulphate input into the groundwater only depends on the oxygen supply through the surface of overburden.

Because of the influence of the buffer capacity on the sulphate content, which does not cross the equilibrium concentration of 1200 mg/l, a comparative calculation for the case of a non-buffered weathering reaction is shown in the following. Within the pyrite weathering zone establishing near the surface of the overburden, concentrations of up to 5400 mg SO_4^{2-}/l for the initial distribution of sulphate after buildup of overburden are computed. The calculated sulphate input into the groundwater is shown in Fig. 12.12. Without the buffer the sulphate concentrations are higher than in the case mentioned above. Furthermore, it is shown that the influence of pyrite weathering near the surface of the overburden is stronger (maximum at approx. 30 years).

12.7 Precaution for Reduction of Oxygen Supply

Initially, acceleration of pyrite weathering by microorganisms depends on the supply of oxygen. Therefore it is obvious to search for precautions to reduce the oxygen supply to the reaction zone. The input of oxygen can occur through the gaseous phase as well as through the seepage water by convection and diffusion. The covering of temporary and permanent surfaces with material containing no pyrite-like solidified sand or clay reduces the oxygen supply into the depth of the overburden. If clay is

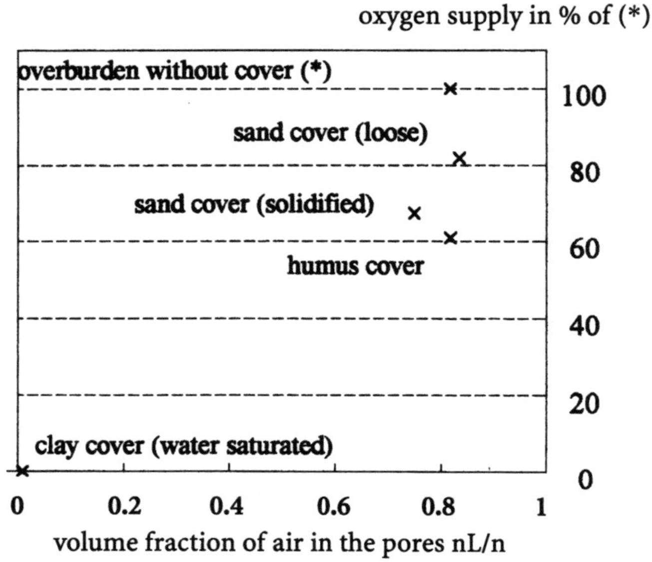

Fig. 12.13. Oxygen supply through the overburden surface depending on the volume fraction of air in the pores of the applied cover

used, the air-filled porosity and the permeability for seepage water are reduced. Besides, it is possible to stimulate the oxygen consumption by covering the overburden with humus. In Fig. 12.13, the relative rates of oxygen supply through the surface of overburden are shown.

The covering of the overburden surface with material containing no pyrite causes a greater diffusion distance and therefore a decrease in the concentration gradient for oxygen. Figure 12.14 shows that the rate of oxygen supply decreases with an increasing cover height. It has to be considered though that the material and labour expenses for covering the overburden will be higher with increasing height of cover relative to the advantage of decreasing oxygen input.

12.8 Conclusions and Prospects

The limitation of oxygen supply creates a restricted zone of pyrite weathering. It is possible to differentiate between two time periods of sulphate input into the groundwater. The first period is determined by the way the overburden is built up; the second one decisively depends on the oxygen supply through the overburden surface. Because of the displacement of the pyrite-weathering zone into the depth and the related

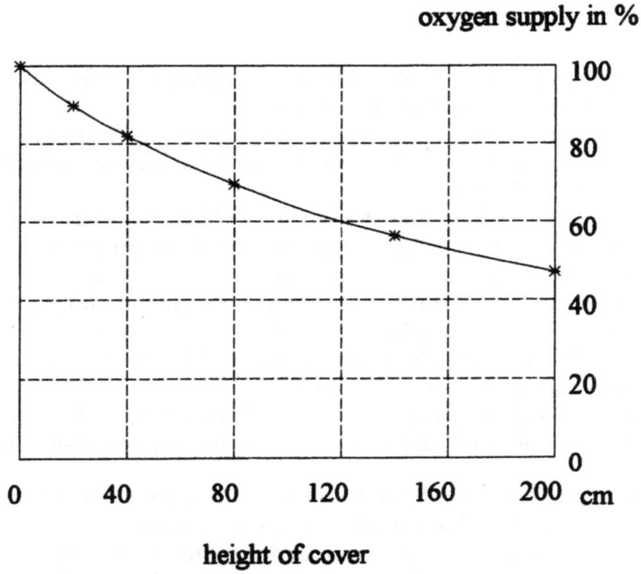

Fig. 12.14. Oxygen supply through the overburden surface depending on the cover height

increase of the diffusion distance, a decrease in sulphate release is typical for the second period.

During the formation of the overburden, temporary surfaces are created. The number and time of existence of these surfaces decisively influence sulphate release. During this time, the vertical distribution of sulphate in the overburden is establishing and determines the sulphate input into the groundwater later on. This sulphate input depends on the leakage rate of seepage water and the influence of buffering by calcium carbonate for precipitation of sulphate as calcium sulphate.

After formation of parts of the overburden, the design of the surface is especially important for the further development of sulphate input into the groundwater. For covering the overburden surface, low permeable material containing no pyrite should possibly be used, to effectively limit the oxygen supply into the depth of overburden.

In order to forecast the sulphate concentration in seepage water reaching the groundwater, detailed knowledge of the distribution of pyrite in the undisturbed rock and the overburden is necessary. Furthermore, the location and the duration of existence of the temporary surfaces should be known to determine the range in which weathering is possible and the reaction rate of weathering, respectively. Therefore the transport and displacement of the overburden should be comprehensible.

References

Ahonen L, Tuovinen OH (1989) Microbiological oxidation of ferrous iron at low temperatures. Appl Environ Microbiol 2:312–316

Beveridge G, Southam G (1992) Enumeration of *Thiobacilli* within pH-neutral and acidic mine tailings and their role in the development of secondary mineral soil. Appl Environ Microbiol 58:6

Brierley CL (1978) Bacterial leaching. CRC Crit Rev Microbiol 6:207–262

Curutchet G (1992) Effect of iron(III) and its hydrolysis products of *Thiobacilli ferrooxidans*. Biotechnol Lett 14(4):329–334

Farrier DF, Massmann J (1992) Effects of atmospheric pressure on gas transport in the vadose zone. Water Resour Res 28(3):777–791

Karavaiko GI (1988) Biotechnology of metals. UNEP, Centre for International Projects, GNKT, Moscow, pp 70–87, 126–155

Kölling M (1990) Modellierung geochemischer Prozesse im Sickerwasser und im Grundwasser. Berichte Fachbereich Geowissenschaften, Universität Bremen, 135, Bremen

Näveke R (1991) Sauerwässer in Braunkohleabraumkippen: Bakterielle Genese und mikrobielle Behandlung, Braunschweig, 25 S (unpublished)

Prein A (1993) Sauerstoffzufuhr als limitierender Faktor für die Pyritverwitterung in Abraumkippen von Braunkohlentagebauen. Mitteilungen des Instituts für Wasserwirtschaft, Hydrologie und landwirtschaftlichen Wasserbau der Universtität Hannover, Hannover, Nr. 79, S. 3–126, 74 Abb., 8 Tab.

Singer PC, Stumm W (1970) Acidic mine drainage: the rate-determining step. Science 167:1121–1123

Stumm W, Morgan J (1984) Aquatic chemistry. Wiley, New York, p 780 pp

Tuovinen OH, Kelly DP (1973) Studies on the growth of *Thiobacillus ferrooxidans* II. Arch Microbiol 98:51–364

13 Effects of Superficial Tertiary Dump Substrates and Recultivation Variants on Acid Output, Salt Leaching and Development of Seepage Water Quality

J. Katzur and F. Liebner

Forschungsinstitut für Bergbaufolgelandschaften e.V., Brauhausweg 2, 03238 Finsterwalde, Germany

13.1 Introduction

With 750 km² of dumps, waste piles and residual holes, the Lausitz brown-coal district is the largest mining area in Germany. The extraction of brown coal generally occurs by surface mining. As a result, the surface rock layers are encallowed and dumped as overburden either on waste piles in the environs of the open mine or on dumps in worked-out open mine areas. Besides Quaternary surface rock layers, Tertiary overburden also appeared on the surface of the dumps and waste piles in the past owing to unspecific accretion.

Quaternary dump substrates can be easily cultivated. In contrast, the quantitatively dominating Tertiary dump substrates are extremely acid and philistine and cause serious problems in recultivation. Hence the reason for many investigations on agricultural (Katzur and Hanschke 1990; Haubold-Rosar et al. 1993) and forest (Preußner and Kilias 1992; Katzur 1993) recultivation of Tertiary dump substrates.

Because of missing information about the substance loads, especially the often toxic heavy metal contents of percolates out of Tertiary dump substrates, the problems of groundwater protection in brown-coal mining areas have been totally underestimated up till now. Unlike the calculability of quantitative water balances, a prediction of the future water quality in the Lusatian brown-coal district is very difficult. Particularly the strongly increasing groundwater table as well as the low acid neutralisation capacity of Tertiary dump substrates have led to strong acidification of residual lakes in the last years (see also Geller, Klapper and Schultze, this Vol.). The groundwater increase in brown-coal mining areas is decisively influenced by seepage of atmospheric water. Since the resulting quality of seepage water is a reflection of the geochemical composition of superficial dump substrates, Tertiary and Quaternary final dumps have an important influence on groundwater quality. Thus, long-term investigations have been carried out on groundwater recharge by seepage and on translocation processes in the water-unsaturated zone of dump substrates since 1992. [The initiating research programme "Substance loads of percolates and development of the disposal potential of agriculturally used dump soils" was funded by the German Federal Ministry for Education, Science, Research and Technology (No. 033 93 93 C).] Test factors for these investigations are the type of dump substrate, the kind of substrate layering and melioration agent (lime marl and brown-coal ash), the intensity of cultivation (intensive and extensive agricultural recultivation) and the recultivation period. The aim of the investigations is to enable conclusions to be made concerning both the amount of groundwater recharge by seepage and the mobility, translocation and leaching of individual species by balancing the corresponding input (dump substrate, melioration agent, fertiliser, precipitation) and output amounts (percolates, plants).

Lysimeters are considered to be an important link between laboratory investigations and field trials in the investigation of groundwater recharge and translocation processes. Thus, groundwater-free lysimeters without subpressure were chosen as the experimental basis for the described investigations. Lysimeters with a depth of 300 cm were used because the distance between rhizosphere and seepage impounding sphere influences the amount of seepage water and evaporation in gravity lysimeters (Katzur 1986).

13.2 Experimental Procedure

Two dump soils typical of the Lusatian brown-coal district were chosen as experimental soils. The variants differ in the method of melioration and in the manner of stratification:

- Variant Q: oligotrophic Pleistocene dump sand (mSl).
- Variant Tk: Tertiary sulphurous weak loamy dump sand (xSl), lime-meliorated.
- Variant Ta: Tertiary sulphurous weak loamy dump sand (xSl), ash-meliorated.
- Variant Q/T: Quaternary casing layer (100 cm) over Tertiary subsoil (200 cm).

The meliorative ash and lime dressings for the Tertiary substrates were calculated from the lime requirement (acid–base balance; Illner and

Raasch 1966) and the soil effective base content of the melioration agents (uncombined plus hydrolysable lime). Lime dressings for the variants Q and Q/T, however, were determined according to Schachtschabel (1971). The aimed pH value was uniformly 5.5. The melioration depths of Tertiary dump substrates were 1 m, while in the Quaternary substrates only the 60-cm-thick upper soil layer was limed. The lysimeters (epoxide-resin-coated steel tanks, 3 m × 1 m²) were filled stepwise with 10-cm deep layers of dump substrate according to the natural compactness. To avoid impounding effects on boundaries of the bed, the surface was roughened between the individual substrate layers. Basic N–P–K fertiliser was added to all lysimeters uniformly. The incorporation depth was 60 cm for P and K; N was added only superficially.

Within the dump substrate/melioration variants (six lysimeters each), four lysimeters were managed intensively (crop rotation: winter rye, forage rye, 3 years of lucerne) and two lysimeters extensively (grassland). Two of the intensively managed lysimeters were irrigated additionally depending on the soil moisture (Table 13.1).

Table 13.1. Dump substrates, melioration agents and recultivation variants

<table>
<tr><td colspan="4">Substrates</td></tr>
<tr><td>Tk</td><td>Ta</td><td>Q</td><td>Q/T</td></tr>
<tr><td colspan="3">3 m troughout</td><td>1 m Q over 2 m T</td></tr>
<tr><td colspan="4">Melioration</td></tr>
<tr><td colspan="2">Acid–base balance</td><td colspan="2">According to Schachtschabel (1971)</td></tr>
<tr><td>Lime marl</td><td>Power plant ash</td><td colspan="2">Lime marl</td></tr>
<tr><td colspan="2">1 m</td><td colspan="2">60 cm</td></tr>
<tr><td colspan="4">Final pH value 5.5
Recultivation</td></tr>
<tr><td colspan="4">Intensive (crop rotation: winter rye, forage rye, 3 years lucerne</td></tr>
<tr><td colspan="2">Without irrigation</td><td colspan="2">Two lysimeters each</td></tr>
<tr><td colspan="2">With irrigation</td><td colspan="2">Two lysimeters each</td></tr>
<tr><td colspan="4">Extensive (typical grassland)</td></tr>
<tr><td colspan="2">Without irrigation</td><td colspan="2">Two lysimeters each</td></tr>
</table>

13.3 Analysis and Evaluation

The analysis of the dump substrates at the beginning of the experiment included the investigation of soil physical parameters (compactness, pore volume, moisture content, grain size distribution, water retention curve) and the determination of lime requirement, potential cation exchange capacity and the nitrogen-binding forms (ammonium, nitrate, nitrite, nitrogen in organic compounds). Furthermore, the contents of macronutrients (digestion with 10 % hydrochloric acid, double-lactate method), heavy metals, trace elements (pulping with aqua regia) as well as total and volatile sulphur (sulphates of monovalent or trivalent cations, sulphides) were determined. The melioration agents were analysed for their contents of total carbon, nitrogen, sulphur, plant-available macro- and micronutrients and the potential cation exchange capacity.

For all rotation periods, the fresh and dry weight yields were determined. Representative plant samples of each lysimeter were analysed for their contents of heavy metals (Cr, Ni, Cu, Zn, Cd, Pb, Mn, Fe) as well as Al, Ca, Mg, K and As.

Three times per week (since January 1992), the amounts of seepage water were registered and aliquots were frozen in collecting flasks. At the end of the month, representative composite samples from each lysimeter were analysed, together with the atmospheric water, for:

- pH, EC (electrochemical conductivity), Eh (reduction potential), COD (chemical oxygen demand), evaporating residue, C_{total}, N_{total}
- NH_4^+, NO_3^-, NO_2^-, Cl^-, PO_4^{3-}, SO_4^{2-}
- Mg, Ca, K, Fe(III), Fe(II), Mn, Al
- Cr, Ni, Cu, Zn, Cd, Pb, As

The monthly species concentrations of the percolates were evaluated by unifactorial variance analysis (test factor dump substrate/melioration agent). In case of variance homogeneity (Cochran test, $\alpha = 5\,\%$), the mean differences within the substrate variants were analysed for significance with the F-test and the two-sided multiple *t*-test procedure.

13.4 Results and Discussion

According to the low geogenic and anthropogenic (melioration) load, the percolates out of the Quaternary dump substrate reach almost drinking water quality. With the exception of iron and manganese, the heavy metal concentrations did not exceed the limits of the European Community Water Quality Norm and the German Drinking Water Decree during the whole experimental period. The concentrations of ammonium and nitrite as well as the values of Kjeldahl-nitrogen were only slightly over the limits at the beginning of the experiment. Only the high concentrations of potassium, which were caused by the low nutrient storage capacity of the sands, exceeded the limits distinctly in the first 2 years; the same applies to the concentrations of iron, aluminium, sulphate and the chemical oxygen demand (COD) values. In the summer, increased species concentrations were usually measured because of lower amounts of seepage water.

The geogenic load of the carboniferous weak loamy dump sand is as low as the Pleistocene dump sand (Table 13.2). However, the species concentrations of the percolates out of the Tertiary substrate were significantly higher compared with those of the Quaternary variants (Table 13.3). The extremely high concentrations of iron, aluminium, sulphate and heavy metals are typical for percolates out of aerated Tertiary dump substrates and may be explained by intensive sulphide and silicate weathering processes. In the first year, the iron and aluminium limits of the German Drinking Water Decree were exceeded by a factor of 15,000, for example. The limits for sulphate, arsenic, nickel and chromium were exceeded by a factor of 100. The COD and Kjeldahl values and the total carbon and ammonium concentrations also exceeded the limits. The species concentrations in the Tertiary percolates decreased distinctly (K by 87%, Al 72%, Zn 70%, sulphate 63%) during the investigations, but they were still over the limits at the end of the fourth year. Reasons for this are the progressive leaching of water-soluble sulphates, the wearing off of weath-

Table 13.2. Average heavy metal contens (ppm) in topsoils of fields and allotments in agglomeration areas (Brüne 1985) compared with the Tertiary and Quaternary dump substrates and the limits of the Eikmann-Kloke List for multifunctionally useable soils

	Topsoils		Experimental substrates		Limits of Eikmann-Kloke list
	Fields	Allotments	Quaternary	Tertiary	
Pb	25	225	1.7	4.1	100
Cd	0.1	0.7	0.03	0.03	1
Cr	39	51	2.8	6.1	50
Cu	18	88	18.8	21	50
Ni	38	28	4.4	5	40
Zn	66	324	10.3	12.8	150

Table 13.3. Yearly averages of selected species concentrations in seepage water of intensively recultivated substrate variants

Iron 0.2 mg/l[a]	1992	1993	1994	1995
Tk	2952	1922	1242	1136
Ta	2132	1827	991	793
Q	0.76	0.21	0.01	0.01
Q/T	1435	1116	907	548

Aluminium 0.2 mg/l[a]	1992	1993	1994	1995
Tk	3191	1556	999	882
Ta	2939	1599	879	679
Q	3.40	0.29	0.08	0.03
Q/T	2656	1758	1604	1065

Sulphate 0.24 g/l[a]	1992	1993	1994	1995
Tk	23.72	15.06	13.36	8.80
Ta	20.29	14.54	10.40	6.47
Q	0.26	0.18	0.30	0.06
Q/T	18.38	13.42	14.05	7.01

Chloride 250 mg/l[a]	1992	1993	1994	1995
Tk	35.70	39.87	49.23	51.00
Ta	40.06	37.92	51.15	57.63
Q	57.87	49.15	64.75	38.29
Q/T	50.73	68.95	39.77	68.50

Calcium 400 mg/l[a]	1992	1993	1994	1995
Tk	345	247	249	276
Ta	395	264	285	281
Q	62	57	143	30
Q/T	377	301	292	293

Magnesium 50 mg/l[a]	1992	1993	1994	1995
Tk	532	540	611	475
Ta	269	266	279	240
Q	18	32	29	4
Q/T	185	121	132	111

Manganese 0.05 mg/l[a]	1992	1993	1994	1995
Tk	37.5	20.5	17.4	13.7
Ta	26.9	16.4	11.5	8.8
Q	0.07	0.02	0.002	0.001
Q/T	26.9	14.2	17.9	17.7

Zinc 5 mg/l[b]	1992	1993	1994	1995
Tk	29.2	12.5	8.5	6.4
Ta	19.0	10.8	6.2	4.1
Q	0.69	0.07	0.005	0.012
Q/T	20.6	11.0	12.3	8.6

Copper 3 mg/l[a]	1992	1993	1994	1995
Tk	1.25	0.72	0.58	0.48
Ta	1.11	0.78	0.50	0.36
Q	0.012	0.005	0.004	0.006
Q/T	0.62	0.56	0.56	0.23

Cadmium 5 µg/l[b]	1992	1993	1994	1995
Tk	107	69	35	21
Ta	79	57	23	11
Q	0.7	0.4	0.1	0.0
Q/T	73	45	28	15

Table 13.3 (continued)

Chromium 0.05 mg/l[a]	1992	1993	1994	1995	Lead 40 µg/l[a]	1992	1993	1994	1995
Tk	4.47	2.09	1.52	1.44	Tk	47.8	42.7	17.8	16.6
Ta	3.73	2.08	1.39	1.24	Ta	40.8	35.4	18.2	25.9
Q	0.002	0.001	0.001	0.001	Q	1.7	1.0	1.1	4.1
Q/T	2.40	1.59	1.28	1.12	Q/T	69.0	37.1	13.4	15.0
Potassium 12 mg/l[a]	**1992**	**1993**	**1994**	**1995**	**Ammonium 0.5 µg/l[a]**	**1992**	**1993**	**1994**	**1995**
Tk	21.5	82.0	2.6	2.7	Tk	8.3	7.9	12.4	17.0
Ta	67.7	1.7	1.4	1.2	Ta	10.0	10.1	15.3	19.3
Q	208.1	52.0	64.4	13.5	Q	0.94	0.11	0.11	0.04
Q/T	82.0	7.9	6.4	4.8	Q/T	11.9	10.1	11.1	8.8
Nitrate 50 mg/l[b]	**1992**	**1993**	**1994**	**1995**	**Phosphate 5 mg/l[b]**	**1992**	**1993**	**1994**	**1995**
Tk	1.17	1.22	0.48	0.54	Tk	6.72	2.97	1.40	0.73
Ta	1.65	0.76	0.41	0.31	Ta	6.85	4.24	1.36	0.60
Q	10.96	4.85	0.26	–	Q	0.002	0.026	0.077	0.001
Q/T	1.20	1.78	0.34	0.24	Q/T	3.37	2.04	1.18	0.40
Kjeldahl-N 1 mg/l[a]	**1992**	**1993**	**1994**	**1995**	**COD 5 mg/l[b]**	**1992**	**1993**	**1994**	**1995**
Tk	17.9	7.5	53.1	34.7	Tk	630	198	171	158
Ta	15.5	7.6	48.0	25.2	Ta	615	213	180	150
Q	4.4	<1	<1	<1	Q	28	14	16	5
Q/T	19.2	7.2	37.4	16.0	Q/T	546	163	162	99
EC 2 mS/cm[a]	**1992**	**1993**	**1994**	**1995**	**pH value 6.5–9.5[a]**	**1992**	**1993**	**1994**	**1995**
Tk	12.8	9.2	8.0	7.2	Tk	2.38	2.33	2.47	2.47
Ta	11.1	8.6	6.6	5.6	Ta	2.45	2.28	2.49	2.51
Q	1.11	0.60	0.79	0.24	Q	7.59	7.03	7.24	7.37
Q/T	10.7	8.1	8.0	5.8	Q/T	2.60	2.31	2.45	2.17

EC, Electrical conductivity.

[a] Limit according to the German Drinking Water Decree, 5 December 1990.

[b] Limit according to the EC Water Quality Norm 80/778/EWG; 15 July 1980.

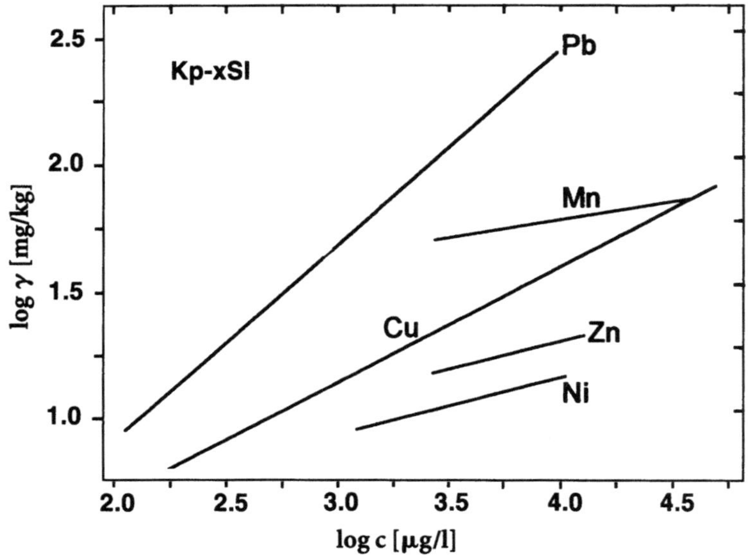

Fig. 13.1. Heavy metal adsorption isotherms (Freundlich; 20 °C, pH 2.5) for Tertiary weak loamy dump sand (Kp-xS1)

ering processes with increasing water saturation, and depth of and the effective neutralisation in the meliorated substrate layers. Exceptions were the species chloride and ammonium as well as the values for Kjeldahl-nitrogen, which increased slightly. The consistently high magnesium concentrations in variant Tk have to be put down to the lime fertiliser (Mg-containing marl) and exchange processes caused by calcium.

The reason for the up till now still extremely high substance loads and unchanged low pH values of the percolates is, firstly, the seasonally fluctuating but in general unhindered oxygen supply to the unmeliorated substrate layers. In the untreated subsoil, the sulphuric acid formed by pyrite oxidation cannot be neutralised. Secondly, the developing acid conditions cause strong silicate weathering and cation mobility. Because of lower water-saturation, increased oxygen supply and microbial activity, the pyrite oxidation rates were high in the summer months and low in the winter.

The species concentrations of the percolates are obviously much more affected by transformation and translocation processes in the untreated Tertiary dump substrate zone than by the applied melioration agents. Strongly acidic conditions owing to progressive weathering processes and very low adsorption capacity of the sandy substrates (Fig. 13.1) cause high species mobility in the unmeliorated dump substrate layers. In contrast, the species mobility in the melioration horizon is low because of effective neutralisation of sulphuric acid. Thus, because of the generally high substance loads of Tertiary percolates, a relatively different species input owing to variation of the melioration agent hardly affects the seepage water quality. So, the substantially higher input of Pb, Zn and Cd with the lime melioration had only slightly increased the Cd and Zn concentrations of the percolates. Pb is relatively immobile (Fig. 13.1), so that despite an additional input of about 21,500 mg Pb with the lime the output was only about 4 mg per lysimeter higher compared with variant Ta.

In the percolates of variant Ta, most pollutant concentrations were even up till now significantly lower than in variant Tk, although the amounts of some heavy metals (Cr, Cu, Ni) applied to the substrate by ash melioration were distinctly higher. This could be explained with heavy metal immobilisation by various sesquioxides (Fe-, Al- and Mn-oxides) of the brown-coal ashes. Furthermore, the melioration of Tertiary dump substrates with brown-coal ash results in better internal humidification and neutralisation of sulphuric acid formed in the melioration horizon. Thus, the pH values in the melioration horizon are more homogeneous in variant Ta than in variant Tk, resulting in higher heavy metal immobilisation.

Compared with the Quaternary dump substrate, the nitrate concentration of the Tk and Ta percolates was significantly lower in the first 2 years. In the average of the years and both Tertiary variants, only 2% of the inorganic nitrogen (84% in variant Q) was leached as nitrate nitrogen, because the largest part of nitrate translocated in the depth is reduced to ammonium. Recent investigations (Liebner 1997) of geochemical processes in water-saturated unmeliorated Tertiary dump substrates (same substrata as in the lysimeter experiment) showed that the reductions of nitrite and nitrate already occur at redox potentials of about 750 mV (Eh). These conditions are realised in water-saturated substrate layers of the Tertiary variants. Whereas in the unsaturated upper substrate layers redox potentials of about 830 mV were measured, the average redox potentials of the percolates at a depth of 3 m was about 690 mV. These high values are normal in Tertiary percolates and must be due to high concentrations of oxidised species and extremely low pH values. Altogether, these effects explain the high concentrations of ammonium and Kjeldahl-nitrogen that were above the limits of the German Drinking Water Decree and significantly higher than in variant Q. The large share of organic nitrogen in these percolates is mainly the result of translocation of water-soluble humus-like materials (fulvic acids, fulvates).

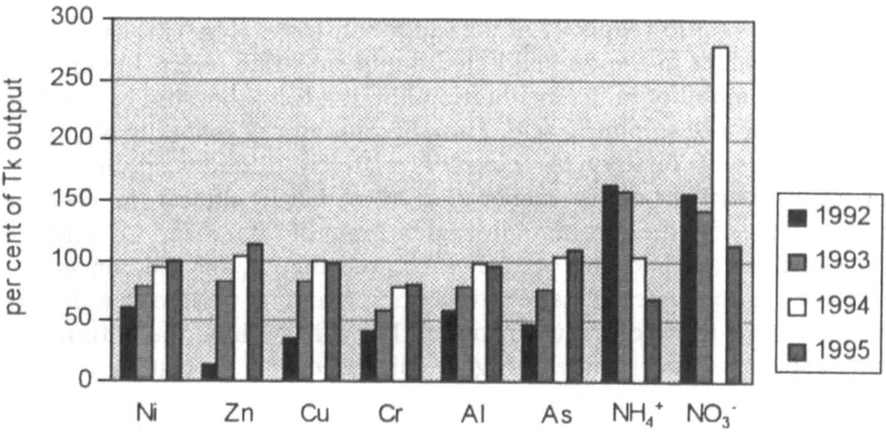

Fig. 13.2. Substance output of variant Q/T in percent of Tk (66% corresponds to the share of Tertiary dump substrate)

The species concentrations of the seepage waters out of variant Q/T were in proportion approximately to the share of Tertiary dump substrate (66%) in the first year. In the following years, most species concentrations decreased too, but less than in the variants Tk and Ta. While in the Tertiary dump substrates (Tk and Ta) distinctly lower redox potentials are ensued with increasing depth and water saturation, the redox potentials in variant Q/T remained almost unchanged (about 820 mV). This effect is probably caused by the Quaternary casing layer which enables an increased oxygen input in the unmeliorated Tertiary dump substrate. The more intensive weathering processes, stronger heavy metal mobilisation and higher nitrogen release in the subsoil lead to proportionately higher species concentrations in variant Q/T compared with Tk and Ta. Despite significantly lower amounts of seepage water (Table 13.4), the species leaching was in most cases higher too (Fig. 13.2).

The amount of salt leaching and groundwater contamination generally depends on the amount of seepage water. The height of seepage water run-offs or rather the extent of groundwater recharge is largely determined by the dump substrate type and the accessible rooting depths (Table 13.4). Highest amounts of seepage water were determined for the xSl-substrate. In the variants Tk and Ta the extremely philistine substrate under the melioration horizon limits the exploitation of the stored water in deeper soil layers by the plants. This applies to both extensive and intensive recultivation. The flat soil depth does not apply to the Quaternary variant Q. The greater rooting depths in Quaternary substrates

Table 13.4. Average quarterly and annual amounts of atmospheric (P) and seepage water (mm) of intensively (Int) and exensively (Ext) recultivated dump substrates (1992–1995)

	Tk		Ta		Q		Q/T		P
	Int	Ext	Int	Ext	Int	Ext	Int	Ext	
First quarter	113.9	121.2	113.4	115.93	109.8	127.32	114.0	129.83	119.0
Second quarter	87.7	103.3	80.1	102.47	49.2	92.03	43.8	97.91	162.2
Third quarter	72.0	90.2	70.9	87.30	4.6	43.36	18.2	59.32	188.8
Fourth quarter	73.1	86.0	70.7	81.81	9.9	28.86	31.0	57.49	116.1
Annual average	346.6	400.7	335.2	387.5	173.4	291.6	206.9	344.5	586.1

result in higher plant yields (30–50% more dry matter), higher evaporation losses and lower amounts of seepage water (average annual difference 133 mm) compared with the intensively recultivated Tertiary variants. Also compared with these variants, the Quaternary casing layer over Tertiary subsoil (Q/T) reduces the amount of seepage water on average by 77 mm/year. This may be explained by impounding effects on boundaries of the bed caused by water-bearing menisci in the pores. Hence larger amounts of water are available for the plants. The results are higher plant yields (for lucerne up to 100%), higher evaporation losses and lower amounts of seepage water compared with the Tertiary variants. In addition to substrate type and kind of layering, the groundwater recharge by seepage is also influenced by the intensity of land use. The amounts of seepage water were highest for extensive recultivation, because of lower evapotranspiration losses. This applies to all substrate/melioration variants.

A comparison of the cumulative anionic output by seepage water illustrates the extreme (up to 50-fold higher than in variant Q) salt loads of the Tertiary variants. While in the percolates of variant Q chloride and nitrate were found in larger amounts (24 and 1%, respectively), in all variants with Tertiary dump substrates about 99.5% of the salts was translocated as sulphates (Table 13.5). Considering the high sulphate content, the quantitative composition of the percolates as well as the total amounts of leached salts could be determined relatively exactly for the Tertiary variants (Table 13.6).

Table 13.5. Cumulative anionic output of lysimeters [substance amounts (val) and percentages (%) of total anionic output] over 4 years

	Tk		Ta		Q		Q/T	
	(mol_c)	(%)	(mol_c)	(%)	(mol_c)	(%)	(mol_c)	(%)
Sulphate	419.96	99.579	351.14	99.476	5.93	74.951	226.05	99.347
Chloride	1.64	0.388	1.69	0.480	1.90	24.081	1.41	0.619
Phosphate	0.12	0.029	0.14	0.040	0.001	0.010	0.06	0.027
Nitrate	0.018	0.004	0.016	0.005	0.076	0.958	0.07	0.007
Total	421.7	100	353.0	100	7.9	100	227.5	100

Table 13.6. Total amounts and percentages of the most important leached salts in the variants Tk, Ta and Q/T (1992–1995)

Compound	Tk		Ta		Q/T	
	(g/lys)	(%)	(g/lys)	(%)	(g/lys)	(%)
$Al_2(SO_4)_3$	13495	48.86	12355	54.21	8566	62.29
$Fe_2(SO_4)_3$	7709	27.91	6162	27.04	2814	20.46
$FeSo_4$	792	2.87	605	2.66	723	5.26
$MnSO_4$	79	0.39	55	0.24	33	0.24
$CaSO_4$	1324	4.79	1388	6.09	860	6.25
$MgSO_4$	3692	13.37	1726	7.57	436	3.17
K_2SO_4	81	0.29	36	0.16	35	0.26
$(NH_4)_2SO_4$	59	0.21	69	0.30	38	0.27
$ZnSO_4$	45	0.16	31	0.14	17	0.13
$NiSO_4$	23	0.09	19	0.08	13	0.10
H_2SO_4	319	1.15	342	1.50	216	1.57

lys, Lysimeter.

The main compounds in all percolates of Tertiary variants were aluminium sulphate as silicate weathering product (with a share of about 50%) and iron(III) sulphate as pyrite weathering product (20–27%). For variant Tk, 13.5 kg aluminium sulphate, 7.7 kg iron(III) sulphate and 3.7 kg magnesium sulphate have been leached with the seepage water up till now, for example. This amounts to about 33.7 t aluminium sulphate and 19.2 t iron(III) sulphate per ha per year. Due to lower species concentrations and slightly lower amounts of seepage water, the salt output at variant Ta was about 16% (17.2 val) lower than at variant Tk (intensive recultivation). In the Tertiary variants, the heavy metal species were

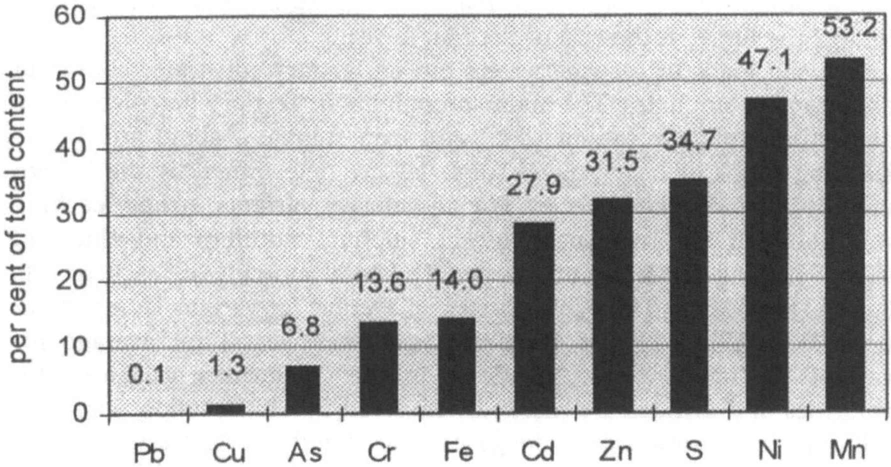

Fig. 13.3. Total heavy metal, arsenic and sulphur leaching in variant Tk (intensive cultivation) after 4 years (in percent of the basic contents)

leached almost exclusively as sulphates, too, and the leaching ratio is determined by the mobility of the respective species. In both Tertiary variants, the highest leaching ratios (percent of basic content) were obtained for Mn, Ni and S and the lowest for copper and lead (Fig. 13.3).

Considering the exponential decay of the monthly sulphate concentrations and leachings as well as the total sulphur output of about 34.7% related to the basic content within 4 years, at a rough guess the period of extremely high acid and pollutant output with the seepage water will last for more than 20 years.

13.5 Conclusions

The effects of typical dump substrates (Quaternary, ash- and lime-meliorated Tertiary, Quaternary casing layer over Tertiary subsoil) and recultivation variants (intensive and extensive agricultural utilisation) on groundwater recharge and the development of seepage water quality were investigated using large-scale lysimeters. Contrary to the near drinking water quality of seepage waters out of Quaternary dump substrates, the percolates out of aerated Tertiary substrates contain extremely high salt concentrations. The limits for drinking water were greatly exceeded in the case of iron, aluminium, sulphate, nickel, zinc, chromium, arsenic and cadmium because of strong weathering processes. Due to the progressive leaching and wearing off of weathering processes, the substance loads and pollutant concentrations decreased with time, but were still distinctly over the limits at the end of the fourth year.

The salt loads of seepage waters out of Tertiary substrates consist of about 99.5% sulphates. The main compounds in Tertiary percolates were aluminium sulphate (about 50%) and iron sulphate (about 30%). The leaching ratios of heavy metal sulphates are determined by the known mobility of individual species. For all Tertiary variants, strong leaching was obtained for manganese, nickel, sulphur, cadmium and zinc and slight leaching for lead and copper. The total amount of leached salts (1992–1995) was 27.6 kg per lysimeter on average for variant Tk (276 t/ha/4 years) and 22.8 kg per lysimeter (228 t/ha/4 years) on average for variant Ta (intensive recultivation). In the Quaternary variants, also chlorides and nitrates are markedly involved in translocation. The total salt output was substantially lower compared with the Tertiary variants.

The low pH values of Tertiary percolates have remained almost unchanged up till now since the acid release in the unmeliorated subsoil is predominant. Thus, because of very low adsorption capacity, the salt concentrations in the Tertiary percolates are much more affected by transformation and translocation processes in the untreated dump substrate zone than by the applied melioration agents. Compared with lime melioration, the use of ash as a melioration agent for Tertiary dump substrates results in slightly lower salt leaching, obviously by sesquioxide immobilisation.

The amount of salt leaching and groundwater contamination generally depends on the amount of seepage water. The extent of groundwater recharge is largely influenced by the dump substrate type and the accessible rooting depths. The highest amounts of seepage water were determined for the xSl variants since the evaporation losses are significantly lower than with the Quaternary substrate. Philistine conditions in the acidic subsoil of Tertiary variants cause significantly lower plant yields by limiting the rooting depth. This applies in the case of both extensive and intensive recultivation.

The expected positive influence of the Quaternary casing layer over Tertiary subsoil (variant Q/T) has not yet been proved. Although the covering layer reduced the amount of seepage water by impounding effects on boundaries of the bed, the substance output was very high. Because of the almost unhindered oxygen supply in deeper, unmeliorated substrate layers, the pyrite weathering was hardly reduced. Compared with the Tertiary variants, this will result in high species concentrations over a longer period. At the same time, the output of organic nitrogen increased. This would suggest that the high oxygen permeability of the Quaternary casing layer leads to an increasing oxidative deamination in the Tertiary subsoil. The long-term effects of Quaternary casing layer, melioration agents and recultivation variants on groundwater recharge and seepage water quality from superficial dump substrate layers cannot be finally judged at the present time. For this, a longer experimental period is needed.

References

Brüne H (1985) Kongreßband 1985. VDLUFA-Schriftenreihe 16:85–102

Haubold-Rosar M, Katzur J, Schröder D, Schneider R (1993) Bodenentwicklung in grundmeliorierten tertiären Kippsubstraten der Niederlausitz. Mitt Dtsch Bodenkundl Ges 72(II):1197–1202

Illner K, Raasch H (1966) Zur Bestimmung des Kalkbedarfes für die Melioration von schwefelhaltigen Tertiärkippen. Z Landeskultur 7:285–290

Katzur J (1986) Einfluß der Tiefe grundwasserfreier Lysimeter ohne Unterdruck auf Sickerwasserablauf und Verdunstung. Arch Acker Pflanzenbau Bodenkd 30:227–233

Katzur J (1993) Bodenmelioration und forstliche Rekultivierung auf den Kippen und Halden des Braunkohlenbergbaues unter besonderer Berücksichtigung der extrem sauren schwefel- und kohlehaltigen Kippböden. Jahrestagung 1993, Regionalverband Lausitz, SDW

Katzur J, Hanschke L (1990) Pflanzenerträge auf meliorierten schwefelhaltigen Kippböden und die bodenkundlichen Zielgrößen der landwirtschaftlichen Rekultivierung. Arch Acker Pflanzenbau Bodenkd 34:35–43

Liebner F (1997) Reduktionsprozesse in der gesättigten Zone belüfteter kohlehaltiger Tertiärsubstrate. Mitt Dtsch Bodenkundl Ges 83:175–178

Preußner K, Kilias G (1992) Erfahrungen bei der forstlichen Rekultivierung in der Lausitz. AFZ 18:982–985

Schachtschabel P (1971) Methodenvergleich zur pH-Bestimmung von Böden. Z Pflanzenernährung Düngung Bodenkd 130:37–43

Part 4
Remediation Concepts and Case Studies

14 Microbial Processes for Potential in Situ Remediation of Acidic Lakes

K. Wendt-Potthoff and T.R. Neu

Department of Inland Water Research, UFZ Centre for Environmental Research, Leipzig-Halle, Am Biederitzer Busch 12, 39114 Magdeburg, Germany

14.1 Introduction

Natural acidification originates, for example, from humic substances which are of significance in a number of water bodies. However, freshwater systems are anthropogenically polluted by atmospheric acid deposition and even more by mining activity. The water bodies exposed to acid precipitation usually have a pH value in the range of 4–5, whereas acid mine drainage may cause a drop in pH value down to 2. Both types of pollution result in increased concentrations of acidity, sulfate and metals. Nevertheless, there are a number of chemical reactions that can moderate acidification. These include the buffering capacity of the water body as well as ion exchange reactions with colloidal materials. In addition, there are several biological processes with the potential of reducing the acidity of contaminated water bodies. These reactions may temporarily or permanently be responsible for neutralization. The biological reactions which may be of significance are oxygen reduction (photosynthesis), nitrate reduction (denitrification), manganese and iron reduction and sulfate reduction. In addition, there are others having minor or indirect effects, such as amino acid fermentation or methanogenesis. These reduction processes increase alkalinity and thus lead to the neutralization of the water (Mills et al. 1989).

For further discussion it may be necessary to distinguish between acidity/alkalinity and pH. The acidity is the difference between the equivalent parts of the anions and cations of strong acids. If the acidity becomes negative the term alkalinity is used (Stumm 1992). The contribution of different ions to water alkalinity is summarized in the following equation:

$$\text{Alkalinity} = [Na^+] + [K^+] + 2[Ca^{2+}] + [NH_4^+] + 2[Mg^{2+}] + 2[Fe^{2+}] + 3[Fe^{3+}] - [Cl^-] - 2[SO_4^{2-}] - [NO_3^-] - [org^-]. \quad (1)$$

The pH, as an intensity factor, is defined as the negative logarithm (log) of the hydrogen (H^+) ion concentration. It is important to note that a change in acidity/alkalinity must be accompanied by a change in anions and cations other than hydrogen to assure electroneutrality.

In the following section, the various reduction processes are briefly discussed in the same order as the bacteria may use them as the energetically most favourable electron acceptors. In addition, the available literature is discussed to show applied aspects of acidic lake restoration.

14.2 Theoretical Considerations

Equations used in this chapter are varied from Mills et al. (1989).

14.2.1 Oxygen Reduction (Photosynthesis)

Photosynthesis can be written as follows:

$$nCO_2 + nH_2O = (CH_2O)_n + nO_2 . \quad (2)$$

According to this equation, there is no change in acidity, but due to the removal of CO_2 from solution as carbonate or bicarbonate, the local pH will increase. The situation will be different if photosynthesis is combined with the assimilation of nitrate according to:

$$106CO_2 + 16NO_3^- + HPO_4^{2-} + 122H_2O + 18H^+ = (C_{106}H_{263}O_{110}N_{16}P) + 138O_2 . \quad (3)$$

The alkalinity will then increase due to the removal of hydrogen ions, nitrate and phosphate ions.

14.2.2 Nitrate Reduction (Denitrification)

For nitrate (NO_3^-) reduction an assimilative (plants and bacteria, aerobic) and a dissimilative pathway (bacteria only, anaerobic) have to be distinguished. The anaerobic route will lead via nitrite (NO_2^-) to ammonia (NH_4^+) or to dinitrogen gas (N_2) as a product. It may be important to mention at this point that the intermediate of this reaction, nitrite (NO_2^-), is a very potent mutagen under acidic conditions as it will form nitrous acid (HNO_2).

Denitrification is the reduction of nitrate to nitrogen gas, which can be lost from the environment. During this reaction, nitrate and hydrogen ions will be consumed as follows:

$$\text{Glucose} + 4.8NO_3^- + 4.8H^+ = 6CO_2 + 2.4N_2 + 8.4H_2O . \quad (4)$$

If nitrate is reduced to ammonia, nitrate and hydrogen ions are also removed according to:

$$\text{Glucose} + 3NO_3^- + 3H^+ = 6CO_2 + 3NH_4^+ + 3OH^- . \quad (5)$$

Both reactions will result in an increase in alkalinity due to the removal of nitrate and hydrogen ions.

14.2.3
Iron and Manganese Reduction

It has been known for several years that bacteria can catalyse the reduction of Fe(III) and Mn(IV) (Lovley and Phillips 1986; Ehrlich 1996). Nevertheless, iron and manganese reduction in the environment was believed to be largely an abiotic process (Zehnder and Stumm 1988) and direct catalysis via respiratory-linked metal reduction has only recently been proved with pure cultures (Lovley and Phillips 1988). Bacteria are able to use either iron or manganese as an electron acceptor during anaerobic respiration. Ferric iron, (Fe^{3+}) in the form of poorly crystalline Fe(III) oxides, is the most abundant form in many suboxic microbial habitats and can be reduced by a variety of bacteria to ferrous iron (Fe^{2+}) (Lovley 1991). As the reduction potential of Fe^{3+}/Fe^{2+} is very high, it can be coupled to the oxidation of a wide variety of organic and inorganic electron donors. If organic acids are used as a carbon source the following reactions can be proposed:

$$FeOOH + CH_3CHOHCOO^- + H_2O$$
$$= Fe^{2+} + CH_3COO^- + CO_2 + 1.5H_2 + 2OH^- . \quad (6)$$

$$MnO_2 + CH_3CHOHCOO^- + H_2O$$
$$= Mn^{2+} + CH_3COO^- + CO_2 + 2OH^- + H_2 . \quad (7)$$

Both reactions reveal a gain in alkalinity if the metal ions are removed from the system.

14.2.4
Amino Acid Fermentation

Fermentation reactions do not require external electron acceptors. Thus the oxidation-reduction reactions are internally balanced. If pairs of amino acids are used (Stickland fermentation), one may serve as an electron donor and the other may serve as the electron acceptor:

$$\text{Alanine } [CH_3CH(NH_2)CO_2H] + 2 \text{ glycine } [H_2NCH_2CO_2H] + 2H_2O$$
$$= 3NH_4^+ + 3CH_3COO^- + CO_2 + (H). \quad (8)$$

The production of organic acids and ammonium will decrease the acidity. Alkalinity can also increase from the fermentation of a single amino acid such as:

$$\text{Serine } [HOCH_2CH(NH_2)CO_2H] = CH_3COCOO^- + NH_4^+. \quad (9)$$

Many fermentation products represent an energy source for other fermentative organisms. Ultimately these products lead to acetate, H_2 and CO_2 which serve as substrates for methanogenic archaea.

14.2.5 Sulfate Reduction

Sulfate-reducing bacteria use sulfate as terminal electron acceptor under anaerobic conditions. The electron donor may be H_2, organic acids, fatty acids, alcohols or hydrocarbons. Sulfate is a much less favourable electron acceptor than O_2 or NO_3^-; therefore growth yields of sulfate-reducing bacteria are much lower. Sulfate reduction can be represented by two equations which may be combined to form a general equation [Eq. (12)] according to:

$$2CH_2O + SO_4^{2-} = S^{2-} + 2CO_2 + 2H_2O. \quad (10)$$

$$S^{2-} + 2CO_2 + 2H_2O = H_2S + 2HCO_3^{2-}. \quad (11)$$

$$2CH_2O + SO_4^{2-} = H_2S + 2HCO_3^{2-}. \quad (12)$$

As indicated by Eq. (12) the removal of sulfate results in a decrease in acidity.

If sulfate-reducing bacteria will oxidize organic acids and if the H_2S gas can escape from the system, the equation may become:

$$SO_4^{2-} + CH_3CHOHCOO^- + 2H_2 = H_2S\,(g) + CO_2 + CH_3COO^- + 2OH^- + H_2O. \quad (13)$$

The metabolism of *Desulfovibrio*, one of the best studied and most versatile representatives of the sulfate-reducing bacteria, has been the subject of a recent review (Voordouw 1995). From this article it becomes clear that *Desulfovibrio* is not only involved in sulfate reduction but also in the reduction of various other metals.

14.2.6 Methanogenesis

Methanogenesis is carried out by a highly specialized group of archaea, the methanogens. The overall reaction is as follows:

$$4H_2 + CO_2 = CH_4 + 2H_2O. \quad (14)$$

There is no change in acidity during methanogenesis, but the local pH may increase due to the reduction of CO_2. As only very few substrates (CO_2-type substrate, methyl substrates, acetoclastic substrate) can be converted to methane, the methanogenesis is dependent on the production of these few carbon compounds by other organisms.

14.2.7
Factors Controlling the Increase in pH/Alkalinity

Few data sets are available to calculate separate alkalinity budgets for lakes and their watersheds. These were analysed and reviewed by Schindler (1986) who found that in-lake production of alkalinity is an environmentally significant process. The factors controlling in-lake alkalinity generation are physical and chemical processes. The physical factors include the volume of the hypolimnion, light penetration, temperature, the degree of mixing during lake turnover and transport processes which are dependent on the size of the terrestrial catchment. The two major chemical factors which are of significance for sulfate reduction are the available organic substrates and the sulfate concentration.

The redox processes mediated by bacteria depend on the redox potential. The bacteria will use from the pool of electron acceptors one which will gain the most energy from an available substrate. At pH 7 the sequence of reactions may have the following order: denitrification, manganese reduction, nitrate reduction to ammonia, iron reduction, sulfate reduction and methanogenesis (Sigg and Stumm 1994; see also Fig. 14.1). The situation is complicated by the fact that in reality the cycle of each element is dependent on or influenced by other elements, and that due to limited organic substrate availability the reactions are competitive (Kelly et al. 1982). Dependent on the availability of exogenous electron acceptors, different fermentations go on parallel to anaerobic respiration. Under circumneutral conditions, some anaerobic phototrophic bacteria can oxidize ferrous iron, thereby generating reducing power for CO_2 fixation (Widdel et al. 1993). This unusual bacterial process permits a complete iron cycle under anaerobic conditions. In acidic mining lakes the most significant electron acceptors may be MnO_2, $Fe(OH)_3$ and SO_4^{2-}. Due to the redox potential, manganese reduction is energetically favoured over iron reduction which again is favoured compared with sulfate reduction. This reveals the order in which certain electron acceptors are used by bacteria. Thus the biogeochemical cycles of sulfur and of iron and to a certain degree of manganese have a strong impact in acidic mining lakes. Very recently, the general significance of metal reduction in se-

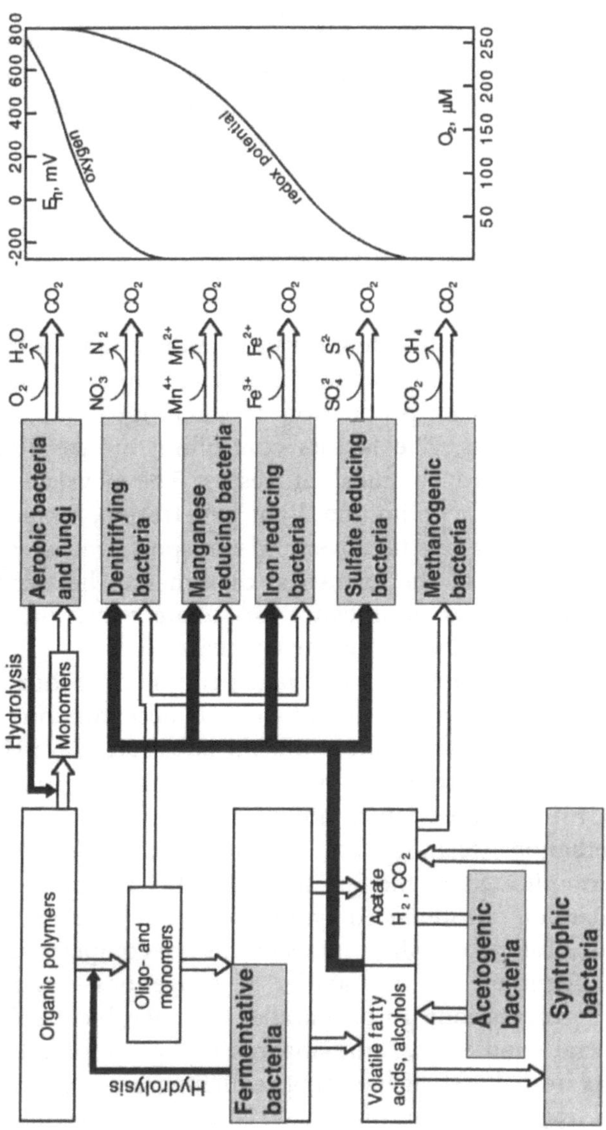

Fig. 14.1. Sequence of redox processes during degradation of organic matter, e.g. in lake sediments. The redox potentials represent standard conditions at pH 7 for the various reactions, and will often be lower in nature. (Westermann 1993, with permission; redox potentials from Zehnder and Stumm 1988)

diments and aquatic systems has been discussed in detail (Lovley 1993; Nealson and Saffarini 1994).

A central role in acidic lakes can be attributed to iron reduction which consumes hydrogen ions. In addition, iron can deliver sulfate to anaerobic regions in a lake. This transport in the form of a coprecipitate of oxyhydroxide and sulfate may be responsible for retention of sulfate in sediments. Iron is also of importance as a compound which will precipitate with the end products of sulfate reduction to FeS. The precipitated iron monosulfides have to be retained in the reduced form, otherwise the reverse process of sulfide oxidation may occur which will liberate protons.

14.2.8 Applied Aspects

The method of choice for restoration of acidic lakes may be to support sulfate-reducing bacteria (Tuttle et al. 1969b). The problem with this approach is the necessity of the responsible bacteria for anaerobic conditions. These conditions could be established at different scales and locations within the acidic lake habitat.

Firstly, anaerobic conditions will be found in the sediment of the lakes. At the aerobic/anaerobic interface within the sediment the desired process of sulfate reduction will occur naturally together with the reverse process of sulfide oxidation. In typical mining lakes the reverse process is favoured by continuous mixing due to wind exposure and low depth (Herlihy et al. 1987).

Secondly, in deeper and wind-protected stratified lakes, the aerobic/anaerobic interface may be located within the water column. In stratified environments the pH is usually higher in the anaerobic zone compared with the aerobic zone (Blowes et al. 1991; Fortin et al. 1995). The duration of stratification determines the influence of alkalinity-generating anaerobic processes on the lake. This may imply a method by which this aerobic/anaerobic interface is gradually lifted towards the surface of an artificially meromictic lake, thereby creating a pH neutral lake. Practically, this will be impossible and a lake treated in such a way must necessarily be treated by a second methodology to re-establish a natural lake environment. Nevertheless, it implies the establishment of anaerobic interfaces within the lake.

Thirdly, an anaerobic interface could be established within microbial biofilms as demonstrated in several investigations (Nielsen 1987; Kühl and Jørgensen 1992). Biofilms at root interfaces of plants have been successfully used for wastewater treatment in artificial wetlands (Gray and Biddlestone 1995). Biofilms used for lake neutralization may be located on natural surfaces, e.g. on the roots of certain plants, or on artificial interfaces in the form of an in-lake or out-of-lake biofilm reactor (see Table 14.1). An example for an in-lake biofilm reactor could be the addition of a solid substrate which also serves as the substratum for the microbial film.

Table 14.1. Overview, presence and location of aerobic/anaerobic interfaces within natural and artificial systems which may be of significance for controlling microbial redox processes of acidic lakes

Location of interface	Natural and artificial examples
Lake bottom	Aerobic/anaerobic interface within the sediment of a lake
Water column	Aerobic/anaerobic interface in stratified lakes
Shoreline	Aerobic/anaerobic interface within the biofilm associated with the root area of plants
Lake surface	Aerobic/anaerobic interface within the biofilm associated with the root area of floating plant islands
Substratum	Aerobic/anaerobic interface on artificial substratum at lake bottom
Biofilm reactor	Aerobic/anaerobic interface within a microbial biofilm grown in a special reactor
Anaerobic reactor	Operation under reducing conditions

Anaerobic conditions may be generated by adding organic substrates to the lake. The activity of facultative anaerobic bacteria then will lower the redox potential and at the same time supply degradation intermediates which can be utilized by sulfate-reducing bacteria. However, this approach would have to be planned carefully to avoid permanent eutrophication of the lake.

14.3 Recent Developments

14.3.1 Sulfate Reduction in Acidic Environments

The most widely used approach for neutralization of natural waters polluted by acid mine drainage is to stimulate bacterial sulfate reduction, since sulfate is abundant in mining waters. As most sulfate-reducing bacteria are heterotrophs, organic carbon sources have to be added. Sulfate-reducing bacteria are taxonomically and physiologically diverse. They are found among the Gram-positive and Gram-negative bacteria as well as in the archaeal domain (Devereux and Stahl 1992), and new types are continuously being detected (Rueter et al. 1994; Kawaguchi et al. 1995). These organisms can use a wide variety of carbon substrates (reviewed by Colleran et al. 1995; see also Table 14.2), so one can choose something rea-

Table 14.2. Energy substrates for sulfate-reducing bacteria. (Modified after Colleran et al. 1995)

Aliphatic mono-carboxylic acids	Formate, acetate, propionate, butyrate, isobutyrate, 2 methylbutyrate, 3 methyl-butyrate, 3 methylvalerate, fatty acids up to 20 carbon atoms, pyruvate, lactate
Dicarboxylic acids	Succinate, fumarate, malate, oxalate, maleinate, glutarate, pimelate
Alcohols	Methanol, ethanol, propanol-1 and 2, butanol-1 and 2, isobutanol, pentanol-1, ethylene glycol, 1-2 propanediol, 1-3 propanediol, glycerol
Amino acids	Glycine, serine, alanine, cysteine, cystine, threonine, valine, leucine, isoleucine, aspartate, glutamate, phenylalanine
Sugars	Fructose, glucose, mannose, xylose, rhamnose
Aromatic compounds	More than 35 different substances, for example benzoate, phenol, indole, resorcinol, catechol, p-cresol, quinoline, nicotinic acid, phenylacetate, vanillin, syringaldehyde, trimethoxybenzoate
Miscellaneous	Betaine, choline, furfural, acetone, cyclohexanone, etc.
Inorganic compounds	CO_2 as sole carbon source and H_2 as electron donor

dily available in bulk quantities for technical application. Materials like mushroom compost (Dvorak et al. 1992) and sawdust, sometimes in combination with limestone, have been applied with some success (Tuttle et al. 1969a). On the other hand, some monomeric carbon sources may be inhibitory to sulfate reduction in certain microbial communities (Tuttle et al. 1969b). Therefore, substrate testing before large-scale field application is advisable. The finding that the highest sulfate reduction occurred in a wood dust pile and not in the water body indicates that biological neutralization may be improved by solid substrates that allow biofilm formation. The sulfate-reducing bacteria themselves do not utilize complex polymers, but these are degraded by several other physiological types of bacteria, thereby consuming oxygen, creating anaerobic conditions and providing substrate for the sulfate-reducing bacteria. Anaerobiosis is necessary for sulfate reduction, although many sulfate-reducing bacteria are no longer believed to be strictly anaerobic. In fact, *Desulfovibrio* strains have recently been shown to carry out aerobic respiration coupled to ATP formation (Dilling and Cypionka 1990). In addition, the highest numbers of sulfate-reducing bacteria are often found at the oxic/anoxic interface where they are likely to be exposed to molecular oxygen (Jørgensen and Bak 1991).

The pH optima of sulfate reducers in pure culture are near neutrality, and no truly acidophilic sulfate-reducing bacteria have been isolated in pure culture yet. Nevertheless, sulfate reduction has been detected in sediments of coal strip mining lakes and in primary enrichments from this habitat at pH < 4, and sulfate-reducing bacteria isolated from acid streamers could be grown in special acidified media at pH 2.9 (Johnson et al. 1993). Comparisons between two lakes with different influence of acid mine drainage (reservoir 29 and lake B: lake water pH 2.7 and 3.2; sediments pH 3.8 and 6.2) gave pH optima for sulfate reduction of 5 and 6.2, respectively (Gyure et al. 1990). Mixed bacterial cultures could reduce sulfate at pH 3 in the laboratory with sawdust as the only nutrient (Tuttle et al. 1969a). These results demonstrate that sulfate reduction is possible in situ at low pH and that microbial communities adapt to the extreme environmental conditions of mining waters.

Most studies on acid mine drainage neutralization have been conducted in experimental bioreactors or in constructed wetlands. Dvorak et al. (1992) treated coal mine water in anaerobic reactors (Pittsburgh model) with spent mushroom compost, thereby raising the pH from 3.2 to 6.4. Alkalinity was generated by sulfate reduction and limestone dissolution. H^+, Al and Mn acidity had to be neutralized by the limestone to produce a H_2S-free effluent. Béchard et al. (1993) used flow-through bioreactors filled with straw or hay for treatment of artificial acid mine drainage. The pH was raised by microbial processes from 3.5 to 6.5 in their reactors, but addition of a soluble carbon source, such as sucrose, was necessary to maintain a stable system. A recent study by Brugam et al. (1995) used polyethylene in-lake enclosures enriched with wheat straw in different quantities. Sulfide concentrations, pH and acid neutralizing capacity increased during the experiment, but returned to the level of the surrounding lake after some time. This was believed to be due to enclosure leakage. These studies show that despite some problems, organic additions to acidic mine waters cause the production of acid neutralizing capacity via bacterial sulfate reduction.

14.3.2
The Role of Iron in Microbiological Neutralization of Acidity

Relatively few studies have investigated the role of iron in mine water neutralization, but those that have indicate that iron may be more important than generally believed until now. Iron influences biological neutralization in several ways. An important but unwanted reaction of iron is the so-called armoring (Henrot and Wieder 1990). When limestone is applied for chemical neutralization, its surface is covered with iron oxide deposits so that it is no longer reactive. Therefore, liming, which might seem easier to handle and more predictive than biological neutralization, is not feasible in iron-rich waters. Furthermore, the presence of iron may stimulate sulfate reduction. Herlihy and Mills (1989) found significantly higher sulfate reduction rates in Lake Anna (Virginia, USA) sediment microcosms in the presence of iron and hypothesized that adsorption of sulfate onto precipitating iron hydroxide flocs would enhance sulfate transport to the sediment. This could lead to more rapid sulfate removal in metal-rich (acid mine drainage-polluted) systems compared with metal-poor systems such as those influenced by acidic deposition.

The potential of Fe(III) reduction itself for alkalinity generation still has to be studied. Fortin et al. (1995) found that sulfate reduction could not be entirely responsible for pH increase in mine tailings and suggested that limestone dissolution and iron hydroxide dissolution could also play important roles. Vile and Wieder (1993) tested the contribution of Fe(III) reduction for alkalinity generation in constructed wetlands and flask experiments with sawdust, straw/manure, mushroom compost and sphagnum peat as organic substrates. They found that biologically mediated Fe(III) reduction was the most important process for generating alkalinity whereas alkalinity generation by sulfate reduction was not observed in any treatment. Bell et al. (1990) studied sediment recovery in Lake Anna after deposition of aerobic, acidic material by a storm event. They observed that pH and alkalinity increased before products of sulfate reduction accumulated and assumed that a succession of proton adsorption, iron reduction and fermentative activity was the responsible process. Chapelle and Lovley (1992) found that Fe(III)-reducing bacteria can even mediate competitive exclusion of sulfate reduction. In contrast, co-occurrence of sulfate reduction and Fe(III) reduction was found in Lake Anna sediments.

In recent years some knowledge has accumulated about the organisms involved in ferric iron reduction, which could help us to understand the observations listed above. Iron-reducing bacteria have been found in different groups, such as marine bacilli, magnetic bacteria and Gram-negative heterotrophic bacteria (Beveridge and Doyle 1989). Certain sulfur-reducing bacteria can also reduce Fe(III) (Roden and Lovley 1993; Coates et al. 1995). A well-characterized bacterium that couples the complete oxidation of organic compounds to ferric iron reduction is *Geobacter metallireducens* (Lovley et al. 1993). Recently *Geovibrio*, a phylogenetically distinct genus of dissimilatory iron-reducing bacteria, has been discovered (Caccavo et al. 1996). Fungal iron reduction has also been reported (Ottow and von Klopotek 1969), but is probably of less environmental significance. Johnson et al. (1993; Johnson and McGinness 1991) found that approximately 40% of the mesophilic heterotrophic acidophiles isolated from "acid streamers" and sediments of an acid mine drainage stream were able to reduce ferric iron under microaerophilic and anoxic conditions. Crystalline, amorphous and soluble ferric irons were reduced.

Coleman et al. (1993) could show with chemotaxonomic methods that sulfate-reducing bacteria were responsible for iron reduction in salt marsh sediments, but this environment was not acidic. This implies that the heterotrophic acidophiles may play an important role in biological alkalinity generation as they tolerate oxygen and respire efficiently under acidic conditions. Their activity may be a prerequisite for sulfate reduction in mining environments, an assumption which is supported by the redox sequence shown in Fig. 14.1.

These findings indicate that both Fe(III) reduction and sulfate reduction contribute to the biological neutralization of mining waters, but quantitative relationships between sulfate reduction, pyrite accumulation and alkalinity generation have not yet been determined (Vile and Wieder 1993). There is some evidence that sulfate reduction only generates alkalinity when pyrite or organic sulfur compounds are formed, but not when FeS is produced (Anderson and Schiff 1987, cited in Vile and Wieder 1993). The process by which pyrite is formed from metastable iron monosulfides is still poorly understood, but it is known that high sulfide concentrations and reactive detrital iron minerals are needed, whereas crystalline forms of iron are less effective. Even under these conditions not all the sedimentary iron is converted to pyrite and a major part of the sulfides is reoxidized (Berner 1984).

14.3.3 Alkalinity Generation by Other Biological Reduction Processes

The major factors that determine the neutralizing capacity of whole lakes are organic matter input and lake hydrology (Herlihy and Mills 1989). It has been shown that the anaerobic routes of carbon and electron flow do not differ from neutral eutrophic lakes, but are markedly slowed by low pH. They also do not function at high carbon flux rates (Goodwin and Zeikus 1987).

If iron or sulfur reduction is monitored by accumulation of Fe^{2+} or S^{2-} these processes may be falsely underestimated in the presence of manganese oxides (Beveridge and Doyle 1989). Fe(II) does not accumulate in the presence of MnO_2 because this acts as an external electron sink and pH buffer. This reduction of manganese oxides occurs even faster at lower pH values. Microbes catalyse manganese reduction both directly and indirectly. Indirect manganese reduction is mediated by excretion of metabolic end products. At pH values below 5, nearly all organic acids catalyse manganese reduction. Some obligate anaerobic manganese reducers have also been isolated, but have not been taxonomically classified (Lovley and Phillips 1988). These bacteria also reduce Fe(III) and use butyrate, propionate or ethanol as electron donors.

There are no experimental data on photosynthesis as a potential neutralization procedure. McConathy and Stahl (1982) studied the occurrence of different protozoa and rotifers in acidic lakes (pH 2.4–3.2) and found that individual and species numbers were higher in clumps of filamentous algae than in plankton. This is supported by our own observations (Packroff et al., in prep.); in addition, we consistently found higher pH values in algal clumps than in lakewater without algae. Nevertheless, no causality can be inferred from these preliminary data.

14.4 Conclusions

Microbial processes as a potential approach to in situ acidic lake remediation include not only sulfate reduction but also other reduction processes. Of the processes discussed, iron reduction seems to be the most significant, which has been neglected in nearly all previous studies. In environments where Fe(III) and Mn(IV) both occur at high concentrations, manganese reduction may also be a significant process. The most promising techniques in acidic lake restoration take advantage of microbial activities at interfaces. Within biofilms anaerobic conditions, necessary for biological reduction processes, are easily established. The crucial point in biological neutralization of acidic mining lakes is the permanent trapping of metal sulfides to prevent reoxidation and regeneration of acidity. Another critical point may be the choice of a suitable electron donor to support heterotrophic activities. Therefore, the environmental application of one or of a combined approach has to be tested first in a pilot plant. In addition, the economic feasibility of a particular restoration approach for a whole lake has yet to be established.

Acknowledgements. We thank Martin Schultze for computer processing of the figure.

References

Béchard G, Rajan S, Gould WD (1993) Characterization of a microbiological process for the treatment of acidic drainage. In: Torma AE, Apel ML, Brierley CL (eds) Biohydrometallurgical technologies. The Minerals, Metals and Materials Society, pp 277–286

Bell PE, Herlihy AT, Mills AL (1990) Establishment of anaerobic, reducing conditions in lake sediment after deposition of acidic, aerobic sediment by a major storm. Biogeochemistry 9:99–116

Berner RA (1984) Sedimentary pyrite formation: an update. Geochim Cosmochim Acta 48:605–615

Beveridge TJ, Doyle RJ (eds) (1989) Metal ions and bacteria. Wiley, New York

Blowes DW, Reardon EJ, Jambor JL, Cherry JA (1991) The formation and potential importance of cemented layers in inactive sulfide mine tailings. Geochim Cosmochim Acta 55:965–978

Brugam RB, Gastineau J, Ratcliff E (1995) The neutralization of acidic coal mining lakes by additions of natural organic matter: a mesocosm test. Hydrobiologia 316:153–159

Caccavo F Jr, Coates JD, Rossello-Mora RA, Ludwig W, Schleifer KH, Lovley DR, McInerney MJ (1996) *Geovibrio ferrireducens*, a phylogenetically distinct dissimilatory Fe(III)-reducing bacterium. Arch Microbiol 165:370–376

Chapelle FH, Lovley DR (1992) Competitive exclusion of sulfate reduction by Fe(III)-reducing bacteria: a mechanism for producing discrete zones of high-iron ground water. Ground Water 30:29–36

Coates JD, Lonergan DJ, Phillips EJP, Jenter H, Lovley DR (1995) *Desulfuromonas palmitatis* sp. nov., a marine dissimilatory Fe(III) reducer that can oxidize long-chain fatty acids. Arch Microbiol 164:406–413

Coleman ML, Hedrick DB, Lovley DR, White DC, Pye K (1993) Reduction of Fe(III) in sediments by sulfate-reducing bacteria. Nature 361:436–438

Colleran E, Finnegan S, Lens P (1995) Anaerobic treatment of sulfate-containing waste streams. Antonie van Leeuwenhoek 67:29–46

Devereux R, Stahl DA (1992) Phylogeny of sulfate-reducing bacteria and a perspective for analyzing their natural communities. In: Odom JM, Singleton R Jr (ed) The sulfate-reducing bacteria: contemporary perspectives. Springer, Berlin Heidelberg New York, pp 131–160

Dilling W, Cypionka H (1990) Aerobic respiration in sulfate-reducing bacteria. FEMS Microbiol Lett 71:123–12

Dvorak DH, Hedin RS, Edenborn HM, McIntire PE (1992) Treatment of metal-contaminated water using bacterial sulfate reduction: results from pilot-scale reactors. Biotechnol Bioeng 40:606–616

Ehrlich HE (1996) Geomicrobiology, 3rd edn. Marcel Dekker, New York

Fortin D, Davis D, Southam G, Beveridge JT (1995) Biogeochemical phenomena induced by bacteria within sulfidic mine tailings. J Ind Microbiol 14:178–185

Goodwin S, Zeikus JG (1987) Ecophysiological adaptations of anaerobic bacteria to low pH: analysis of anaerobic digestion in acidic bog sediments. Appl Environ Microbiol 53:57–64

Gray KR, Biddlestone AJ (1995) Engineered reed-bed systems for wastewater treatment. Trends Biotechnol 13:248–252

Gyure RA, Konopka A, Brooks A, Doemel W (1990) Microbial sulfate reduction in acidic (pH 3) strip-mine lakes. FEMS Microbiol Ecol 73:193–202

Henrot J, Wieder RK (1990) Processes of iron and manganese retention in laboratory peat microcosms subjected to acid mine drainage. J Environ Qual 19: 312–320

Herlihy AT, Mills AL, Hornberger GM, Bruckner AE (1987) The importance of sediment sulfate reduction to the sulfate budget of an impoundment receiving acid mine drainage. Water Resour Res 23:287–292

Herlihy AT, Mills AL (1989) Factors controlling the removal of sulfate and acidity from the waters of an acidified lake. Water Air Soil Pollut 45:135–155

Jørgensen BB, Bak F (1991) Pathways and microbiology of thiosulfate transformations and sulfate reduction in a marine sediment (Kattegat, Denmark). Appl Environ Microbiol 57:847–856

Johnson DB, McGinness S (1991) Ferric iron reduction by acidophilic heterotrophic bacteria. Appl Environ Microbiol 57:207–211

Johnson DB, Ghauri MA, McGinness S (1993) Biogeochemical cycling of iron and sulfur in leaching environments. FEMS Microbiol Rev 11:63–70

Kawaguchi R, Burgess JG, Sakaguchi T, Takeyama H, Thornhill RH, Matsunaga T (1995) Phylogenetic analysis of a novel sulfate-reducing magnetic bacterium, RS-1, demonstrates its membership of the δ-Proteobacteria. FEMS Microbiol Lett 126:277–282

Kelly CA, Rudd JWM, Cook RB, Schindler DW (1982) The potential importance of bacterial processes in regulating rate of lake acidification. Limnol Oceanogr 27:868–882

Kühl M, Jørgensen BB (1992) Microsensor measurement of sulfate reduction and sulfide oxidation in compact microbial communities of aerobic biofilms. Appl Environ Microbiol 58:1164–1174

Lovley DR (1991) Dissimilatory Fe(III) and Mn(IV) reduction. Microbiol Rev 55: 259–287

Lovley DR (1993) Dissimilatory metal reduction. Annu Rev Microbiol 47:263–290

Lovley DR, Phillips EJP (1986) Organic matter mineralization with reduction of ferric iron in anaerobic sediments. Appl Environ Microbiol 51:683–689

Lovley DR, Phillips EJP (1988) Novel mode of microbial energy metabolism: organic carbon oxidation coupled to dissimilatory reduction of iron or manganese. Appl Environ Microbiol 54:1472–1480

Lovley DR, Giovannoni SJ, White DC, Champine JE, Phillips EJP, Gorby YA, Goodwin S (1993) *Geobacter metallireducens* gen. nov. sp. nov., a microorganism capable of coupling the complete oxidation of organic compounds to the reduction of iron and other metals. Arch Microbiol 159:336–344

McConathy JR, Stahl JB (1982) Rotifera in the plankton and among filamentous algal clumps in 16 acid strip-mine lakes. Trans Ill State Acad Sci 75:85–90

Mills AL, Bell PE, Herlihy AT (1989) Microbes, sediments, and acidified water: the importance of biological buffering. In: Rao SS (ed) Acid stress and aquatic microbial interactions. CRC Press, Boca Raton, pp 1–19

Nealson KH, Saffarini D (1994) Iron and manganese in anaerobic respiration: environmental significance, physiology, and regulation. Annu Rev Microbiol 48: 311–343

Nielsen PH (1987) Biofilm dynamics and kinetics during high-rate sulfate reduction under anaerobic conditions. Appl Environ Microbiol 53:27–32

Ottow JCG, von Klopotek A (1969) Enzymatic reduction of iron oxide by fungi. Appl Microbiol 18:41–43

Roden EE, Lovley DR (1993) Dissimilatory Fe(III) reduction by the marine microorganism *Desulfuromonas acetoxidans*. Appl Environl Microbiol 59:734–742

Rueter P, Rabus R, Wilkes H, Aeckersberg F, Rainey FA, Jannasch HW, Widdel F (1994) Anaerobic oxidation of hydrocarbons in crude oil by new types of sulfate-reducing bacteria. Nature 372:455–458

Schindler DW (1986) The significance of in-lake production of alkalinity. Water Air Soil Pollut 30:931–944

Sigg L, Stumm W (eds) (1994) Aquatische Chemie. Teubner, Stuttgart

Stumm W (ed) (1992) Chemistry of the solid water interface. Wiley, New York

Tuttle JH, Dugan R, MacMillan CB, Randles CI (1969a) Microbial dissimilatory sulfur cycle in acid mine water. J Bacteriol 97:594–602

Tuttle JH, Dugan PR, Randles CI (1969b) Microbial sulfate reduction and its potential utility as an acid mine water pollution abatement procedure. Appl Microbiol 17:297–302

Vile MA, Wieder RK (1993) Alkalinity generation by Fe(III) reduction versus sulfate reduction in wetlands constructed for acid mine drainage. Water Air Soil Pollut 69:425–441

Voordouw G (1995) The genus *Desulfovibrio*: the centennial. Appl Environ Microbiol 61:2813–2819

Westermann P (1993) Wetland and swamp microbiology. In: Ford TE (ed) Aquatic microbiology – an ecological approach. Blackwell Scientific, Oxford, pp 215–238

Widdel F, Schnell S, Heising S, Ehrenreich A, Assmus B, Schink B (1993) Ferrous iron oxidation by anoxygenic phototrophic bacteria. Nature 362:834–835

Zehnder AJB, Stumm W (1988) Geochemistry and biogeochemistry of anaerobic habitats. In: Zehnder AJB (ed) Biology of anaerobic microorganisms. Wiley, New York, pp 1–38

15 Biological Abatement of Acid Mine Drainage: The Role of Acidophilic Protozoa and Other Indigenous Microflora

D.B. Johnson

School of Biological Sciences, University of Wales, Bangor LL57 2UW, UK

15.1 Introduction

Extremely acidic, metal-rich environments such as acid mine drainage and industrial leachate liquors may be populated by a considerable diversity of obligate acidophilic microorganisms. These include the familiar metal-mobilising chemolithotrophic bacteria (*Thiobacillus ferrooxidans*, *Leptospirillum ferrooxidans*, etc.) as well as other bacteria and eukaryotes, some of which have received relatively little attention, but which have considerable potential either in controlling the production of acidic mine effluents, or in treating this form of pollution once it has formed. Several isolates of acidophilic protozoa have been shown to graze mineral-oxidising and other acidophilic bacteria and, in some cases, thereby to decrease the rate of pyrite oxidation in coal samples. The predator–prey relationship which exists between acidophilic bacteria and grazing protozoa has been found to suppress rather than to eliminate metal-mobilising bacteria and, at present, prospects for biological control of acid mine drainage using protozoa would appear to be somewhat remote. Other indigenous acidophilic microflora include some heterotrophic bacteria which essentially reverse the reactions of pyrite oxidation, by inducing either the dissimilatory reduction of ferric iron or of sulphate. These reactions generate net alkalinity, and also cause chalcophilic metals present in acidic effluents to precipitate as highly insoluble sulphides; they therefore have considerable potential in the development of novel bioremediation schemes for acid mine water pollution.

15.2 Genesis of Acid Mine Drainage and Conventional Control Measures

Acid mine drainage (AMD) is one of the more serious forms of pollution afflicting industrialised societies, particularly those currently or historically involved in mining (coal and metals) activities. A central reaction in the genesis of AMD is the oxidation of sulphide minerals, such as pyrite and marcasite, by ferric iron, as:

$$FeS_2 + 14Fe^{3+} + 8H_2O \rightarrow 15Fe^{2+} + 2SO_4^{2-} + 16H^+ .$$

This reaction can occur abiotically, and also under anoxic conditions (such as exist in the cores of mineral waste heaps). However, the oxidant (ferric iron) is consumed in the reaction, and its regeneration does require the presence of oxygen, as:

$$4Fe^{2+} + O_2 + 4H^+ \rightarrow 4Fe^{3+} + 2H_2O.$$

Whilst this reoxidation reaction can occur spontaneously, it does so relatively slowly in highly acidic, sterile solutions. However, the rate of ferrous iron oxidation may be accelerated by a factor of between 10^4 and 10^6 by some obligate acidophilic chemolithotrophic bacteria, such as *Thiobacillus ferrooxidans* and *Leptospirillum ferrooxidans*. This mechanism of accelerated sulphide oxidation, which is generally known as the 'indirect' mechanism, has recently been proposed to be the single mode by which acidophilic iron bacteria (including those attached to mineral surfaces) bring about mineral dissolution (Sand et al. 1995), though electrochemical interactions between minerals during ore leaching can also lead to selective accelerated sulphide oxidation (Mustin et al. 1992). It follows, therefore, that removal or inhibition of these particular acidophilic bacteria in a mineral-leaching environment would effectively limit the rate of AMD formation, particularly as pure cultures of other acidophilic microorganisms (such as the sulphur-oxidiser *Thiobacillus thiooxidans*) are unable to oxidise sulphide minerals.

Various approaches may be adopted to control AMD pollution. One is to minimise the oxidation of sulphide minerals. Limiting the exposure of sulphide-rich strata and mine spoils to oxygen and/or moisture may be achieved mechanically by sealing mine shafts and adits, or by underwater disposal of mine tailings, or by the use of 'dry covers' on spoil heaps. The feasibility of any of these approaches depends on the size, location and nature of the sulphidic deposit in question. An alternative approach to controlling the activities of iron-oxidising acidophiles in mine wastes involves the application of surfactants, such as sodium dodecyl sulphate, to which these bacteria are highly sensitive (Dugan 1987). However, whilst short-term control may be achieved in this way, reports of longer-term problems with using surfactants (e.g. Johnson 1991a) have placed doubt on the effectiveness of using chemical controls. An alternative approach, as yet untried, is to consider biological control of metal-mobilising bacteria using naturally occurring antagonistic microflora.

Conventional treatment of AMD, once formed, involves addition of neutralising chemicals such as sodium hydroxide or calcium carbonate, thereby causing the pH of the acidic waters to increase, and the precipitation of many potentially highly toxic soluble metals (such as aluminium and lead) as their hydroxides. Bulky iron-rich precipitates are formed, which require collection and disposal, thereby increasing considerably the overall cost of the treatment. Also, it is essential to ensure that any residual ferrous iron in AMD is oxidised at or before the point of alkali addition, as any ferrous iron still in solution beyond this will oxidise downstream, generating further acidity. Ferrous iron oxidation may be induced by adding chemical oxidants such as hydrogen peroxide, or biologically using bacteria such as *T. ferrooxidans* immobilised onto rotating contactors (Unz and Deitz 1986); aeration of AMD may be necessary to facilitate complete iron oxidation (Hustwit et al. 1992).

An alternative to chemical treatment of AMD, which has found particular favour in North America, is the use of wetlands. These may be natural or, more often, constructed features. This approach uses biological mechanisms to ameliorate the acidity and metal contents of AMD and, although the microbial ecology of AMD-receiving wetlands has not received detailed study, bacteria which induce the dissimilatory reduction of iron and sulphur (as described in Sect. 15.4) are generally considered to be of primary importance in wetland ecosystems (Kalin et al. 1991). Biological mitigation of AMD using constructed wetlands is appealing since these systems tend to cost less to set up and maintain than conventional chemical treatment plants. However, their effectiveness at mitigating AMD pollution is not always as consistent as is desired (Wieder 1994).

Given the scale of the AMD pollution, the variability and distribution of potentially polluting sites, and the problems encountered with existing remediation technologies, it is worth considering alternative and novel approaches to the problem. Biological control is one such approach. Bacteria such as *T. ferrooxidans* and *L. ferrooxidans* are not unique in being tolerant of the extreme conditions (acidity, metal and sulphate concentrations, etc.) of mineral leachate liquors and AMD. Indeed, the microbiology of these environments is surprisingly complex, both in terms of the diversity of indigenous microbial populations and the interactions that occur between them (Tables 15.1 and 15.2; Fig. 15.1). Certain acidophilic microorganisms appear to have at least some potential for controlling AMD pollution, either by essentially reversing the reactions of mineral sulphide oxidation, or else by controlling populations of AMD-generating bacteria in situ.

Table 15.1. Acidophilic microorganisms indigenous to acid mine drainage waters

Prokaryotic microorganisms

1. Chemolithotrophs
   a) Iron-oxidisers
   Thiobacillus ferrooxidans[a], *Thiobacillus prosperus*, *Leptospirillum ferrooxidans*
   b) Sulphur-oxidiser
   Thiobacillus thiooxidans[a]
2. Mixotrophs
   Thiobacillus acidophilus[a], *Thiobacillus cuprinus*
3. Heterotrophs
   a) Iron-oxidisers: '*T-21*' group[a]
   b) Iron-reducers: various *Acidiphilium* spp.
   c) Sulphate-reducers: *Desulfotomaculum*-like isolates
   d) Others: various *Acidiphilium* spp. (also *Acidomonas* and *Acidobacterium* spp.)

Eukaryotic microorganisms

1. Yeasts/fungi
   Rhodotorula spp., *Trichosporon* spp., *Aspergillus* spp. and others
2. Algae
   Chlamydomonas, *Navicula viridis* and *Chlorella vulgaris* have been identified
3. Protozoa
   Isolates of phyla *Ciliophora*, *Sarcodina* and *Mastigophora* have been identified
4. Others
   Rotifera (*Philodina*-like) observed in 'acid streamer' growths; cattails (*Typha latifolia*) may colonise AMD-impacted sites

This list excludes moderately thermophilic bacteria (e.g. *Sulfobacillus* spp.) and extremely thermophilic archaea (e.g. *Acidianus* and *Sulfolobus* spp.) which are found in thermal acidic waters.

[a] These bacteria have also been reported to reduce ferric iron under appropriate conditions.

Table 15.2. Microbial interactions in extremely acidic, metal-rich environments

Type of interaction	Examples
Competition	Ferrous iron oxidation by *T. ferrooxidans* and *L. ferrooxidans*
Predation	Grazing of acidophilic bacteria by protozoa and rotifera
Commensalism	Utilisation by acidophilic heterotrophs of organic material leaked by chemolithotrophic iron-oxidising bacteria, thereby 'detoxifying' the environment for organic-sensitive strains of the latter
Synergism	Oxidation of pyrite by mixed cultures of *T. thiooxidans* and heterotrophic iron-oxidising bacteria, neither of which are capable of mineral dissolution in pure culture

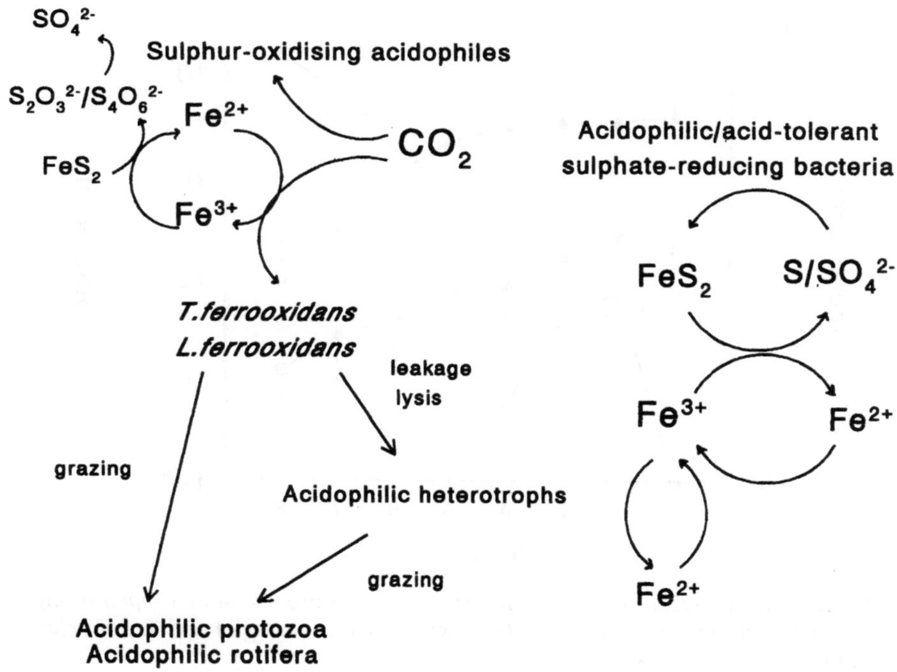

Fig. 15.1. Food web based on oxidation of pyrite (FeS_2) by acidophilic bacteria, illustrating the various dissimilatory transformations of iron and sulphur that occur in acidic, metal-rich environments

15.3
Control of Populations of Metal-Mobilising Bacteria by Acidophilic Protozoa

The presence of eukaryotic microorganisms in AMD streams was noted by some of the first researchers who worked in this area. Many yeasts and fungi are tolerant of moderate or extreme acidity, and some appear to be adapted to the extreme conditions which are characteristic of AMD, and are readily isolated from such on solid heterotrophic media (Cooke 1966). Protozoa were observed in AMD by both Lackey (1938) and Joseph (1953), the former describing green colonies of *Euglena mutabilis* on stones within streams, and the latter identifying non-photosynthetic Protozoa (*Amoeba proteus* and *Paramecium caudatum*) as well as *Euglena* spp. However, it was Ehrlich (1963) who first described an acidophilic or acid-tolerant protozoan (a biflagellate, tentatively identified as a *Eutreptia* sp.) growing in enrichment cultures of acidophilic iron-oxidisers. He also found amoeba present in stored (49-day) AMD samples, though he did

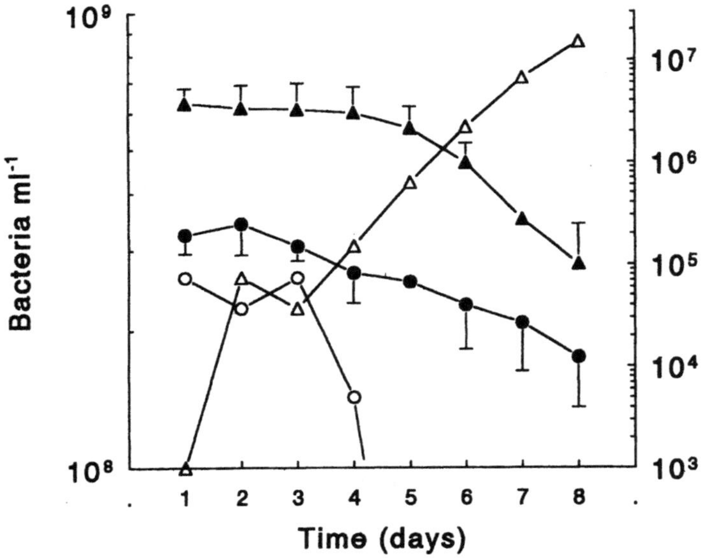

Fig. 15.2. Relationships between numbers of flagellate protozoan PR1 (*open triangle, open circle*) and viable counts of *T. ferrooxidans* (*solid triangle*) and *T. thiooxidans* (*solid circle*) in shake flask cultures

not observe these protozoa in fresh AMD samples. Ehrlich (1963) concluded that the protozoa were grazing bacteria in these cultures, though no evidence was presented in support of this. Johnson (1991b) described a food chain based on the chemolithotrophic oxidation of ferrous iron by *T. ferrooxidans* and *L. ferrooxidans*. Protozoa and rotifera (both observed grazing native 'acid streamer' bacterial growths) formed the highest trophic levels on the food web. It was possible to set up and maintain such a system in batch and continuous cultures in the laboratory, with a biflagellate as the single grazing acidophile. Later, McGinness and Johnson (1992) studied this biflagellate (coded isolate PR1, and also identified as a *Eutreptia* sp.) in greater detail, examining, amongst other things, its pH range and metal tolerance, and also the range of acidophilic bacteria and yeasts that it consumed. Both *T. ferrooxidans* and *L. ferrooxidans* were found to be grazed by PR1, though the protozoan displayed a preference for grazing *T. ferrooxidans* when grown in mixed culture with both iron-oxidisers. *Acidiphilium cryptum* and various unclassified acidophilic isolates were also grazed, though grazing of an acidophilic yeast was limited, presumably because of the relatively small size of the biflagellate (some 8 by 1 μm). Interestingly, it appeared that neither *T. thiooxidans* nor the mixotroph *T. acidophilus* were grazed by PR1. Figure 15.2 shows how increases in the total numbers of PR1 (from direct microscopic counts) correlated inversely with viable counts (using specific solid media) of *T. ferrooxidans*, but not with those of *T. thiooxidans*. By examining changes in bacterial and flagellate populations with time, it was possible to evaluate the ratio of numbers of acidophilic bacteria grazed to numbers of protozoa produced; this varied from 19 to 39, depending on the biovolume of the particular bacterium being grazed.

The flagellate protozoan PR1 was isolated from acid streamer growths found within a pyrite mine (Cae Coch) in North Wales which had been abandoned for some 70 years. Detailed observation of streamers taken from different locations in the main AMD stream within this mine revealed that other flagellates and ciliates (of varying sizes and morphologies) were actively grazing the bacteria which make up the gelatinous streamer growths. By careful manipulation and selection of enrichment cultures, it was possible to obtain cultures that were 'pure' with regard to protozoa (i.e. they contained only one morphological type), though they were invariably 'mixed' with regard to other acidophiles (i.e. they contained various strains of iron-oxidising chemolithotrophs, acidophilic heterotrophic bacteria and, occasionally, yeasts and fungi). Communities could be maintained by subculturing in ferrous sulphate-basal salts medium, and 'protozoan-free' control cultures (i.e. with identical bacterial populations, but devoid of protozoa) could be obtained by filtering young cultures through 3-µm membranes; full details of the procedures are contained in Johnson and Rang (1993). In this way, cultures of three other protozoa from Cae Coch mine were established: PR3, a ciliate identified as a *Cinetochilium* sp.; PR4, a second biflagellate which was slightly larger than PR1 but also probably a *Eutreptia* (or *Bodo*) sp.; PR5, an amoeboid protozoan which was considered to be a *Vahlkampfia* sp. A fifth protozoan (another biflagellate similar to PR1) was isolated from an experimental coal processing pilot plant at the Idaho National Engineering Laboratory (INEL), USA, courtesy of Dr. Graham Andrews. Microscope examination of the acidic leachate liquor at the plant had shown that the normally high (10^7–10^8/ml) numbers of metal-mobilising bacteria had shown a sudden and dramatic decline, which coincided with the appearance of a small biflagellate (PR2). This was subsequently isolated by the author in the same manner as other acidophilic protozoa. Further details of the five protozoan isolates (and of a *Euglena* sp. isolated from a derelict copper mine) are given in Table 15.3.

One apparent anomaly found during early experimental work with protozoan PR1, and also noted with subsequent work with other acidophilic Protozoa, was seemingly an inability to grow in liquid media containing either finely ground pyrite or pyritic coal. Detailed analysis of the

Table 15.3. Characteristics of some acidophilic protozoa

Isolate code	Site of origin	Morphological characteristics	Identification	Notes
PR1	Cae Coch sulphur mine (North Wales)	Small biflagellate (~ 8 × 1 μm)	*Eutreptia* sp.	Obligate acidophilic (pH optimum 3–4)
PR2	INEL coal processing pilot plant	Small biflagellate (~ 7 × 1 μm)	*Eutreptia* sp.	Less tolerant of ferric iron than other isolates
PR3	Cae Coch sulphur mine	Small ciliate (~ 12 × 7 μm)	*Cinetochilium* sp.	
PR4	Cae Coch sulphur mine	Small biflagellate (~ 10 × 1 μm)	*Eutreptia* or *Bodo* sp.	
PR5	Cae Coch sulphur mine	Cylindrical limax amoeba (~ 5 × 10 μm)	*Vahlkampfia* sp., possibly *V. lacustris*	
PR6	Parys copper mine (North Wales)	Photosynthetic flagellate (3–40 μm long); cleavage monsters produced	*Euglena mutabilis*	Non-bactivorous; highly tolerant of ferric iron (> 200 mM) and more copper-tolerant (10 mM) than other isolates

INEL, Idaho National Engineering Laboratory, USA.

leachate liquor in oxidising pyrite cultures showed that measured physico-chemical parameters (pH, metal concentrations, etc.) were all within the known ranges of those tolerated by PR1, and there was an abundance of the self-same acidophilic bacteria that were readily grazed by the protozoan in ferrous sulphate liquid medium. The pyrite ore (80% FeS_2) itself originated from Cae Coch mine, and was ground (> 95% at < 200 μm) but not graded further. However, in media containing pyrite ore or pyritic coal that had been sorted (as 61- to 200-μm and < 61-μm fractions) it was found that whilst protozoa did not grow (< 10^2/ml) in cultures containing finer-grain minerals, they became highly abundant in cultures containing coarse-grained minerals (Johnson and Rang 1993). It was inferred that inadvertent ingestion of bacteria-sized pyrite or coal particles resulted in the inactivation or death of the protozoa.

In the light of the above observation, leaching experiments were conducted using coarse-grain (61- to 200-μm) pyritic coal, in which the

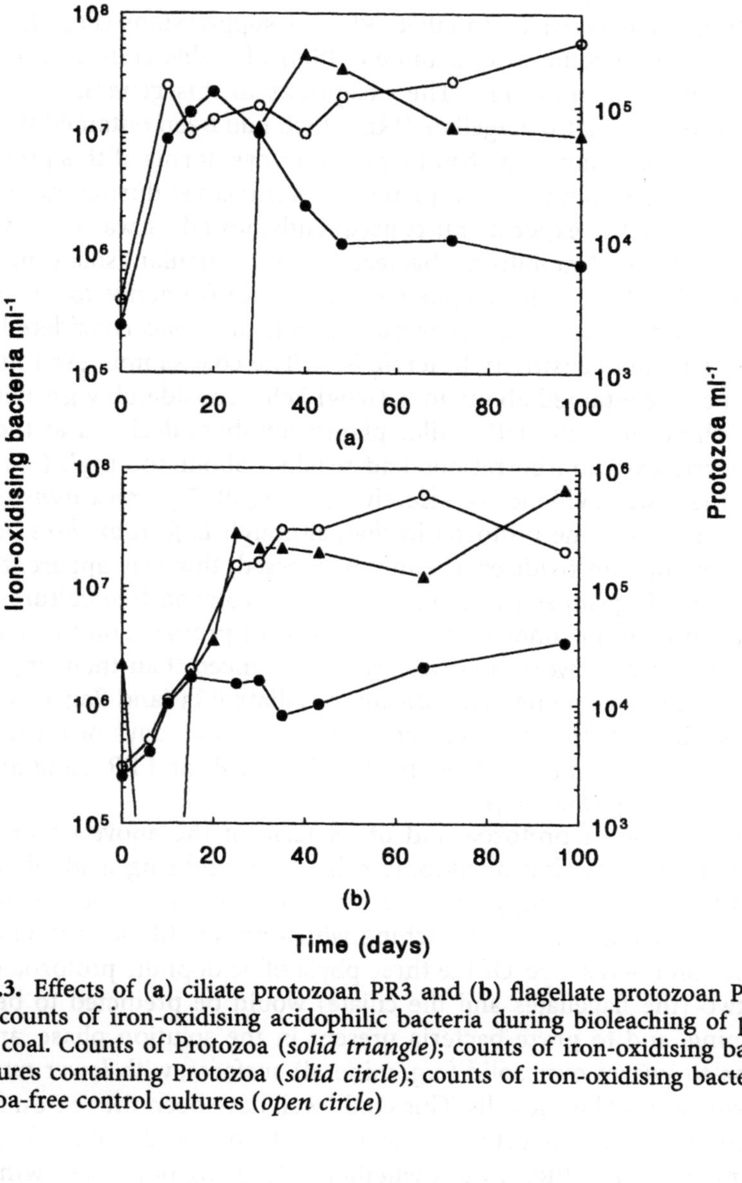

Fig. 15.3. Effects of (**a**) ciliate protozoan PR3 and (**b**) flagellate protozoan PR4 on viable counts of iron-oxidising acidophilic bacteria during bioleaching of pyritic (20%) coal. Counts of Protozoa (*solid triangle*); counts of iron-oxidising bacteria in cultures containing Protozoa (*solid circle*); counts of iron-oxidising bacteria in protozoa-free control cultures (*open circle*)

effects of individual protozoan cultures on populations of metal-mobilising bacteria and pyrite solubilisation were monitored (Johnson and Rang 1993). Some of the data obtained are shown in Figs. 15.3 and 15.4. Differences between the five protozoan isolates were apparent. Flagellates were detected (>10^2/ml) in cultures much earlier than the ciliate PR3, or the amoeba PR5, though replicate cultures of the latter displayed considerable disparity. Increases in protozoan populations corresponded with dramatic declines (with ciliate PR3) or suppressions (flagellates PR1 and PR4 and, to some extent, amoeba PR5) of viable counts of *T. ferrooxidans* and *L. ferrooxidans*. The exceptions to this general trend were cultures containing the flagellate PR2, which had been obtained from the INEL coal processing plant. Numbers of non-cyst forms of this protozoan went into sharp decline at day 40 and did not recover during the remaining 60 days of the experiment; consequently, no effect was observed on numbers of metal-mobilising bacteria. This particular isolate had also been found to have a more ephemeral existence (in active form) in iron sulphate medium, and the behaviour in both cases was considered to be due to its higher sensitivity to ferric iron. The coal sample used in these experiments contained about 20% (w/w) FeS_2 (considerably greater than that of the coal in the INEL pilot plant); numbers declined at the time when ferric iron concentrations had reached about 500 mg/l (~9 mM). Again, there was evidence of selective grazing of *T. ferrooxidans* over *L. ferrooxidans* by some protozoa in that, although *L. ferrooxidans* became the dominant iron-oxidiser as time progressed, this was apparent at an earlier stage in protozoa-containing than in protozoa-free cultures, with the exception of the amoeba PR5. The effects of protozoa on the net solubilisation of pyrite were, however, less pronounced than their impact on bacterial numbers, being insignificant for ciliate PR3 (and flagellate PR2), and lowering soluble iron concentrations (after 100 days of incubation) by 15, 25 and 12% (mean values) for flagellates PR1 and PR4, and amoeba PR5, respectively (Fig. 15.4).

Counts made of protozoa and of bacteria in the above experiments were of free-swimming planktonic cells. Iron-oxidising acidophiles are known to adhere strongly and selectively to sulphide phases in mineral matrices, causing patterns of etching which are readily identifiable with the electron microscope. Of the three phyla of acidophilic protozoa examined, two (the flagellates and the ciliate) would be predicted to be free-swimming, and to graze bacteria present in the solution phase, and the amoeba to swarm over solid (e.g. mineral) surfaces, with lower numbers occurring as planktonic cells. This could well have been the reason for the variation between replicate cultures (in protozoan and bacterial counts) containing amoeba PR5. To test whether, by grazing both free-swimming and attached metal-mobilising bacteria, a mixed community of acidophilic protozoa would be more effective at limiting bacterially promoted pyrite solubilisation than pure cultures, a mixed population (coded PR-M) was assembled. This contained the flagellate PR1, ciliate PR3 and amoeba PR5. A corresponding protozoan-free control culture containing the same bacteria but no protozoa was also prepared; other control cultures were those containing either PR1, PR3 or PR5. Pyritic (20% FeS_2) coal was then

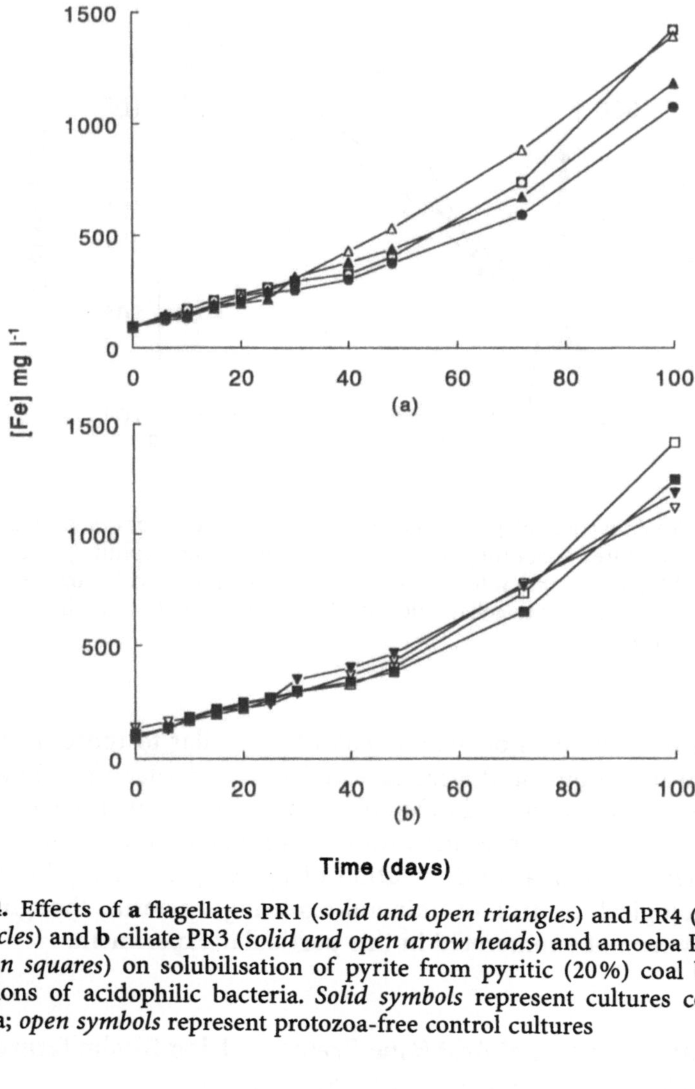

Fig. 15.4. Effects of **a** flagellates PR1 (*solid and open triangles*) and PR4 (*solid and open circles*) and **b** ciliate PR3 (*solid and open arrow heads*) and amoeba PR5 (*solid and open squares*) on solubilisation of pyrite from pyritic (20%) coal by mixed populations of acidophilic bacteria. *Solid symbols* represent cultures containing protozoa; *open symbols* represent protozoa-free control cultures

subjected to leaching by these systems, using the methods described by Johnson and Rang (1993), and changes in pyrite solubilisation were compared. Results, shown in Fig. 15.5, compare data of protozoan counts and pyrite solubilisation in cultures containing the mixed acidophilic protozoan population with those from cultures containing either the flagellate PR1 or the ciliate PR3, over a relatively small time scale (28 days). Acidophilic protozoa were far more numerous in the mixed culture than in either PR1- or PR3-containing cultures between days 7 and 21; the dominant eukaryote in the mixed cultures throughout was the flagellate,

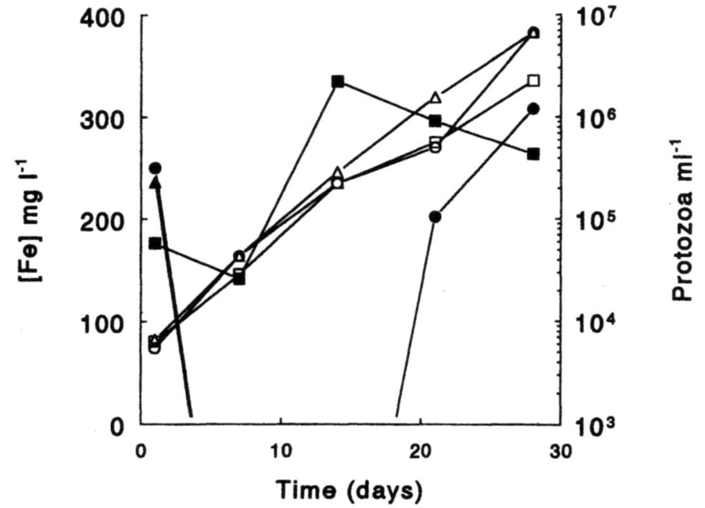

Fig. 15.5. Solubilisation of pyrite (*open symbols*) and numbers of protozoa (*solid symbols*) in cultures containing a mixed community of acidophilic protozoa (PR1, PR3 and PR5; *open and solid squares*) and comparison with similar bacterial communities containing flagellate PR1 (*open and closed circles*) or ciliate PR3 (*open and closed triangles*)

though small numbers of the amoeba PR5 (similar to those in the pure culture of this protozoan) were detected at day 28 (data not shown). No ciliates were found either in the mixed or pure cultures beyond day 0. Again, there was evidence that protozoa were able to limit, indirectly, to some extent the oxidative dissolution of pyrite, presumably by exerting some biological control of mineral-oxidising bacteria, though viable bacteria were not enumerated in this particular experiment.

15.4 Biological Mitigation of Acid Mine Drainage Using Dissimilatory Iron- and Sulphate-Reducing Acidophilic Bacteria

As described in Section 15.3, the indigenous microflora of AMD and mineral leachate liquors include bacteria other than the familiar acidophilic *Thiobacilli;* some of these may be involved also in promoting, either directly or indirectly, mobilisation and precipitation of metals. Some acidophiles bring about the dissimilatory reduction of ferric iron to ferrous (Pronk and Johnson 1992) in which ferric iron is used as an electron sink. Whilst this trait is particularly a characteristic of heterotrophic acidophiles (e.g. many species of *Acidiphilium*), even the most well-known iron-oxidiser (*T. ferrooxidans*) will reduce iron under appropriate conditions (for example, when oxidising elemental sulphur under anoxic conditions). The ability to reduce ferric iron appears also to be ubiquitous amongst thermo-tolerant ('moderately thermophilic') iron-oxidising bacteria (e.g. *Sulfobacillus* spp.; Ghauri and Johnson 1991). Acidophilic iron-reducers tend to be facultative anaerobes (though ferric reduction is often more rapid in micro-aerobic rather that strictly anoxic conditions) and, whilst the reduction of soluble ferric iron does not decrease solution acidity, the reduction of solid-phase (crystalline and amorphous) ferric iron compounds does, as:

$$Fe(OH)_3 + 3H^+ + e^- \rightarrow Fe^{2+} + 3H_2O,$$

where e^- originates from an organic electron donor (*Acidiphilium* spp., and heterotrophically grown 'iron-oxidisers').

Considerably less is known about acidophilic bacteria which reduce sulphate to sulphide. Observations that sulphate reducing bacteria (SRBs) were active in anaerobic zones in AMD streams were made by Tuttle et al. (1969) and by Johnson et al. (1993). However, isolates thus far described appear to be acid-tolerant rather than truly acidophilic. For example, a pure SRB culture isolated from a pH-3.38 pond by Tuttle et al. (1969) would not reduce sulphate in media poised below pH 5.5 in the laboratory, and an isolate (possibly a *Desulfotomaculum* sp.) from a pH-2.4 stream had a lower pH limit of 2.9 (Johnson et al. 1993). However, both research teams noted that mixed, undefined cultures of SRBs from AMD sites were significantly more tolerant to extreme acidity (to pH 3.0 and pH 2.4; Tuttle et al. and Johnson et al., respectively) than pure cultures. The significance of acidophilic/acid-tolerant SRBs in AMD mitigation is two-fold. Firstly, sulphate reduction is an acid-consuming reaction, as:

$$SO_4^{2-} + 8H^+ + 8e^- \rightarrow S^{2-} + 4H_2O.$$

Culture pH values in acidic SRB media rise by up to 2–3 units during incubation. Secondly, many metals found in AMD (such as copper, lead and zinc) are chalcophilic (i.e. they have high affinities for S^{2-}) and form highly insoluble sulphides. Sulphide genesis is therefore a potential method of removing many toxic heavy metals from solution.

A scheme which utilises acidophilic bacteria to ameliorate AMD pollution by, essentially, reversing the reactions of sulphide mineral oxidation has been proposed by Johnson (1995). In brief, the proposed system involves in-line bioreactors containing, in sequence, immobilised populations of acidophilic ferric iron-reducing bacteria and SRBs, through which raw AMD would flow. Initial conditioning of AMD by iron-reducing heterotrophs (immobilised, for example, onto carrier materials which could serve both as support matrix and carbon source) would condition the water to facilitate more efficient processing by immobilised SRBs. This the activities of the obligate anaerobic SRBs), eliminating soluble and particulate ferric iron which would otherwise react rapidly with nascent sulphides formed by the SRBs and, ideally, providing organic substrates (as secondary metabolites) for SRBs. Sulphide deposits accumulating in SRB bioreactors would be collected and disposed of, their bulk being considerably less than those of iron sludges produced via conventional chemical treatment of AMD.

15.5 Pros and Cons of Biological Abatement of Acid Mine Drainage

Acid mine drainage pollution is essentially a bacterially driven phenomenon, and the prospect of using indigenous microorganisms to inhibit its formation and/or mitigate the problem once formed is an attractive alternative to chemical remediation techniques. The use of biological agents to control metal-mobilising acidophilic bacteria could, theoretically, include bacteriophage and bacteriocins. However, although a phage which infected some strains of *Acidiphilium* has been described (Ward et al. 1990), there have been no reports of any which infect either *T. ferrooxidans* or *L. ferrooxidans*. Exploratory work in the author's laboratory also failed to isolate phage which was active against iron-oxidising acidophiles. Likewise, bacteriocins which inhibit or kill metal-mobilising acidophiles may exist, but are yet to be described. On the other hand, acidophilic protozoa that graze pyrite-oxidising bacteria are known, and appear to be well distributed in AMD environments. It is patently obvious, however, that any control they exert on AMD production is limited. For example, the derelict Cae Coch mine, which served as the source for four of the protozoa describedin Section 15.3, contains an estimated 100 m^3 + of acid streamer growths, which are composed of mixed communities of (predominantly) acidophilic heterotrophs and chemolithotrophs.

Grazing by protozoa and rotifera has apparently made little impact on net streamer biomass. Some of the reasons for this, and also for the limited impact that protozoa had on pyrite solubilisation (in contrast to their effects on bacterial numbers) in the experiments described in Section 15.3, are readily appreciated. Firstly, the relationship between acidophilic protozoa and bacteria is one of predator and prey, so that reductions in the numbers of bacteria are followed by reductions in protozoan numbers; this has been demonstrated with acidophilic bacteria and protozoa grown in continuous culture (Johnson 1991b). Populations of metal-mobilising bacteria tend, therefore, to be subjected to periodic suppression rather than being eliminated by grazing protozoa. Secondly, bacterial aggregation may serve as a defence mechanism for some acidophiles. This is thought to be the reason for limited acid streamer grazing, and also why *T. ferrooxidans* is grazed selectively over *L. ferrooxidans*, since the latter acidophile characteristically produces bacterial aggregates which may become visible to the naked eye. Likewise, attachment of bacteria to mineral surfaces may protect them from free-swimming protozoa. Additionally, it appears that even if populations of metal-mobilising acidophiles are lowered significantly by protozoan grazing, this may not necessarily produce an appreciable slowing down of pyrite oxidation. This is probably due to the fact that ferric iron regeneration is sufficiently rapid, even with relatively small populations of iron-oxidisers, to prevent this reaction becoming the rate-limiting step in pyrite oxidation. In this regard, it is interesting to note that the ciliate PR3 produced the greatest reduction in numbers of metal-mobilising bacteria in the coal leaching experiments referred to in Section 15.3, but this effect was delayed relative to the cultures containing flagellates PR1 and PR4, and there was consequently no detectable effect on net pyrite oxidation in cultures of this protozoan.

Other constraints on the effectiveness of protozoan grazing of acidophilic bacteria might be various physico-chemical factors. Extreme acidity appears not to be a bar to protozoa; ciliate PR3 had been observed to be active in pyrite cultures measured at pH 1.6, and flagellate PR1 has also been shown to graze an *Acidiphilium* sp. in media poised at pH 1.8 (McGinness and Johnson 1992). Certain heavy metals, however, appear to be much more toxic to some acidophilic protozoa than to iron-oxidising bacteria such as *T. ferrooxidans*. Copper (II), added at 1–5 mM, has been used to control the activities both of flagellate PR1 in laboratory cultures and of flagellate PR2 in the INEL coal processing plant. However, increased copper tolerance by both protozoa, presumably by adaptation or selection of tolerant variants, has been observed (unpubl. data). The extremely high copper tolerance of the non-grazing *Euglena* isolated from a copper mine (Table 15.3) suggests that protozoa tolerant to various heavy metals may be readily obtained from appropriate sites.

Acidophilic iron- and sulphate-reducing bacteria (SRBs) are probably important contributors to the neutralisation and detoxification of AMD that occurs in efficient natural and constructed wetland sites. However, given the heterogeneity of wetland ecosystems (particularly at the microsite level) it is probably the case that neutrophilic iron-reducers and SRBs also contribute significantly to dissimilatory iron and sulphur transformations. Whether net sulphate reduction is a characteristic of all but newly commissioned wetland ecosystems receiving AMD has been questioned by Vile and Wieder (1993). These researchers concluded that ferric iron reduction was a far more important source of alkalinity in most established wetlands, and that most of any sulphide formed was rapidly oxidised by ferric iron in the in-flowing AMD. Displacing biological ferric iron and sulphate reduction, such as in the bioremediation scheme proposed by Johnson (1995), would overcome this. Other perceived advantages of this system are the flexibility it allows in scaling bioreactors to facilitate treatment of small-scale AMD effluents, where wetland construction may not be feasible or economical, and the possibility of recovering metals from the accumulated sulphide precipitates.

Increased awareness of the complexity of the microbial ecology of highly acidic, metal-rich environments has resulted in the indication that certain acidophilic microflora may be of considerable use in controlling the production, or in mitigating the impact, of AMD. More extensive basic research, coupled with studies of the application of such approaches, will be necessary to evaluate whether biological control of AMD, as described in this chapter, could become a real alternative means of pollution abatement.

References

Cooke WB (1966) The occurrence of fungi in acid mine drainage. Proc Ind Waste Conf 21:258–274

Dugan PR (1987) Prevention of formation of acidic drainage from high-sulfur coal refuse by inhibition of iron-oxidizing and sulfur-oxidizing bacteria. 2. Inhibition in run of mine refuse under simulated field conditions. Biotechnol Bioeng 29:49–54

Ehrlich HL (1963) Microorganisms in acid drainage from a copper mine. J Bacteriol 86:350–352

Ghauri MA, Johnson DB (1991) Physiological diversity amongst some moderately thermophilic iron-oxidising bacteria. FEMS Microbiol Ecol 85:327–334

Hustwit CC, Ackman TE, Erickson PE (1992) The role of oxygen transfer in acid mine drainage (AMD) treatment. Water Environ Res 64:179–186

Johnson DB (1991a) Biological desulfurization of coal using mixed populations of mesophilic and moderately thermophilic acidophilic bacteria. In: Dugan PR, Quigley DR, Attia YA (eds) Processing and utilization of high sulfur coals IV. Elsevier, New York, pp 567–580

Johnson DB (1991b) Diversity of microbial life in highly acidic, mesophilic environments. In: Berthelin J (ed) Diversity of environmental biogeochemistry. Elsevier, Amsterdam, pp 225–238

Johnson DB (1995) Acidophilic microbial communities: candidates for the bioremediation of acidic mine effluents. Int Biodeter Biodegrad 35:41–58

Johnson DB, Rang L (1993) Effects of acidophilic Protozoa on populations of metal-mobilizing bacteria during the leaching of pyritic coal. J Gen Microbiol 139:1417–1423

Johnson DB, Ghauri MA, McGinness S (1993) Biogeochemical cycling of iron and sulphur in leaching environments. FEMS Microbiol Rev 11:63–70
Joseph JM (1953) Microbiological study of acid mine waters; preliminary report. Ohio J Sci 53:123–127
Kalin M, Cairns J, McCready RGL (1991) Ecological engineering methods for acid mine drainage treatment of coal wastes. Resour Conserv Recycl 5:265–275
Lackey JB (1938) Aquatic life in waters polluted by acid mine waste. Public Health Rep 54:740–746
McGinness S, Johnson DB (1992) Grazing of acidophilic bacteria by a flagellated protozoan. Microbial Ecol 23:75–86
Mustin C, de Donato P, Berthelin J (1992) Quantification of the intergranular porosity formed in bioleaching of pyrite by *Thiobacillus ferrooxidans*. Biotechnol Bioeng 39:1121–1127
Pronk J, Johnson DB (1992) Oxidation and reduction of iron by acidophilic bacteria. Geomicrobiol J 10:153–171
Sand W, Gehrke T, Hallmann R, Schippers A (1995) Sulfur chemistry, biofilm, and the (indirect) attack mechanism – a critical evaluation of bacterial leaching. Appl Microbiol Biotechnol 43:961–966
Tuttle JH, Dugan PR, Randles CI (1969) Microbial dissimilatory sulfur cycle in acid mine water. J Bacteriol 97:594–602
Unz RF, Deitz JM (1986) Biological applications in the treatment of acid mine drainages. Biotechnol Bioeng Symp 16:163–170
Vile MA, Wieder RK (1993) Alkalinity generation by Fe(III) reduction versus sulfate reduction in wetlands constructed for acid mine drainage treatment. Water Air Soil Pollut 69:425–441
Ward TE, Rowland ML, Bruhn DF, Watkins CS, Roberto FF (1990) Studies on a bacteriophage which infects members of the genus *Acidiphilium*. In: Salley J, McCready RGL, Wichlacz PL (eds) Biohydrometallurgy 1989. Int Symp Proc, Canmet, Canada, pp 159–169
Wieder RK (1994) Diel changes in iron(III)/iron(II) in effluent from constructed acid mine drainage treatment wetlands. J Environ Qual 23:730–738

16 Treatment of Mine Drainage by Anoxic Limestone Drains and Constructed Wetlands

R. L. P. Kleinmann, R. S. Hedin and R. W. Nairn***

US Dept. of Energy, P. O. Box 10940, Pittsburgh, Pennsylvania 15236-0940, USA

16.1 Introduction

As contaminated mine drainage flows into receiving streams, rivers, and lakes, its toxic characteristics decrease naturally as a result of chemical and biological reactions and by dilution with uncontaminated water. During the last decade, the possibility that these processes might be used to treat mine water passively has developed from an experimental concept to full-scale implementation at hundreds of sites. Ideally, passive treatment systems require no input of chemicals and very little operation and maintenance. However, passive treatment systems use contaminant removal processes that are slower than that of conventional treatment and thus require longer retention times and larger areas in order to achieve similar results.

Passive systems incorporate at least one of three basic techniques: aerobic wetland systems, wetlands that contain an organic substrate (compost wetlands) and anoxic limestone drains (ALDs). Each technology is most appropriate for a particular type of mine water problem. Often, they are most effectively used in combination with each other. In this chapter a model is presented that is useful in deciding whether a mine water problem is suited to passive treatment and, also, in designing and constructing effective passive treatment systems.

16.2 Design and Sizing of Passive Treatment Systems

Two sets of sizing criteria have been developed based on our observations and measurements of contaminant removal at existing constructed

* *Present address:* Hedin Environmental, 634 Washington Rd, Mt Lebanon, Pennsylvania, USA 15228.

** *Present address:* Univ. of Oklahoma, Norman, OK, USA 73019.

Table 16.1. Recommended sizing for passive treatment systems. (Hedin et al. 1994)

	AML criteria g m^{-2} day^{-1}		Compliance criteria g m^{-2} day^{-1}	
	Alkaline	Acid	Alkaline	Acid
Fe	20	NA	10	NA
Mn	1.0	NA	0.5	NA
Acidity	NA	7	NA	3.5

AML, abandoned mined land; NA, not applicable.

wetlands (Hedin et al. 1994 a, b; Table 16.1). The "abandoned mined land (AML) criteria" will cost-effectively decrease contaminant concentrations, but will not consistently achieve a specific effluent concentration. A more conservative sizing value is provided for systems where the effluent must meet US regulatory guidelines of 3 mg Fe l^{-1} and 2 mg Mn l^{-1}. We refer to these as "compliance criteria".

The Mn removal rate used for compliance, 0.5 g m^{-2} day^{-1}, is based on the performance of five treatment systems, three of which consistently lower Mn concentrations to compliance levels. A higher removal value, 1 g m^{-2} day^{-1}, is suggested for AML sites. Because the toxic effects of Mn at moderate concentrations (<50 mg l^{-1}) are generally not significant, except in very soft water (Kleinmann and Watzlaf 1988), and the size of a wetland necessary to treat Mn-contaminated water is so large, AML sites with Fe problems should receive a higher priority than those with only Mn problems.

These criteria are appropriate at most coal mines, which typically discharge low concentrations of metals such as Cd, Cu and Zn. Efforts to extend passive treatment to sites where such metals are more of a problem must consider the potential for food-chain effects and whether the long-term accumulation of such metals, as sulfides, in substrate sediment is desirable. Research is continuing; no recommendations can be made at this time.

The acidity removal rate is based on observed performance of compost wetlands and is influenced by seasonal variations in sulfate reduction (Hedin et al. 1991). This is not a problem for mildly acidic water, where the wetland can be sized in accordance with winter performance, nor should it be a major problem in warmer climates. In northern Appalachia, however, we know of no compost wetland that consistently transforms highly acidic water (>300 mg l^{-1} acidity) into alkaline water. One of our study sites, which receives water with an average of 600 mg l^{-1} acidity and does not need to meet a Mn standard, has discharged water that only required

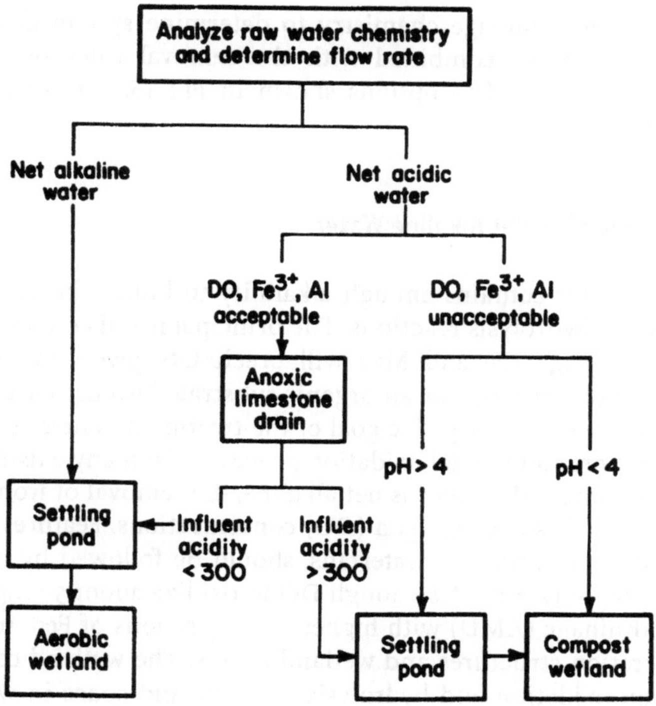

Fig. 16.1. Flow chart showing chemical determinations necessary for the design of passive treatment systems. *DO* Dissolved oxygen

chemical treatment during winter months. While considerable cost savings are realized at the site due to the compost wetland, the passive system must be supported by conventional treatment during a portion of the year.

Passive systems constructed at active sites need not be sized according to the compliance criteria provided in Table 16.1. Sizing must balance available space and system construction costs vs. influent water quality and chemical treatment costs. Mine water can be treated passively before the water is chemically treated to reduce water treatment costs or as a potential part-time alternative to full-time chemical treatment. In those cases where a combination of passive and chemical treatment methodologies have been utilized, many operators find that they recoup the cost of the passive system in less than a year by using less expensive chemical treatment systems and/or by decreasing the amount of chemicals used.

The design and sizing model is summarized in Fig. 16.1. The empirical basis of this model has been presented elsewhere (Hedin et al. 1994). The model uses mine drainage chemistry to determine system design, and contaminant loadings combined with the removal rates of Table 16.1 to define system size. The options shown in Fig. 16.1 are discussed in Section 16.2.1.

16.2.1 Passive Treatment of Net Alkaline Water

Net alkaline water contains enough alkalinity to buffer the acidity produced by metal hydrolysis reactions. The principal metal contaminants in coal mine drainage (Fe and Mn) will precipitate given enough time; incorporation of limestone or an organic substrate into the passive treatment system is unnecessary. The goal of the treatment system is to aerate the water and promote metal oxidation processes. In many existing treatment systems where the water is net alkaline, the removal of iron appears to be limited by dissolved oxygen (DO) concentrations. Features that can aerate the drainage, such as waterfalls, should be followed by quiescent areas. Aeration only provides enough DO to oxidize about 50 mg Fe^{2+} l^{-1}. Acid mine drainage (AMD) with higher concentrations of Fe^{2+} requires a series of aeration structures and wetland basins. The wetland cells allow time for iron oxidation and hydrolysis to occur and space in which the iron floc can settle out of suspension. The entire system can be sized based on the Fe removal rates shown in Table 16.1. If Mn removal is desired, the system is sized based on the Mn removal rates in Table 16.1. Removal of Fe and Mn occurs sequentially. If both Fe and Mn removal are necessary, the two wetland sizes are added together.

A typical aerobic wetland is constructed by planting *Typha* rhizomes in soil or non-acidic spoil. Some systems have been planted by simply spreading *Typha* seeds, with good plant growth attained after 2 years. The depth of the water in a typical aerobic system is 10–50 cm. Ideally, a cell should not be of a uniform depth, but should include both shallow and deep marshy areas. It is recommended that the freeboard of aerobic wetlands constructed for the removal of iron is at least 1 m. The water level and flow can be controlled by the downstream dam spillway or adjustable riser pipes.

Spillways should consist of wide cuts in the dike with side slopes no steeper than 2 H (horizontal): 1 V (vertical), lined with non-biodegradable erosion control fabric, and coarse rock if high flows are expected (Brodie 1991). Proper spillway design can preclude future maintenance costs due to erosion and/or failed dikes. If pipes are used, small diameter (< 30-cm) pipes should be avoided because they can plug with litter and FeOOH deposits. Pipes should be made of plastic coated for long-term stability.

The geometry of the wetland site as well as flow control and water treatment considerations may dictate the use of multiple wetland cells. The intercell connections may also serve as aeration devices. If there are elevation differences between the cells, the interconnection should dissipate kinetic energy and be designed to avoid erosion and the mobilization of precipitates. More details on construction of aerobic systems can be found elsewhere (Hammer 1992).

Some of the aerobic systems that have been constructed to treat alkaline mine water have little emergent plant growth. Metal removal rates in these plantless, aerobic systems appear to be similar to those observed in aerobic systems containing plants. However, plants may provide values that are not reflected in measurements of contaminant removal rates. For example, plants can facilitate the filtration of particulates, prevent flow channelization and provide wildlife benefits that are valued by regulatory and environmental groups.

16.2.2
Passive Treatment of Net Acid Water

Treatment of acidic mine water requires the generation of enough alkalinity to neutralize the excess acidity. In passive treatment systems where mine water flows through anaerobic environments, its chemistry is affected by chemical and biological processes that generate bicarbonate and hydrogen sulfide.

16.2.2.1
Pretreatment of Acidic Water with an ALD

A major source of bicarbonate in anaerobic environments is the dissolution of carbonate minerals, such as calcite. The absence of Fe^{3+} limits the formation of FeOOH coatings that would otherwise armor carbonate surfaces and inhibit further carbonate dissolution in aerobic environments (US EPA 1983). Also, anaerobic mine water environments commonly contain high CO_2 partial pressures, which, in turn, increases dissolution of carbonate minerals (Butler 1991).

Since Turner and McCoy described their anaerobic limestone treatment system in 1990, dozens of limestone treatment systems have been constructed (Brodie et al. 1991; Nairn et al. 1991; Skousen and Faulkner 1992). These systems have become known as anoxic limestone drains or ALDs. In an ALD, mine water is made to flow through a bed of limestone gravel that has been buried to limit inputs of atmospheric oxygen. The containment caused by the burial also traps CO_2 within the treatment

Table 16.2. Chemistry (mg l^{-1}) of mine water flowing through the Howe Bridge, western Pennsylvania, anoxic limestone drain, 23 January 1992

Ion	In	Well 1	Well 2	Well 3	Well 4	Eff
pH	5.9	6.1	6.4	6.5	6.5	6.3
Alkalinity	39	75	141	179	183	176
Ca	140	150	183	201	206	198
Fe^{2+}	249	237	246	246	245	244
Fe^{3+}	<1	<1	<1	<1	<1	<1
Mn	34	33	34	34	34	34
Al	<1	<1	<1	<1	<1	<1
Mg	90	87	91	91	90	90
Na	11	11	11	11	11	11
SO_4	1175	1175	1200	1150	1200	1200
CO_2	6.3	4.0	4.7	4.3	4.7	NA

Water flows linearly from the influent (In) through wells 1, 2, 3 and 4 and out the effluent (Eff). CO_2 values are the partial pressure (atm) of gas samples collected from the headspace within the sampling wells. No gas sample could be collected for the effluent because it is an open pipe. NA, not applicable.

system (Nairn et al. 1992). An equilibrated open system would only produce alkalinity in the range of 50–60 mg l^{-1}, and increase calcium concentrations by 4–8 mg l^{-1}.

Water quality data from an ALD in western Pennsylvania are shown in Table 16.2. This ALD is a rectangular bed of limestone gravel that is 31 m long by 12 m wide by 1.5 m deep. The limestone bed is covered with filter fabric and 1 m clay. No organic matter was incorporated into the limestone system. Water samples were collected from the ALD influent and effluent and at four locations within the ALD. The influent mine water contained high concentrations of Fe^{2+} and Mn^{2+} and a small amount of alkalinity. As the mine water flowed through the ALD, pH and concentrations of Ca and alkalinity increased while other measured parameters were unchanged. Between the influent and effluent locations, changes in concentrations of alkalinity (137 mg l^{-1}) and Ca (58 mg l^{-1}) were in stoichiometric agreement with those expected from $CaCO_3$ dissolution. Concentrations of alkalinity and Ca changed little between the third well and the ALD effluent. Similar observations of solubility-limited alkalinity generation have also been made at a second site (Nairn et al. 1992).

ALDs produce alkalinity at a lower cost than do compost wetlands. However, not all water is suitable for pretreatment with ALDs. The primary chemical factors that limit the utility of ALDs are Fe^{3+}, Al^{3+} and DO. When acidic water containing any Fe^{3+} or Al^{3+} contacts limestone, metal hydroxide particulates [FeOOH or $Al(OH)_3$] will form. No oxygen is necessary. Ferric hydroxide can armor the limestone, limiting its further dissolution. Aluminium hydroxide fills the void space in the ALD. The buildup of both precipitates within the ALD can eventually decrease the drain permeability and cause plugging. The presence of DO in mine water will promote the oxidation of Fe^{2+} to Fe^{3+} within the ALD, and thus cause armoring and plugging. The short-term performance of ALDs that receive water containing elevated levels of Fe^{3+}, Al^{3+} or DO can be spectacular. ALDs constructed to treat mine water contaminated with Fe^{3+} and Al and having acidities greater than 1000 mg l^{-1} have discharged net alkaline water. However, the long-term prognosis for these metal-retaining systems is not good (Nairn et al. 1991). Systems have been known to clog in less than 6 months.

In some cases, the suitability of mine water for pretreatment with an ALD can be evaluated based on the type of discharge and measurements of field pH. Mine waters that seep from flooded underground coal mines and have a field $pH > 5$ characteristically have very low concentrations of DO, Fe^{3+}, and Al. Such sites are generally good candidates for pretreatment with an ALD. Mine waters that discharge from open drift mines or have $pH < 5$ must be analysed for Fe^{3+} and Al. Mine waters with $pH < 5$ can contain dissolved Al; mine waters with $pH < 3.5$ can contain dissolved Fe^{3+}. The long-term effects of other dissolved metals, such as Cu and Zn, on ALD performance are currently being investigated.

It is important to use limestone with a high $CaCO_3$ content. The limestones used in most successful ALDs contain 80–95% $CaCO_3$. Most effective systems have used very coarse limestone. Some systems constructed with limestone fines and small gravel have failed, apparently because of plugging problems (Hedin et al. 1994). The ALD must be sealed so that inputs of atmospheric oxygen are minimized and the accumulation of carbon dioxide within the ALD is maximized. This is usually accomplished by burying the ALD with clay. Plastic is commonly placed between the limestone and clay as an additional gas barrier. In some cases, the ALD has been completely wrapped in plastic before burial (Skousen and Faulkner 1992). The ALD should be designed so that the limestone is inundated at all times. To accomplish this, riser pipes can be placed at the outflow of the ALD.

The dimensions of ALDs vary considerably. Most older ALDs were constructed as long narrow drains, approximately 0.6–1.0 m wide (Nairn et al. 1991). At sites where linear ALDs are not possible, limestone beds have been constructed that are 10–20 m wide. These bed systems have produced alkalinities similar to those produced by the more conventional systems.

The mass of limestone required to neutralize a certain discharge for a specified period of time can be calculated. Approximately 12 h of contact time between mine water and limestone is necessary to achieve a maximum concentration of alkalinity (Hedin et al. 1994a, b). In order to achieve this, ~3000 kg limestone rock is required for each litre of mine water flow per minute. An ALD that produces 275 mg alkalinity l^{-1} (the maximum sustained concentration thus far observed for an ALD), dissolves ~1600 kg limestone a decade for each litre of mine water flow per minute. To construct an ALD that contains sufficient limestone to insure a 12-h retention time for 30 years, the limestone bed should contain ~7800 kg limestone for each litre of flow per minute. The calculation assumes that the ALD is constructed with 90% $CaCO_3$ limestone rock that has a porosity of 50%. The calculation also assumes that original mine water does not contain Fe^{3+} or Al.

ALDs are often used to improve the performance of existing constructed wetlands. At many poorly performing wetlands that receive acidic water, the wetland was built too small to treat an acidic, metal-contaminated influent, but is large enough for an alkaline, metal-contaminated influent. For example, one site that we have studied, the Morrison wetland (Pennsylvania, USA), was undersized for the highly acidic water that it received. As a result, the wetland effluent required supplemental treatment with chemicals. Since construction of an ALD, and the addition of 275 mg bicarbonate alkalinity l^{-1} to the water, the discharge of the wetland has been alkaline, low in dissolved metals, and does not require any supplemental chemical treatment (Hedin et al. 1994). Similar enhancements in wetland performance using ALDs have been reported elsewhere (Turner and McCoy 1990; Brodie 1991).

The anoxic limestone drain is one component of a passive treatment system. When the ALD operates ideally, its only effect on mine water chemistry is to raise pH to circumneutral levels and increase concentrations of calcium and alkalinity. Dissolved Fe and Mn should be unaffected by flow through the ALD. In general, the water should be aerated as soon as it exits the ALD and then directed into a settling pond. An aerobic wetland should follow the settling pond. The total post-ALD system should be sized according to the criteria provided earlier for net alkaline mine water. Mine waters with acidities less than 150 mg l^{-1} are readily treated with an ALD and aerobic wetland system.

If the mine water is contaminated with only Fe^{2+} and Mn, and the acidity exceeds 300 mg l^{-1}, it is unlikely that an ALD will discharge net alkaline water. When this partially neutralized water is aerated, much of the iron will precipitate, but the absence of sufficient buffering will result in a discharge with low pH. Building a second ALD, to recharge the mine water with additional alkalinity after it flows out of the aerobic system is not feasible because of the high DO of water flowing out of aerobic systems. If the treatment goal is to neutralize all of the acidity passively, then a compost wetland should be built so that additional alkalinity can be generated. Such a treatment system thus contains all three passive technologies: the mine water flows through an ALD, into a settling pond and an aerobic system, and then into a compost wetland. Systems have recently been constructed that repeat the sequence, using the compost wetland to remove the DO and reduce the Fe^{3+} so that water drained from the bottom can be directed into another ALD/wetland sequence (Kepler and McCleary 1994). Since the compost wetland provides hydrostatic head, such a system has been used to force water through an ALD where plugging with AL is known to be a potential problem.

When a mine water is contaminated with Fe^{2+} and Mn and has an acidity between 150 mg l^{-1} and 300 mg l^{-1}, the ability of an ALD to discharge net alkaline water will depend on the concentration of alkalinity produced. An experimental method has been developed that results in an accurate assessment of the amount of alkalinity that will be generated when a particular mine water contacts a particular limestone (Watzlaf and Hedin 1993).

16.2.2.2
Sulfate Reduction

When mine water flows through an anaerobic environment that contains an organic substrate, bacterial sulfate reduction can occur. In this process, bacteria oxidize organic compounds using sulfate as the terminal electron sink, and release hydrogen sulfide and bicarbonate. Bacterial sulfate reduction requires the presence of sulfate, suitable concentrations of low-molecular-weight carbon compounds, pH greater than 4, and the absence of oxidizing agents such as oxygen, Fe^{3+} and Mn^{4+} (Postgate 1984). These conditions are commonly satisfied in treatment systems that receive coal mine drainage and contain organic matter. The oxygen demand of organic substrates causes the development of anoxic conditions and an absence of oxidized iron or Mn. The low-molecular-weight compounds that sulfate-reducing bacteria utilize are common end products of microbial fermentation processes in anoxic environments. The pH requirements can be satisfied by alkalinity generated by microbial activity and carbonate dissolution.

Bacterial sulfate reduction directly affects concentrations of dissolved metals by precipitating them as metal sulfides. The removal of dissolved metals as sulfide compounds depends on pH, the solubility product of the

Table 16.3. Surface and pore water chemistry ($mg\,l^{-1}$) at the Latrobe wetland. (Hedin et al. 1994a)

Parameter	Pore water[a]		Surface water[b]	
	Mean	SD	Mean	SD
Al	1	5	35	5
Ca	467	188	308	29
Fe^{2+}	215	183	3	39
Fe^{3+}	2	9	24	16
H_2S	37	75	<1	0
Mg	175	48	166	9
Mn	24	10	42	2
Na	11	10	5	1
SO_4	1674	532	1967	115
Acidity[c]	439	340	503	86
Alkalinity	885	296	0	0
Net alkalinity[d]	446	NA	-503	NA
pH	6.8	0.8	3.1	0.1

SD, standard deviation; NA, not applicable.

[a] A total of 52 water samples were collected on 25 July and 11 August 1988 by the dialysis tube method. Metals were analysed for every sample. Field pH was measured for 29 samples. Alkalinity was measured for nine samples.

[b] Six samples collected in July and August 1988 from the second wetland cell.

[c] Calculated from pH, Fe^{2+}, Fe^{3+}, Al and Mn and for pore water samples and measured by the H_2O_2 method for surface water samples.

[d] Average alkalinity minus average acidity. The nine pore water samples for which alkalinity was measured had a mean net alkalinity of 653 mg l (SD 590).

specific metal sulfide and the concentrations of the reactants. Table 16.3 shows the chemistry of surface water and substrate pore water samples collected from a wetland constructed with limestone and spent mushroom compost. The compost depth used in most wetlands is 30–45 cm. Typically, 1 t compost will cover about 3.5 m^2 about 45 cm thick. Cattails or other emergent vegetation are planted in the substrate to stabilize it and to provide additional organic matter to "fuel" the sulfate reduction process. As a practical tip, cattail plant/rhizomes should be planted well into the substrate prior to flooding the wetland cell. At the wetland used in this example, 10–15 cm limestone sand was covered with 20–50 cm compost and planted with cattails. Water flowed through the wetland primarily by surface paths; no efforts were made to force the water through the compost.

The data shown in Table 16.3 were collected 15 months after the wetland was constructed. Compared with the surface water, the substrate pore water had higher pH, higher concentrations of alkalinity, Fe^{2+}, Ca and H_2S, and lower concentrations of sulfate, Fe^{3+} and Al. On average, the pore water had a net alkalinity while the surface water had a net acidity. Compared with surface water, the substrate pore water contained elevated concentrations of Fe^{2+}. High concentrations of Fe^{2+} result from the dissolution of FeOOH at the redox boundary. Reduction of FeOOH has no effect on the net acidity of the mine water because the increase in alkalinity is exactly matched by an increase in mineral acidity. If the iron-enriched pore water diffuses into an aerobic zone, the Fe^{2+} content should oxidize, hydrolyse, and reprecipitate as FeOOH.

Because the pore water has circumneutral pH and is strongly buffered by bicarbonate, the iron precipitates rapidly when the water diffuses to the surface. Indeed, during the summer months, when the data in Table 16.3 were collected, comparisons of the wetland influent and effluent indicated that both concentrations of iron and total acidity decreased on every sampling day. The decrease in acidity indicates that alkaline pore water was mixing with surface water and neutralizing acidity. The decrease in concentrations of iron in the surface water indicates that elevated concentrations of Fe^{2+} observed in the pore water were rapidly removed in the surface water environment.

On average, sulfate reduction and limestone dissolution contributed equally to alkalinity generation (54 vs. 46%, respectively). The average sulfate removal rate calculated for sites constructed with a compost substrate, 5.8 g SO_4^{2-} m^{-2} day^{-1}, is equivalent to a sulfate reduction rate of ~200 nmol cm^{-3} day^{-1}. This value is consistent with measurements of sulfate reduction made at constructed wetlands (McIntire and Edenborn 1990) and coastal ecosystems (Skyring 1987).

Compost wetlands in which water flows on the surface of the compost remove acidity (e.g. generate alkalinity) at rates of approximately 2–12 g m^{-2} day^{-1}. This range in performance is largely a result of seasonal variation; rates of acidity removal are lower in winter than in summer (Hedin et al. 1991). Research indicates that supplementing the compost with limestone and incorporating system designs that cause most of the water to flow through the compost can result in higher rates of limestone dissolution and better winter performance.

Compost wetlands should be sized based on the removal rates in Table 16.1. In many wetland systems, the compost cells are preceded with a single aerobic pond in which iron oxidation and precipitation occur. This feature is useful where the influent to the wetland is of circumneutral pH (either naturally or because of pretreatment with an ALD), and rapid, significant removal of iron is expected as soon as the mine water is aerated. Aerobic ponds are not useful when the water entering the wetland system has a pH less than 4. At such low pH, iron oxidation and precipitation reactions are quite slow and significant removal of iron in the aerobic pond would not be expected.

16.3 Operation and Maintenance

Operational problems with passive treatment systems can generally be attributed to inadequate design, unrealistic expectations, pests or poor construction methods. If properly designed and constructed, a passive treatment system can be operated inexpensively. Probably the most common maintenance problem is dike and spillway stability. Reworking slopes, rebuilding spillways and increasing freeboard can all be avoided by proper design and construction.

Pests can plague wetlands with operational problems. Muskrats will burrow into dikes, causing leakage and potentially catastrophic failure problems, and will uproot significant amounts of cattails and other aquatic vegetation. Muskrats can be discouraged by lining dike inslopes with chainlink fence or rip-rap to prevent burrowing (Brodie 1990). Beavers cause water level disruptions due to damming and also seriously damage vegetation. They are very difficult to control once established. Large pipes with 90° elbows on the upstream end have been used as discharge structures in beaver-prone areas (Brodie 1991). Otherwise, shallow ponds with dikes with shallow slopes toward wide, riprapped spillways may be the best design for a beaver-infested system.

16.4 Long-Term Performance

Passive treatment systems cannot be expected to perform indefinitely. In the long term, wetland systems will fill up with metal precipitates or the conditions that facilitate contaminant removal may be compromised. None of the treatment systems that we have studied demonstrated any downward trends in contaminant removal performance. Therefore, estimates of the long-term performance of passive systems must be made by extrapolating available data. Like the design and sizing of passive treatment systems, estimates of long-term performance vary with the chemistry of the mine water. The rapid removal of Fe that occurs in alkaline treatment systems means that such systems will inevitably fill up. Stark (1992) reports that the Fe sludge in a constructed wetland is accumulating at 3–4 cm $year^{-1}$, while measurements at other sites indicate an increase in sludge depth of 2–3 cm $year^{-1}$ (Hellier and Hedin 1992). These measurements suggest that 1 m of freeboard should provide sufficient volume for 25–50 years of performance.

At some surface mines, water quality tends to improve within a decade after regrading and reclamation have been completed (Meek 1991; Ziemkiewicz 1993). At these surface mine sites, 25–50 years of passive treatment may be adequate to mitigate the contaminant problem. At surface mine sites where contaminant production is expected to continue, the system can either be built with greater freeboard or rebuilt when it eventually fills up. Site conditions will determine whether it is more economical to simply bury the wetland system in place and construct a new one, or to excavate and haul away the accumulated solids for proper disposal. Disposal of these excavated sludges is not difficult or unduly expensive for coal mine drainage because the material is not considered hazardous. For metal mine drainage, the accumulation of toxic metals in the substrate and their possible uptake by plants provide some cause for concern. Bioavailability studies, which have been done in wetlands constructed to treat coal mine drainage, have not been undertaken at metal mine drainage sites. Passive treatment of metal mine drainage has generally been limited to sites where metal concentrations have been relatively low, which is probably appropriate.

Wetlands that receive acidic water, and function through the alkalinity-generating processes associated with an organic substrate, may decline in performance as the components of the organic substrate that generate alkalinity are exhausted. The compost wetlands described in this chapter neutralize acidity through the dissolution of limestone and the bacterial reduction of sulfate. A wetland that contains a 40-cm compost depth and 30 kg limestone m^{-3}, and which generates $CaCO_3$-derived alkalinity at a mean rate of 3 g m^{-2} day^{-1} (the average rate measured in this study), will be exhausted of limestone in 11 years. The same volume of compost contains ~40 kg organic carbon. If bacterial sulfate reduction mineralizes 100% of this carbon to bicarbonate at a rate of 5 g m^{-2} day^{-1}, then the carbon will be exhausted in 91 years. This estimate is increased by the carbon input of the net primary production of the wetland system, but decreased significantly by the fact that some of the carbon is mineralized by reactions other than sulfate reduction and some is recalcitrant.

A realistic scenario for the long-term performance of a compost wetland is that sulfate reduction is linked, in a dependent manner, to limestone dissolution. Sulfate-reducing bacteria are inactive below pH 5

(Postgate 1984). Their activity in a wetland receiving lower pH water may depend, in part, on the presence of pH-buffering supplied by limestone dissolution, since limestone dissolution creates alkaline zones in which sulfate reduction can proceed. If this scenario is accurate, it would be advisable to increase the chemical buffering capability of the wetland substrate by adding additional limestone during wetland construction. In fact, this procedure is commonly practised.

The performance of ALDs has many aspects that make long-term expectations uncertain. ALDs function by the dissolution of limestone. Long-term scenarios for ALD performance fail to consider the hydrologic implications of the gradual structural failure of the systems. In large ALDs, most of the limestone dissolution occurs in the upgradient portion of the limestone bed. It is unknown whether this preferential dissolution will produce partial failure of the integrity of the system or whether the permeability will be adversely affected. Another aspect that affects long-term ALD performance is the fact that ALDs retain Fe^{3+} and Al (Nairn et al. 1991; Skousen and Faulkner 1992). This retention can result in armoring of limestone or the plugging of flow paths in as little as 6 months, long before the limestone is exhausted by dissolution (Nairn et al. 1991).

16.5 Summary and Conclusions

Waters that contain high concentrations of bicarbonate alkalinity are most amenable to treatment with constructed wetlands. At circumneutral pH, Fe and Mn precipitation processes are more rapid than under acidic pH conditions. Given the ability of bicarbonate alkalinity to positively impact both the metal precipitation and neutralization aspects of mine water treatment, it is not surprising that the most successful applications of passive treatment have been at sites where the mine water was net alkaline. The most successful wetlands constructed in western Pennsylvania in the early 1980s treated mine waters that contained alkalinity. All of the early successes of the Tennessee Valley Authority (TVA) were, likewise, with waters that were alkaline (Brodie 1990). Similarly, the Simco wetland in Ohio, which has discharged compliance water for several years, receives water containing ~160 mg alkalinity l^{-1} (Stark et al. 1990).

When mine water is acidic, enough alkalinity must be generated by the passive treatment system to neutralize the acidity. The most common method used to passively generate alkalinity is the construction of a wetland that contains an organic substrate in which alkalinity-generating microbial processes occur. If the substrate contains limestone, as spent mushroom compost does, then alkalinity will be generated by both calcite dissolution and bacterial sulfate reduction reactions. The performance of the constructed wetlands that receive acidic water is usually limited by the rate at which alkalinity is generated within the substrate. While wetlands can significantly improve water quality, and have proven to be effective at moderately acidic sites, we know of no wetland systems that consistently and completely transform highly acidic water to compliance quality. Inconsistent or partial treatment indicates undersizing. We believe this is due to a lack of awareness of how much larger wetlands constructed to treat acidic water must be than ones constructed to treat alkaline water. The Fe and acidity removal rates that we have measured indicate that the treatment of 5000 g Fe day^{-1} in alkaline water requires ~250 m^2 aerobic wetland. The treatment of the same Fe load in acidic water (where treatment requires both precipitation of the Fe and neutralization of the associated acidity) requires ~1300 m^2 compost wetland. Thus, wetlands constructed to treat acidic water must be six times larger than ones constructed to treat similarly contaminated alkaline water.

ALDs can lower acidities or actually transform acidic water into alkaline water, and markedly decrease the sizing demands of the wetlands constructed to precipitate the metal contaminants. Because limestone is inexpensive, the cost of an ALD/aerobic wetland passive treatment system is typically much less than the compost wetland alternative. Thus, when the influent water is appropriate, ALDs should be the preferred method for generating alkalinity in passive treatment systems.

Passive treatment, like active treatment with chemicals, requires that the metal contaminants be precipitated and that the acidity associated with these ions be neutralized. By recognizing that these treatment goals need not be accomplished simultaneously, one can focus on optimization of the individual objectives. As a result, the performance and cost effectiveness of passive treatment systems is rapidly improving. Today, most mine operators who install properly designed passive treatment systems rapidly recoup the cost of their investment through decreased water treatment costs. There is no reason to doubt that this technology will continue to improve and that, over time, passive treatment will be used in many applications that are not possible today.

References

Brodie GA (1990) Treatment of acid drainage using constructed wetlands experiences of the Tennessee Valley Authority. In: Graves DH, DeVore, RW (eds) Proc 1990 Natl Symp on Mining. OES Publ, Lexington, Kentucky, pp 77–83

Brodie GA (1991) Achieving compliance with staged aerobic constructed wetlands to treat acid drainage. In: Oaks W, Bowden J (eds) Proc 1991 Natl Meet American Society for Surface Mining and Reclamation, ASSMR, WV, pp 151–174

Brodie GA, Britt CR, Tomaszewski TM, Taylor HN (1991) Use of passive anoxic limestone drains to enhance performance of acid drainage treatment wetlands. In: Oaks W, Bowden J (eds) Proc 1991 Natl Meet American Society for Surface Mining and Reclamation, ASSMR, WV, pp 211–222

Butler JN (1991) Carbon dioxide equilibria and their applications. Lewis, Chelsea, Michigan, 259 pp

Hammer DA (1992) Creating freshwater wetlands. Lewis, Chelsea, Michigan, 298 pp

Hedin RS, Dvorak DH, Gustafson SL, Hyman DM, McIntire PE, Nairn RW, Neupert RC, Woods AC, Edenborn HM (1991) Use of a constructed wetland for the treatment of acid mine drainage at the Friendship Hill national historic site, Fayette County, PA. US BuMines Rep, Interagency Agreement #4000–5-0010, 128 pp

Hedin RS, Nairn RW, Kleinmann RLP (1994a) Passive treatment of coal mine drainage. BuMines IC 9389, 35 pp

Hedin RS, Watzlaf GR, Nairn RW (1994b) Passive treatment of acid mine drainage with limestone. J Environ Qual 23:1338–1345

Hellier WW, Hedin RS (1992) The mechanism of iron removal from mine drainages by artificial wetlands at circumneutral pH. INTECOL'S IV Int Wetlands Conf, Ohio State University, Columbus, Ohio, Abstracts, p 13

Kepler DA, McCleary EC (1994) Successive alkalinity producing systems (SAPs) for the treatment of acidic mine drainage. In: Proc 3rd Int Conf on the Abatement of Acidic Drainage. Bureau of Mines Spec Publ SP 06A-94, 1:195–204

Kleinmann RLP, Watzlaf GR (1988) Should the effluent limits for manganese be modified? In: Proc 1988 Mine Drainage and Surface Mine Reclamation Conf, vol II. BuMines IC 9184, pp 305–310

McIntire PE, Edenborn HM (1990) The use of bacterial sulfate reduction in the treatment of drainage from coal mines. In: Skousen J, Sencindiver J, Samuel D (eds) Proc 1990 Mining and Reclamation Conference and Exhibition, West Virginia University, Morgantown, pp 409–415

Meek FA Jr (1991) Assessment of acid preventative techniques employed at the Island Creek Mining Co. Tenmile Site. In: Proc 12th Annu West Virginia Surface Mine Drainage Task Force Symp, West Virginia University; Morgantown

Nairn RW, Hedin RS, Watzlaf GR (1991) A preliminary review of the use of anoxic limestone drains in the passive treatment of acid mine drainage. In: Proc 12th Annu West Virginia Surface Mine Drainage Task Force Symp, West Virginia University, Morgantown

Nairn RW, Hedin RS, Watzlaf GR (1992) Generation of alkalinity in an anoxic limestone drain. In: Proc 9th Natl American Society for Surface Mining and Reclamation Meeting, ASSMR, WV, pp 206–219

Postgate JR (1984) The sulphate-reducing bacteria, 2nd edn. Cambridge University Press, New York, 208 pp

Skousen J, Faulkner B (1992) Preliminary results of acid mine drainage treatment with anoxic limestone drains in West Virginia. In: Proc 13th Annu West Virginia Surface Mining Drainage Task Force Symp, West Virginia University, Morgantown

Skyring GW (1987) Sulfate reduction in coastal ecosystems. Geomicrobiology 5:295–373

Stark L, Stevens E, Webster H, Wenerick W (1990) Iron loading, efficiency, and sizing in a constructed wetland receiving mine drainage. In: Skousen J, Sencindiver J, Samuel D (eds) Proc 1990 Mining and Reclamation Conf and Exhibition, West Virginia University, Morgantown, vol II, pp 393–401

Stark LR (1992) Assessing the longevity of a constructed wetland receiving coal mine drainage in eastern Ohio. INTECOL'S IV Int Wetlands Conf, Ohio State University, Columbus, Ohio, Abstracts, p 13

Turner D, McCoy D (1990) Anoxic alkaline drain treatment system, a low cost acid mine drainage treatment alternative. In: Graves DH, DeVore, RW (eds) Proc 1990 Natl Symp on Mining, OES Publ, Lexington, Kentucky, pp 73–75

United States Environmental Protection Agency (1983) Design manual: neutralization of acid mine drainage. Office of Research and Development, Industrial Environmental Research Laboratory, Washington, 231 pp

Watzlaf GR, Hedin RS (1993) A method for predicting the alkalinity generated by anoxic limestone drains. In: Proc 14th Annu West Virginia Surface Mine Drainage Task Force Symp, West Virginia University, Morgantown

Ziemkiewicz P (1993) A simple spreadsheet for predicting acid mine drainage. In: Proc 14th Annu West Virginia Surface Mine Drainage Task Force Symp, West Virginia University, Morgantown

17 Biological Polishing of Zinc in a Mine Waste Management Area

M. Kalin

Boojum Research Limited, 468 Queen Street East, Toronto, Ontario, Canada M5A 1T7

17.1 Introduction

Biologically mediated processes have been used for at least 2000 years to treat domestic sewage and agricultural effluent (Wang 1987). These natural water cleansing processes for both organics and inorganics take place not only in sewage and agricultural waste treatment systems, but also in natural systems such as wetlands, rivers, and lakes (Hamilton 1978). Effluents associated with mine waste areas are distinctly different from effluents produced by domestic and agricultural activities. Organic substances required to drive microbial activity are not present. Mine waste effluents are also characterized by low phosphorus levels, elevated concentrations of nitrogen compounds (remnants of blasting), high metal concentrations, and often also extremes in pH. Natural treatment processes based on microbial mineralization of sewage or agricultural waste are therefore not directly applicable to mine effluents.

Organic matter can be supplied as easily degradable biomass and/or generated within the aquatic system though the productivity of species of phytoplankton and periphyton tolerant of mine effluent conditions (Kalin et al. 1989; Smith and Kalin 1991). Studies on passive (biological) treatment systems or constructed wetlands over the past decade indicate that metals are removed through absorption and precipitation of metal hydroxides (Kalin 1992). Under sulfate-reducing conditions, Wieder (1992) found that iron became organically bound and also precipitated as carbonates or hydroxides. Such precipitates remove other metals from the water through adsorption, coprecipitation, and particulate formation (Kalin and Wheeler 1992 a, b).

Biological polishing is a process of metal removal based on continuous algal growth on extensive surface (organic or inorganic) that absorbs metals and traps metal precipitates via extensive colloid production (Kalin 1992; Kalin and Wheeler 1992 a, b). In addition to biological factors causing precipitation and relegation of metals to the sediment, other physico-chemical factors affect the precipitate formation process, such as mechanisms affecting iron oxidation and reduction. All of these processes are essential to the design of treatment systems specific for a particular mine effluent.

Four zinc removal mechanisms were identified or hypothesized for this project. First, iron(III) hydroxides are precipitated as the groundwater enters the oxygenated flooded pit. A fraction of the dissolved zinc is coprecipitated or is adsorbed onto the surface of these precipitates. Second, suspended solids containing zinc adhere to algal biomass surfaces. Third, dissolved zinc is adsorbed onto the algal cell walls. Fourth, a fraction of the zinc is precipitated as zinc carbonate on the algal mats which form in the discharge from the flooded pit. With this last mechanism, it was conjectured that the microenvironment surrounding the algae alters the solution properties (including pH), which leads to precipitation of zinc carbonate on algal surfaces.

This chapter describes the approach used to reduce zinc concentrations in effluents from an abandoned mine. The treatment strategy employed biological polishing. In the wastewater effluent, iron (III) hydroxides are precipitated in circumneutral groundwater discharging from old mine workings. These precipitates are removed from the water through adsorption on algal surfaces which grow attached to brush cuttings placed in the zinc polishing ponds. Although induction of particle formation is not brought about by microbially active sediments in this system, the performance of the biological components in the treatment ponds can be used to derive design criteria for biological polishing ponds in general.

17.2 Buchans, Newfoundland: A Case Study

The abandoned mine site is located close to the town of Buchans, in central Newfoundland, Canada (Fig. 17.1). The waste management system consists of five mines (not shown), three pits, and two tailings ponds (Fig. 17.2). The pits were mined by the gloryhole method, where ore was dropped from the mine seams to the bottom of the pit and hauled to the mill through underground workings. Zinc, gold, copper, and barium were mined from the Buchans site. At mine and mill closure, the pits and the underground workings were force flooded. Groundwater flows into these gloryholes. The resultant overflow has significant concentrations of zinc. The overflow from all of the underground workings averages 20 l/s, contains 20 mg zinc/l, and typically has a pH level of 6.5.

Fig. 17.1. Map of Canada, showing the location of Buchans, Newfoundland

The methodological approach taken in finding a biologically mediated solution consists of four stages:

1. Characterization of the mine waste site and assessment.
2. Formulation of the treatment strategy.
3. Pilot-scale tests.
4. Full-scale treatment.

As this was one of the first programs in which a biologically mediated solution was sought for a site of this type, monitoring and assessment continues to be an ongoing process. This is absolutely essential, since each mine site environment has its own distinct characteristics which vary over the course of the year. Also, there is a significant amount of development work which must be carried out in effecting a successful biologically mediated treatment program.

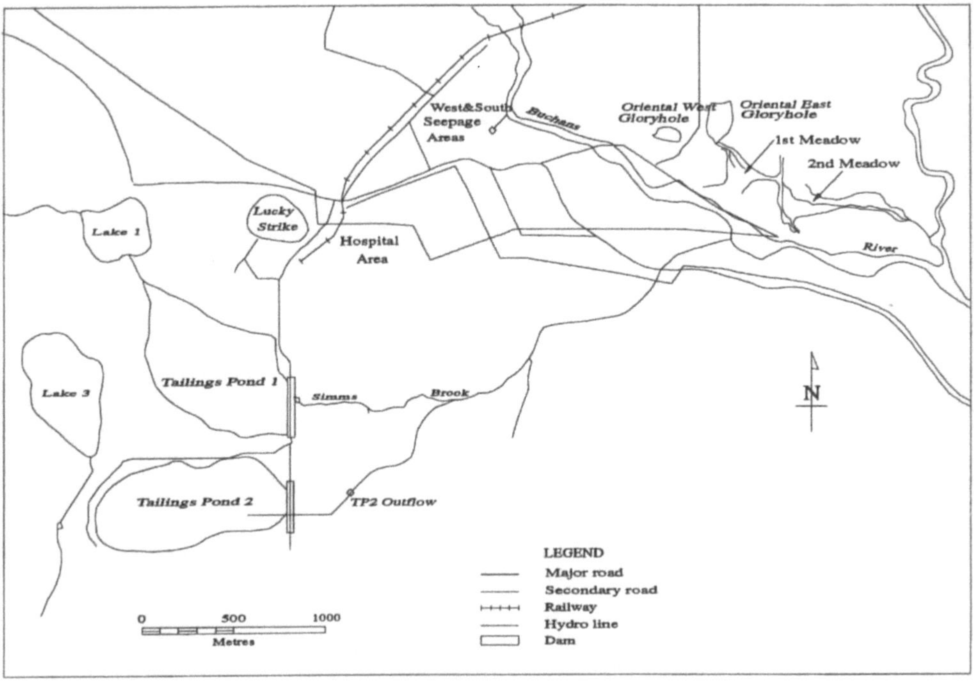

Fig. 17.2. Map of the Buchans mine site, showing locations of gloryholes and wastewater routes

17.3 Characterization of the Mine Waste Site and Assessment

An extensive analysis of the mine site environment was carried out first. This began with a characterization of the hydrological and geochemical conditions to determine the source and mechanism of the continuous release of metal contaminant to the groundwater. Water samples were analyzed for metal and nutrient concentrations, pH, Eh, and temperature throughout the year. It was concluded that zinc was released via groundwater infiltration into the flooded underground workings. Next, taxonomic identification of the indigenous phytoplankton, periphyton, and picoplankton was carried out. Finally, particulate matter, both from the water column and sediment, was collected and analyzed. A number of techniques were employed to analyze the particulates, including elemental analysis, sequential extraction, electron microscopy [secondary electron microscopy (SEM) and energy dispersive X-ray analysis (EDX)] and, in some cases, surface analysis [secondary ion mass spectrometry

(SIMS)]. The large amounts of data collected were used for predicting the long-term behaviour of the waste stream from the mine site.

This information was combined with geochemical simulations, based on water chemistries, to determine optimal precipitation conditions. The literature was then reviewed based on the characterization of the metal concentrations and the algal species found. From these results, and particularly from our experience in developing treatment technologies, the natural processes which remove the metal contaminant (zinc) were defined. The exact removal mechanisms often are not proved at this stage, however a working hypothesis is arrived at. Based upon these removal mechanisms, the treatment strategy was developed which would enhance metal precipitation processes, and continuously (i.e. in a self-sustaining manner) remove contaminant, or at least inhibit its release.

The initial evaluation of the Buchans mine waste site was carried out in 1988. The choice was to treat either the flooded pits directly or the effluent flowing from them. The continuous upwelling of water in the flooded pits meant that establishing a stable sediment was not possible. Thus a treatment strategy using the ARUM (acid reduction using microbiology) process was not optimal for this site. Instead, the decision was taken to employ biological polishing ponds for the pit effluents and enhance the zinc carbonate precipitation on the algal mats. Pilot-scale testing was then the next step.

17.4 Description of Pilot-Scale Biological Polishing Ponds

Previous work conducted on other systems resulted in metal removal rates for Ni, Zn, and As ranging from 0.1 to 3.8 g m^{-3} day^{-1}, iron removal rates from 0.04 to 39 g m^{-3} day^{-1}, and sulfate removal rates from 1.1 to 51 g m^{-3} day^{-1} (Kalin et al., submitted). The relatively large ranges of removal rates were attributed to differences in retention times of the wastewater in the systems, the surface area of microbially active sediment in contact with the wastewater, as well as the chemical characteristics of the wastewaters passing through the systems.

The pilot-scale system was constructed in 1989 (Fig. 17.3) to develop optimal design criteria for biological polishing of acid mine drainage. Each of the six serial ponds had a surface area of 70 m^2 and a volume of 40 m^3. Alder branches were placed in each pond to increase surface area (approximately 3.8 m^2 m^{-3}) for algal growth. Flow through the pond was controlled to provide an overall residence time of between 16 and 79 days.

Data showing the performance of the six ponds are given in Fig. 17.4. The zinc concentration decreases as the effluent passes through the pond

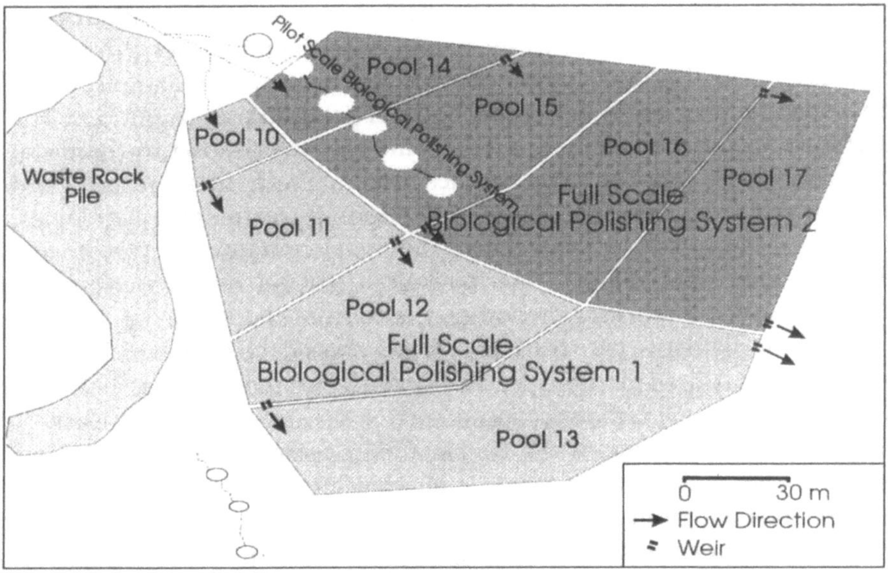

Fig. 17.3. Layout of the pilot-scale and full-scale biological polishing systems for removal of zinc in effluent discharging from flooded underground mine workings

system from an initial level of nearly 16 mg zinc l^{-1} to 1.2 mg zinc l^{-1} (88% removal in water exiting the system). The amount of zinc remaining in each pond was determined in two ways; from the loss in the water, and from the concentration within the biomass lying on the alder branches. The greatest amount of zinc found in the algae or biomass coincided with the largest decrease in the zinc concentration in the pond water (pond 3; not shown). Figure 17.4 illustrates the two approaches which were used to evaluate the performance of the pilot polishing ponds. The first approach determines zinc removal in terms of the difference in zinc concentration in the pond water, flow rates, and residence times. The second approach assesses the algal growth and the concentration of zinc found in the algal biomass.

Using the physical approach (residence time), a linear regression analysis of the percent reduction in zinc concentration versus residence time of wastewater passing through the pond system was performed using all pilot system data available for the period from 1989 to 1993, both winter and summer data (Kalin and Wheeler 1992b). From this analysis it was projected that a zinc concentration decrease of 2.2% per day of wastewater residence time could be expected in the full-scale system. It was observed that the zinc removal rate increased substantially over the

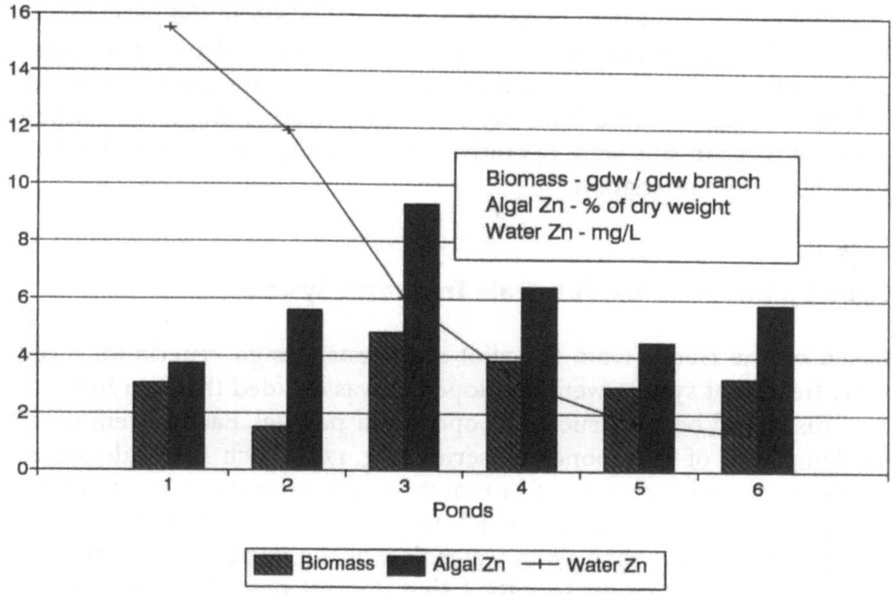

Fig. 17.4. Zinc removal capacity and treatment parameters of the six pilot-scale ponds on 25 August 25 1990; *gdw* grams dry weight

summer months. From a regression analysis using eight sets of data for summer months alone, it was projected that a zinc concentration decrease of 5.2% per day could be expected during summer months (Kalin 1992; Kalin and Wheeler 1992b).

Next, performance based strictly upon algal growth rates and algal zinc content was determined. The measure of performance is in terms of zinc removed per surface area of algae. Algal growth was quantified by suspending netting of a known surface area in the polishing pond and collecting the biomass at different intervals during the growing season (Kalin 1992; Kalin and Wheeler 1992a). Assuming an average growth period of 100 days in the summer, the growth rate of algae on the netting was determined to be 0.75 g m^{-2} day^{-1}. If the growth were averaged over the entire year, a growth rate of 0.20 g m^{-2} day^{-1} would result. Analysis of the algae indicated that 1 kg (dry weight) of this biomass contained 34.7 g zinc.

The effect of fertilizer additions (Plant Products 10–52–10) on algal growth in the pond system was also determined. Addition of fertilizer resulted in growth rates two to three times those of the unfertilized algae. It was anticipated that these increased growth rates would result in increased zinc removal by the biological polishing ponds.

Using both approaches, there was no deterioration in the performance of the polishing ponds over the 5-year study period. If removal mechanisms only involved the ion exchange capacity of the polishing ponds, saturation effects would have been evident. In other words, continuous removal mechanisms were operative, although the mechanisms involved were not fully understood.

17.5 Scale-Up Design to the Full-Scale Treatment System

Based on the results from the pilot-scale tests, design criteria for a full-scale treatment system were developed. It was decided that two full-scale systems would be constructed to operate in parallel. Each system would be comprised of four ponds in series (Fig. 17.3). Each full-scale system would have a surface area of 13,100 m^2 and a capacity of 6000 m^3. The design flow rate was 4.7 l s^{-1}, which would result in a residence time of 15 days. Subsequent changes in water flow at the mine site increased the actual flow rate to closer to 9 l/s. Using the full year pilot projection of 2.2% zinc concentration reduction per day of residence, a 32% reduction in zinc was expected. Using the summer reduction rate of 5.2% per residence day, it was estimated that 78% of the zinc should be removed.

Performance based strictly on algal growth rates and algal zinc content was examined for comparison with that based on percentage zinc removal-residence time (see above). By multiplying the amount of biomass produced per day by the zinc concentration associated with that growth, an estimate of the amount of zinc removed per day was obtained. The total surface area for periphyton growth was calculated from the estimated average surface area of a cutting of alder brush (including leaf area) and the number of brush cuttings used. Using the conservative annualized growth rate and zinc concentration data, it was projected that in a scaled-up system with 51,000 m^2 of substrate surface area provided by alder brush, 1.1 kg zinc would be removed per day by the periphytic algal population in a 6000-m^3 full-scale biological treatment system (Table 17.1). This number would increase in summer to approximately 3.2 kg zinc day^{-1} or greater if the fertilized growth data was used.

17.6 Full-Scale Biological Polishing Pond Performance

Construction of the first section of the full-scale system was completed in late autumn 1993. Alder brush was placed onto the ice-covered ponds in November 1993. An average flow of 8.5 l s^{-1} passed through the system

Table 17.1. Projected performance of polishing ponds 10–13 based on pilot-scale performance using algal growth rates and zinc concentration

	Performance data All year	Performance data Summer	Units
Average algal growth rate without fertilizer (netting, branch data)	0.20	0.75	$g\ m^{-2}\ day^{-1}$
Zinc content (1994–1995 data)	35	35	g zinc kg^{-1} dw algae
Surface area in pond system ($3\ m^2$ per brush cutting with leaves × 17,000 brush cuttings)	51,000	51,000	m^2
Algal biomass growth	33	92	kg day^{-1}
Estimated zinc removed by periphyton	1.1	3.2	kg day^{-1}
Zinc load, [$4.7\ l\ s^{-1}$ flow; $15\ mg\ l^{-1}$, 1994–1995 average (Zn) in OEP discharge]	6.1	6.1	kg day^{-1}
Estimated % zinc removal by periphyton	18	53	%

dw, Dry weight; OEP, Oriental East Pit.

over the next year, which resulted in a calculated residence time of 8.2 days (Table 17.2). During the summer months, the average flow was $4.9\ l\ s^{-1}$, giving a calculated residence time of 14.2 days, close to the design specifications for residence time. During 1995, the second year of testing, the flows averaged 8.8 and $8.5\ l\ s^{-1}$ for the year and summer, respectively. These final flows were higher than originally anticipated.

Despite the higher flows (and lower theoretical residence times), the performance of the scaled-up system was higher than predicted. The annual average of 31% and the summer average of 66% zinc removal indicate that flow and residence time are not the only factors that control zinc removal. During the second year, the percent zinc removed was slightly lower (22 and 53% for annual and summer calculations), due to the much higher flow rates during the summer. However, the total amount of zinc removed during that summer was actually greater during the second year (4.7 vs. 3.5 kg day^{-1}).

Table 17.2. Actual performance of polishing ponds 10–13, 1994–1995

	Performance data All year	Performance data Summer	Units
First year: flow in/out	8.5	4.9	$l\ s^{-1}$
Load in	10.6	5.3	kg day^{-1}
Load out	7.3	1.8	kg day^{-1}
Total removed	3.3	3.5	kg day^{-1}
Percent removed	31	66	%
Second year: flow in/out	8.8	8.5	$l\ s^{-1}$
Load in	9.8	8.8	kg day^{-1}
Load out	7.6	4.1	kg day^{-1}
Total removed	2.2	4.7	kg/day
Percent removed	22	53	%

Performance in the first year was very close to that predicted based on residence time. However, the flow into the system was considerably higher than the design criterion (8.5 $l\ s^{-1}$ compared with 4.7 $l\ s^{-1}$). This greater than predicted performance is attributed to adsorption onto the surfaces of the newly placed alder cuttings. In the second year, the performance dropped to an overall 22% removal which is closer to the prediction based on year-round data (32% based on residence time and 18% based on algal growth). Considering only data for the second summer, actual removal rates (53%) closely matched those predicted based on algal growth (52.5%).

17.7 Discussion

Determining a process which is self-sustaining, proving the process with pilot-scale testing, and then scaling up to a full-scale system is not the end of the development of the biologically mediated solution. Process optimization continues, based upon finding and understanding the mechanisms involved in removing the zinc.

One of the first steps in determining a suitable treatment strategy was defining the removal mechanisms involved. We postulated that the following four mechanisms may be involved. First, iron(III) hydroxides are precipitated as the groundwater enters the oxygenated flooded pit. A fraction of the dissolved zinc is coprecipitated or is adsorbed onto the surface of these precipitates. Second, suspended solids containing zinc adhere to algal biomass surfaces. Third, dissolved zinc is adsorbed onto anion groups on the algal cell walls. Fourth, a fraction of the zinc is precipitated as zinc carbonate on the algal mats which grow in the discharge from the flooded pit. In this last mechanism, it was conjectured that the microenvironment surrounding the algae may be altering the solution properties (including the pH), which in turn leads to the precipitation of zinc carbonate in the vicinity of the algae.

Evidence for the first mechanism is found in the chemical analysis of material collected in sedimentation traps suspended in the flooded pit. Concentration of zinc ranged from 1.1 to 2.5% of sediment dry weight. In addition, SEM/EDX analysis carried out on the precipitates formed in the flooded pit displayed grains which were uniformly coated with what appeared to be precipitation products, having high Fe (40–70%), Si (15–30%), and significant amounts of zinc (approximately 3%). Iron precipitates which remained suspended and flowed into the polishing ponds could be removed by adhering to the periphyton growing in the ponds, the second zinc removal process.

Direct evidence for the second mechanism was obtained through studies of the metal coprecipitation associated with algal/moss growth in wastewater at Buchans and other sites. These populations, because of the close interaction between precipitate and periphyton, have been termed periphyton–precipitate complexes (PPC). The periphyton populations found in the polishing ponds are mainly composed of green filamentous algae. In the absence of precipitates or suspended solids, they contain percentages of zinc of 1% or less. PPC growth rates were quantified in the six pilot-scale polishing ponds in the Buchans wastewater, using either an artificial substrate or alder branches. The artificial substrates, called "peritraps," consisted of an artificial netting structure which housed alder branches. The PPC mass was cleaned off the nets and branches, dried, weighed, and analyzed for elemental content. The percent content of likely compounds of the major elements are presented in Table 17.3. Other compounds accounting for the remaining 33% could include mainly sulfur and silicon compounds. Other determinations, such as loss on ignition, can range from 25 to 42%, through the volatilization of organic and inorganic carbon as CO_2, hydroxides, and water from hydrated compounds (Table 17.3).

Table 17.3. Composition in periphyton–precipitate complexes (PPC) in percent of dry weight

Taxa	Algae	$Fe(OH)_3$	$Zn(OH)_2$	$Mn(OH)_5$	$Al(OH)_3$	$CaCO_3$	Other
Microspora	18.7	21.4	12.7	3.6	3.5	4.1	36.3

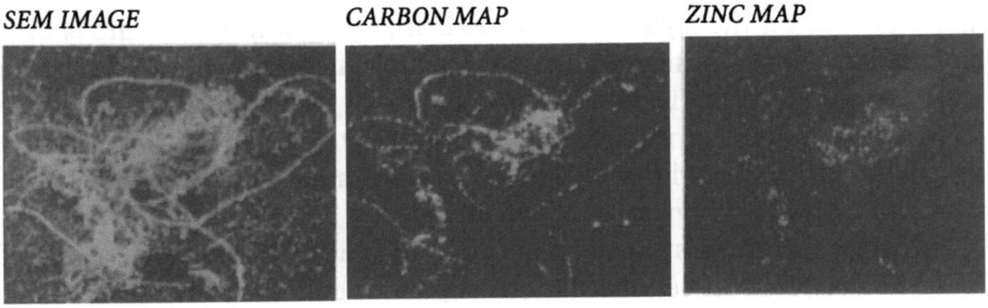

Fig. 17.5. Scanning electron microscope photograph of a leaf surface with filamentous algae and precipitates

The source of the hydroxides is, first, the flooded pit immediately upstream of the polishing ponds, and, second, from further precipitates formed in the ponds themselves. Related work done on PPCs in other wastewater bodies and at Buchans was carried out matching precipitate "growth rate" with the growth rate of the algal portion of the PPCs. The "growth rates" matched well. This suggests that the periphyton surfaces and associated polysaccharides exuded by them are providing a "sticky" surface which appears to remove not only precipitates from the water which have adsorbed zinc, but also some of the dissolved zinc (filterable through 0.45 mm). This provides evidence for the third mechanism of zinc removal.

Evidence for the fourth mechanism was obtained from analysis of surfaces of algae/moss growing in the polishing ponds using a high resolution scanning electron microscope (SEM) and energy dispersive X-ray microanalysis (EDX) using a windowless detector which allowed quantifiable detection of light elements, including carbon and oxygen. Large precipitates were observed among the filamentous algae (Fig. 17.5). The precipitates, which were several micrometres in size, were spheroidal in shape and clustered in large clumps between thin filaments of algae. Elemental image mapping showed that these precipitates were composed of zinc and carbon, with significant amounts of manganese. This provided strong evidence for zinc removal by precipitation as zinc carbonates.

The polishing pond performance during the summer was good. The winter removal rates were significantly lower. There are a number of factors which may account for this. The temperature and pH of the water flowing into the ponds changes significantly over the course of the year (Fig. 17.6). The temperature increases from 3–5 °C during the winter to over 20 °C during the summer. The pH of this water increases from approximately 6.5 during the winter to approximately 7.2 during the summer.

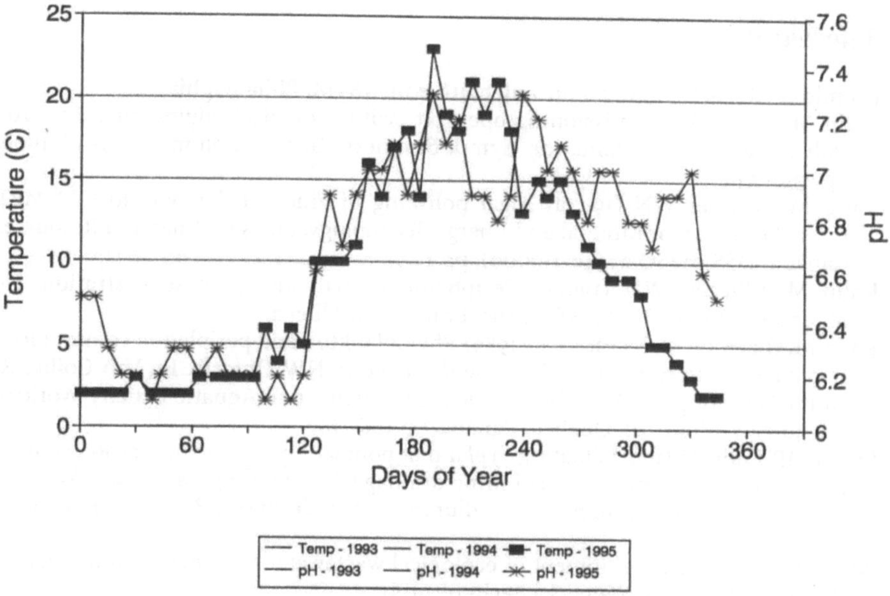

Fig. 17.6. Temperature and pH changes recorded throughout the year at the outflow of the gloryhole

A further factor which affects the metal precipitation conditions is the oxygen concentration in the water. During the winter, the aeration of the pit and the ponds is almost eliminated by the formation of ice on top of the ponds. The combination of the lower temperatures, pH levels, and concentrations of dissolved oxygen of the waters results in reduced iron hydroxide formation. Due to reduced algal growth rates, zinc carbonate precipitates on the periphyton are also reduced.

Zinc removal is affected not only by its coprecipitation (adsorption on iron hydroxide precipitates) but also by its adsorption and adhesion to algal surfaces and extracellular polysaccharides. The increased effectiveness of zinc removal during the summer can also be ascribed to the growth of the algae, since algal surface areas for adsorption, filtration, and coprecipitation are constantly increasing. A decreased effectiveness during the winter may be the result of reduced pond volume in the winter due to ice formation.

Acknowledgements. This work was supported by the ASARCO (American Smelting and Refining Company) – Abitibi Price Joint Working Group at Buchans, Newfoundland, and Natural Resources Canada, Canada Centre for Mineral and Energy Technology (CANMET) and Biotechnology.

References

Hamilton CE (1978) Manual on water, 4th edn. ASTM, Philadelphia

Kalin M (1992) Decommissioning open pits with ecological engineering. In: Proc 16th Annu Mine reclamation Symp, Smithers, British Columbia, 15–18 June, pp 239–246

Kalin M, Wheeler WN (1992a) Algal polishing of zinc. Final report to CANMET (Canada Centre for Mineral and Energy Technology, Energy, Mines and Resources Canada: DSS 034SQ.23440-1-9009), pp 1–57

Kalin M, Wheeler WN (1992b) Periphyton growth and zinc sequestration. In: McCready R (ed) BIOMINET Proc. Edmonton, Alberta

Kalin M, Olaveson M, McIntyre B (1989) Phytoplankton and periphyton communities in a shield lake receiving acid mine drainage in NW Ontario. In: Van Coillie R, Niimi A, Champoux A, Joubert G (eds) Proc 15th Annu Aquatic toxicity Worksh, 28–30 Nov, Montreal, Quebec, Canada, pp 166–187

Smith MP, Kalin M (1991) Floating *Typha* mat populations as organic carbon sources for microbial treatment of acid mine drainage. In: Duarte J, Lawrence RW (eds) Proc IXth Int Symp, Biohydrometallurgy, Quelez de Baixo, Portugal, 9–13 Sept, p 454

Wang B (1987) The development of ecological wastewater treatment and utilization (EWTUS) in China. Water Sci Technol 19:51

Wieder RK (1992) The Kentucky wetlands project: a field study to evaluate man-made wetlands for acid coal mine drainage treatment. Final report, Cooperative Agreement GR-896422 between the US Office of Surface Mining, Reclamation and Enforcement and Villanova University, July 1992

18 Development of a Long-Term Strategy for the Treatment of Acid Mine Drainage at Wheal Jane

H.M. Lamb[1], *M. Dodds-Smith*[1] *and J. Gusek*[2]

[1] Knight Piésold, 35/41 Station Road, Ashford, Kent TN23 1PP, UK
[2] Knight Piésold, Denver, USA

18.1 Background

Wheal Jane is an abandoned underground tin mine in south-west England. Extensive mining began in the area in the early seventeenth century, reaching its peak in the mid-nineteenth century when the region was the largest producer of copper in the world. The industry subsequently went into decline, and, with the closure of Wheal Jane in 1991, there is currently only one tin mine remaining in operation in Cornwall.

Following the closure of Wheal Jane, all underground operations including dewatering ceased. As a consequence, the mine water level rose within the abandoned workings until it reached adit level in late 1991. The failure, in January 1992, of an adit plug allowed the mine water to drain into the local watercourse, the Carnon River. The result was the release of an estimated 30,000 m^3 acidic, metal-rich mine water into the river over an initial 24-h period.

The Carnon River drains an area of some 45 km^2 in a region which has a long history of mining, and the poor river water quality prior to the 1992 incident reflects this history. However, the 1992 incident resulted in concentrations of many metals in the river exceeding Environmental Quality Standards (EQS) by up to two orders of magnitude and a highly visible ochre-coloured plume of polluted water entered the Fal Estuary, an important centre for tourism, commerce and an area of high ecological value.

Knight Piésold was commissioned by the UK National Rivers Authority (NRA) to design and operate a temporary mine water treatment system, and to research the most appropriate and cost-effective long-term treatment strategies for achieving specified water quality objectives in the Carnon River (see Fig. 18.1).

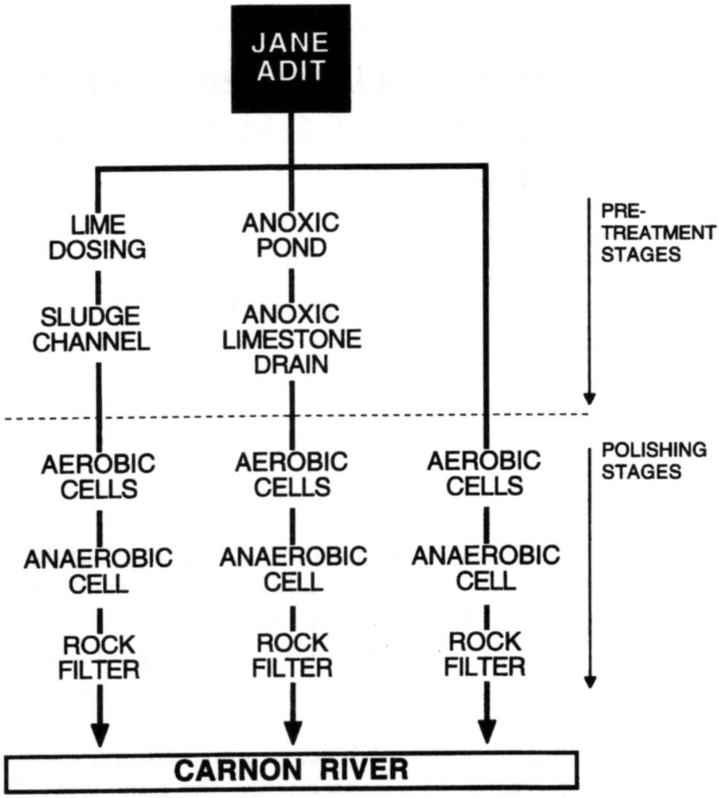

Fig. 18.1. Layout of the Wheal Jane pilot plant, Cornwall, UK, designed by Knight Piésold and Partners on behalf of the National Rivers Authority

18.2
Mine Water Chemistry

The Wheal Jane mine water displays the typical characteristics of acid mine drainage, being highly acidic and having elevated concentrations of a range of metals, principally including iron, zinc, cadmium, copper, arsenic and manganese. Maximum concentrations of metals in the mine water have declined since the initial incident but remain elevated (see Table 18.1).

The decline in metal concentrations has been evaluated to generate predictions of future mine water quality which might form the basis for a long-term treatment strategy. However, attempts to model these changes have indicated that the recorded data does not follow a simple exponen-

Table 18.1. Maximum recorded metal concentrations in mine water

Parameter	Maximum concentration	Mean concentration (Sept. 1994)
pH	2.8 (minimum)	3.5 (minimum)
Arsenic	162	9
Aluminium	190	27
Cadmium	1.7	0.08
Copper	23	1.2
Iron	5070	345
Manganese	27	8
Zinc	2130	132

All metal concentrations are expressed as milligrams of total metal per litre.

tial decline (see Fig. 18.2). Although it is anticipated that concentrations will continue to decline, it is not possible to predict the magnitude of this decline with sufficient accuracy for use in the design of a long-term treatment plant.

18.3 Temporary Treatment

In order to reduce the short-term impacts of the release of mine water, the NRA exercised its statutory powers and implemented a temporary treatment system involving:

- Pumping of mine water from underground.
- Lime-dosing, to raise the pH from 3.5 to 9.5, to precipitate metal hydroxides.
- Flocculation to promote settlement of the precipitate.
- Sedimentation and storage of the resultant metal hydroxide sludge in the mine's existing tailings dam

This system, which has been upgraded since its implementation, continues to form the basis of the current treatment strategy. To date, the system, which can treat up to 300 l s^{-1}, has removed a total of approximately 12,500 t metal from the mine water. The treated mine water discharge is currently of a better quality than that of the receiving river.

The development of a long-term strategy for the treatment of mine water requires:

- The determination of appropriate Water Quality Objectives for the Carnon River, against which the treatment options can be evaluated.

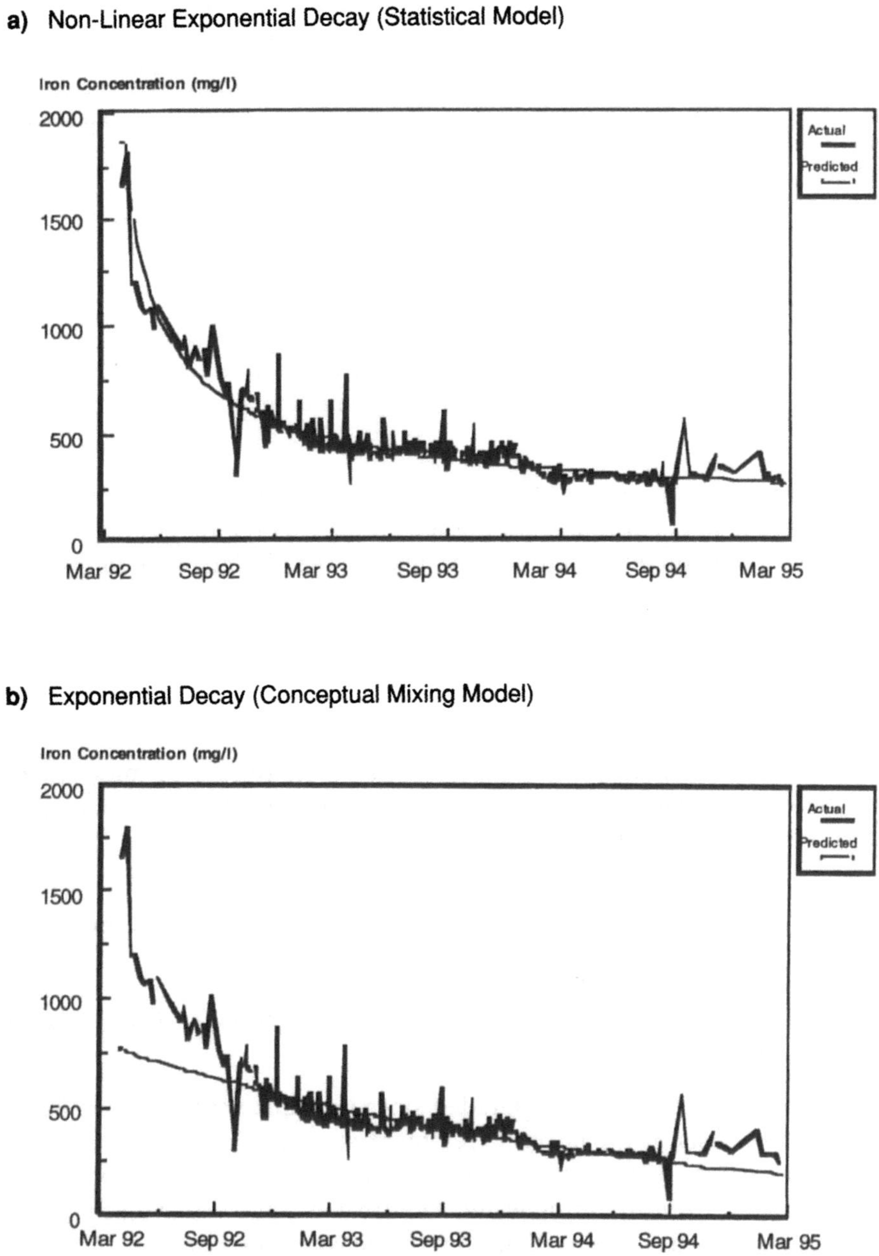

Fig. 18.2. Actual and predicted total iron concentrations in Wheal Jane mine water

- An assessment of the potential economic benefits which might be attained from progressive improvements in river quality.
- A technical and financial review of a range of potential treatment options, including both passive and active systems.

18.4 Water Quality Objectives

Water Quality Objectives (WQOs) are set by the NRA in accordance with both EU standards and national policy. However, modelling of the water quality in the Carnon River indicated that full compliance with all EU and national standards could not be achieved by the treatment of Wheal Jane mine water alone. Consequently, two alternative objectives to assist the development of the treatment strategy were also set, namely:

- The maintenance of the river water quality achieved by the existing temporary treatment system.
- An 80% reduction from the 1985 loadings of zinc, in line with commitments made by the UK to the North Sea Conference.

18.5 Cost Benefit Analysis

The NRA commissioned the specialist consultants Risk and Policy Analysts Ltd. to assess the potential economic benefits which might be derived from the attainment of each of the stated Water Quality Objectives. This was set against a baseline of the disadvantages which would arise in the event that no treatment was undertaken. The following principal areas were considered to be the most vulnerable to the effects of a deterioration in water quality:

- Fisheries (commercial and non-commercial).
- Extraction of maerl (a coralline algae used as a soil conditioner).
- Water sports.
- Boat moorings.
- Aesthetic impacts on property values.

The cost benefit analysis indicated that limited net economic benefit would result from improvements in water quality above that achieved by the temporary treatment system. However, alternative treatment systems were evaluated to assess both the most efficient means of maintaining the existing water quality and the feasibility of achieving a higher quality in the future.

18.6 Passive Treatment Technologies

Despite the current interest in passive treatment technologies, a detailed understanding of the complex physical, chemical and biological processes which operate in these systems is still evolving and many aspects remain poorly understood. In practice, the construction of many types of passive treatment systems is based to a large extent on empirical design parameters, often derived from experience under a wide range of operating conditions of flow, water chemistry, climate and discharge standards. For this reason, it is prudent to construct and operate small pilot plants to test the validity of the design parameters at individual sites.

Following a full chemical characterisation of the mine water, a Pilot Passive Treatment Plant was designed and is being operated by Knight Piésold at Wheal Jane to evaluate the most appropriate treatment processes to suit the particular conditions at the site. The design parameters were based primarily on experience gained by Knight Piésold in the USA, and on the results of bench-scale laboratory tests.

Passive treatment systems are usually based on one of two processes:

1. Aerobic precipitation of metal hydroxides or oxides.
2. Anaerobic precipitation of metal sulphides.

Aerobic systems are best suited to the removal of iron, arsenic and manganese. In contrast, anaerobic systems are not generally configured to remove manganese, but will remove iron, zinc, cadmium and copper.

Given the range of metals present within the Wheal Jane mine water it was necessary to consider the use of both types of system. Consequently, the Wheal Jane Pilot Plant was based upon a combination of:

- An aerobic wetland designed principally to remove iron and arsenic.
- An anaerobic cell designed principally to remove zinc, copper and cadmium.
- An aerobic rock filter utilising algal micro-environments to remove manganese.

The aerobic treatment of large volumes of mine water in particular may require large areas of land. This land requirement may be reduced by "pre-treatment" of the mine water to raise the pH, which allows the subsequent aerobic systems to operate more efficiently. Pre-treatment generally comprises either active lime-dosing or an anoxic limestone drain (ALD). The performance of the ALD may be adversely affected by the presence of dissolved oxygen, ferric iron and aluminium in the mine water. Consequently, the ALD at Wheal Jane has been preceded by an anoxic cell which uses the activity of anaerobic bacteria to promote the removal of dissolved oxygen, the maintenance of iron in the ferrous form and the precipitation of aluminium hydroxide.

Consequently, in order to evaluate the systems with and without pre-treatment, the pilot plant comprises three separate configurations of treatment systems, namely:

1. A "lime-free" system containing an aerobic wetland, an anaerobic cell and a rock filter.
2. A "lime-dosed" system containing an aerobic wetland with pre-treatment by lime-dosing, an anaerobic cell and a rock filter.
3. An "ALD" system, comprising pre-treatment by an anoxic cell and anoxic limestone drain (ALD), followed by an aerobic wetland, an anaerobic cell and a rock filter.

The individual components of the systems at Wheal Jane are described as below.

18.6.1 Aerobic Wetland

The aerobic wetland comprises a shallow reed bed of *Typha*, *Phragmites* and *Scirpus*, planted in a substrate of coarse tailings. This is designed to promote the oxidation of ferrous iron to ferric iron, which is subsequently hydrolysed and precipitated as iron hydroxide. Arsenic will also be removed by co-precipitation with the iron hydroxide.

Experience gained elsewhere has identified the iron loading and the pH of the mine water as the two critical factors in the aerobic wetland design. The sizing of the wetland is based upon the following empirical iron removal rates:

- For an influent pH of <5 (i.e. the lime-free system): 2–4 g Fe removed per m^2 per day.
- For an influent pH of >5 (i.e. the systems with pre-treatment): 5–11 g Fe removed per m^2 per day.

18.6.2 Anaerobic Cell

The anaerobic cell substrate comprises a sawdust and hay mixture, with the addition of cattle manure as an inoculum of sulphate-reducing bacteria. The cell is designed to promote the bacterial reduction of sulphate to hydrogen sulphide, and the subsequent reaction of the hydrogen sulphide with dissolved metal ions to form insoluble metal sulphides. The size of the anaerobic cell is determined by two loading factors:

1. Volumetric: based upon the stoichiometry of the metal sulphide precipitation reactions, and the amount of sulphide which can be generated by the bacteria. A loading factor of 0.3 mol metals m^{-3} of substrate has been used.
2. Surface: determined by the need to moderate the pH of the influent mine water to optimise bacterial activity. Empirical observations suggest that for an influent pH of below around 5, the appropriate surface loading factor is approximately 20 $m^2\ l^{-1}\ min^{-1}$.

18.6.3
Rock Filter

The rock filter is designed to utilise algae to create micro-environments of the high pH required to precipitate manganese. During photosynthesis, the algae remove carbon dioxide from the water and release oxygen, resulting in the creation of a micro-environment with a pH as high as 10. The rock filter comprises shallow cells containing coarse cobbles of granite which provide a substrate for the growth of algae and which maximise the contact of water with the algae. The sizing of the rock filters is based upon experimental data which have shown that manganese can be removed from mine waters at a rate of 2 g Mn per m^2 per day.

18.7
Preliminary Results

Passive treatment plants attain optimum efficiency only when the biological systems reach maturity, which may take several years. Consequently, sufficient data has been collected to date to enable only a preliminary analysis of performance to be undertaken.

18.7.1
Performance of the Anoxic Cell/ALD

The anoxic cell has reduced the dissolved oxygen content of the influent mine water by around 40% to less than 2 mg l^{-1}, and raised the pH from less than 4 to 5. The ALD itself has raised the pH of the mine water to approximately pH 6. A further improvement in the removal of dissolved oxygen may be necessary to fully protect the ALD from the risk of armouring with iron hydroxide. Moreover, aluminium is not being removed fully in the anoxic cell and there is evidence to suggest that clogging of the ALD caused by the precipitation of aluminium hydroxides may become a problem. The operation of the anoxic cell/ALD system is being reviewed to counteract this possibility.

18.7.2 Aerobic Wetlands

At average influent flow rates of between 0.2 and 0.3 l s^{-1}, between 84 and 95% of the iron and between 98 and 99% of arsenic were removed in the aerobic wetlands of the three systems. The iron removal rate is expected to improve further as the system reaches maturity.

18.7.3 Anaerobic Cells

At average influent rates of approximately 0.3 l s^{-1}, the anaerobic cells are currently removing around 82% of the zinc, 99% of the cadmium, 99% of the copper and 33% of the iron which remain following aerobic treatment. The pH of the effluent from the anaerobic cells is currently around 5.5 compared with an influent of less than 3. It is anticipated that the efficiency of zinc removal will improve further and that anaerobic cells will also maintain the removal of other metals at progressively higher flow rates.

18.7.4 Rock Filter

The algal population of the rock filter has not yet reached the level at which significant manganese removal might be expected.

18.7.5 Suitability of Passive Treatment for Wheal Jane Mine water

Due to the preliminary nature of the available data, it would be premature to undertake any detailed evaluation of the performance of the pilot plant. Nevertheless, experience from other sites together with the limited data available from the pilot plant suggest that the Wheal Jane mine water could, in principle, be treated by a system which utilises passive treatment technology. However, in practice, the flow of mine water which requires treatment would necessitate a passive treatment system occupying an area of land which exceeds that currently available to the NRA in the Carnon Valley. This suggests that passive treatment may not be practical as the sole solution at Wheal Jane. Nevertheless, evaluation of passive treatment is continuing to provide information which will be of value in the treatment of polluted waters at other sites, including other smaller sources of acid mine drainage in the Carnon Valley.

18.9 Active Treatment Technologies

The existing treatment operation at Wheal Jane is a form of active treatment which relies on the availability of the tailings dam for both water/sludge separation and sludge disposal. This and alternative forms of active treatment may be feasible in the long term, provided that a suitable method of sludge handling and disposal can be developed.

The active treatment technologies considered comprise three basic stages:

Stage I. Precipitation of dissolved metals from solution by means of either:
- Precipitation of hydroxides
- Precipitation of sulphides
- Selective precipitation

Stage II. Water/metalliferous sludge separation by means of either:
- Thickening
- Hydrocyclones
- Magnetic separation
- Flotation

Stage III. Sludge thickening and dewatering by means of either:
- Rotary or vacuum filters
- Continuous pressure belt presses
- Centrifuges
- Frame and plate presses

A preliminary technical and financial appraisal of the treatment technologies available for each of the stages above was undertaken in order to determine if the method would be a practicable option for Wheal Jane (see Table 18.2).

18.8.1 Economic Appraisal and the Preferred Active Treatment System

Those options with sufficient potential (technically and economically) underwent bench-scale or pilot plant testing, and the results were used in

Table 18.2. Applicability of active treatment technologies

Active treatment technology	Applicable in principle	Practicable option
Stage I		
Precipitation of hydroxides	Yes	Yes
Precipitation of sulphides	Yes	Yes
Selective precipitation	Yes	No
Stage II		
Thickening	Yes	No
Modified thickening	Yes	Yes
Hydrocyclones	No	Yes
Magnetic separation	No	Yes
Flotation	No	Yes
Stage III		
Rotary or vacuum filters	Yes	No
Continuous pressure belt presses	Yes	Yes
Centrifuges	Yes	Yes
Frame and plate presses	Yes	Yes

the evaluation of the preferred active treatment route for the Wheal Jane mine water. As a result of this evaluation, the preferred treatment system comprises two phases:

1. Continued operation of the existing temporary treatment system whilst sludge storage is available in the tailings dam.
2. In the long-term, implementation of active treatment based upon:
   - Lime-dosing to form an hydroxide precipitate
   - Sludge/water separation using thickeners
   - Sludge dewatering using either a centrifuge for on-site sludge storage or a frame and plate filter press for off-site disposal

18.9 Summary and Conclusions

1. Following the closure of Wheal Jane, the water level within the mine rose and, with the failure of an adit plug, there was an initial release of approximately 30,000 m^3 acidic, metal-rich water into the Carnon River.
2. An emergency treatment system was installed, which has successfully removed some 12,500 t metals from the mine water since February 1992.

3. Without treatment, the metal concentrations in the Carnon River would be significantly elevated over the current levels and there would be a risk of widespread and prolonged iron hydroxide discoloration in the Fal Estuary with the possibility of long-term adverse effects on the local environment and economy.
4. Both active and passive treatment systems have been evaluated to assess their ability to both maintain and improve upon current water quality.
5. A full technical and financial evaluation has indicated that the current temporary treatment system is the most appropriate means for the maintenance of the current water quality whilst the capacity to settle and store sludge in the tailings dam remains. Thereafter, an active system utilising sludge thickeners, dewatering and disposal off-site appears to be the preferred option.
6. Further improvements in water quality would have little net national economic benefit.
7. Passive treatment systems, whilst still undergoing full evaluation, are likely to be constrained by the availability of a suitable land area in the Carnon Valley.
8. Full compliance with all national and European water quality standards will require the treatment of sources of contamination in the valley other than Wheal Jane. Passive treatment may have a role in the treatment of these sources.

Acknowledgements. Permission from the National Rivers Authority to publish the details contained in this chapter is gratefully acknowledged.

References

Cambridge M (1995) Use of passive systems for the treatment and remediation of mine outflows and seepages. Proceedings of IMM Conference "Planning for Closure", Minerals Industry International, May 1995

Hamilton RM, Waite RRJ, Postlethwaite NA, Cambridge M (1994) Wheal Jane, its abandonment and treatment of the resultant discharge. 3rd International Conference on Abatement of Acidic Drainage, Pittsburgh, USA, April 1994

Lamb HM, Dodds-Smith ME, Gusek J (1995) The development of a long term strategy for the treatment of acid mine drainage at Wheal Jane. International Workshop on Abatement of Geogenic Acidification in Mining Lakes, Magdeburg, Germany, September 1995

Taberham J, Cambridge M (1994) Developing Treatment Strategies for the Wheal Jane Acidic Minewater. BICS International Conference "Managing Abandoned Mine Discharges and Effluents", London, September 1994

19 Bioremediation of Metals in Acid Tailings by Mixed Microbial Mats

P. Phillips and J. Bender

Clark Atlanta University, Atlanta, Georgia 30314, USA

19.1 Introduction

Mixed microbial mats were used to promote metal removal from mine drainage in Alabama and Colorado, USA. They are a microbial consortium which are highly tolerant of toxic metals and harsh environmental conditions. Under field conditions, these mats are generally modified by the voluntary invasion of local flora, especially filamentous green algae. For bioremediation work, these constructed mats take advantage of the role natural microbial mats play in filtering and transforming metals and metalloids. Constructed microbial mats are synthesized using silage with its associated bacterial flora and the blue-green algae (cyanobacteria) *Oscillatoria.*

In the laboratory, microbial mats removed an array of heavy metals and metalloids, mineralized to carbon dioxide pesticides, PCBs (polychlorinated biphenyls), solvents, oils and explosives and removed mixed waste (radionuclides and heavy metals) in a continuous flow system (Goodroad et al. 1994). In the field, they have removed manganese (Mn), zinc (Zn) and other heavy metals [silver (Ag), chromium (Cr), cadmium (Cd), copper (Cu), lead (Pb), nickel (Ni) and iron (Fe)] from mine drainage and BTEX (benzene, toluene, ethyl-benzene and xylene), a gasoline constituent, from contaminated groundwater (Goodroad et al. 1994; Bender et al. 1995). Landfill leachate is a mixture of toxic organics, metals and ammonia. In collaboration with Waste Management, Inc., a pilot project tested microbial mat reduction of ammonia and metals from sanitary landfill leachate (Goodroad et al. 1995).

Microbial mats offer several advantages for bioremediation. These include low cost, durability, the ability to function in both fresh and salt water and tolerance to high concentrations of metals, metalloids and organic contaminants, as well as mixtures of these.

19.1.1 Natural Microbial Mats

Microbial mats are natural microbial communities dominated by blue-green algae, but also containing a variety of bacteria within the laminated structure of the mat. These systems are resilient microbial communities which self-organize into stratified biofilms arranged for maximum efficiency of interspecies exchanges (Caumette 1989). Their physiological flexibility has been amply documented and includes anoxygenic and oxygenic photosynthesis, survival after desiccation and inclusion of aerobes and anaerobes within the same matrix (Shilo 1989; Stal et al. 1989). During daylight, photosynthesizing mats maintain highly aerobic conditions at the surface while retaining anaerobic zones in lower regions. Thus, the mat ecosystem has aerobic/anaerobic activities in process simultaneously. However, dark periods cause a rapid shift to anaerobic function because there is no light for photosynthesis, and, consequently, for oxygen production. Because microbial mats have evolved under hostile conditions, similar to those expected in highly contaminated environments, survival adaptations of these ecosystems are directly applicable to remediation biotechnology.

19.1.2 Constructed Microbial Mats

Mixed microbial mats can be constructed for specific metal bioremediation tasks by integrating the desired microbial components for bioremediation of metal/metalloid contamination. Among the latter, microbial mats have been found to reduce selenate to elemental selenium, remove Pb, Cd, Cu, Zn, cobalt (Co), Cr, Fe, uranium^{238} (U^{238}) and Mn from water and to remove Pb from sediments. Controlled experiments with one radionuclide, uranium (0.1 mg $U^{238}\,l^{-1}$, spiked in groundwater samples) were successful. Table 19.1 (adapted from Goodroad et al. 1994) presents a summary of metal removal from water and sediments.

Microbial mats are cultured by isolating key microorganisms from the treatment site according to Bender and Phillips (1994). They are developed by enriching a pond or bioreactor surface with ensiled vegetation together with microbial inocula. Silage provides a flora of fermentative bacteria, as well as organic acids that nutritionally support heterotrophic bacteria. Additionally, photosynthetic products of the blue-green algae will support microorganisms that are heterotrophic. The floating silage serves as a secondary structural function for floating microbial mat development. Anaerobic and microaerophilic strains colonize the anoxic zones

Table 19.1. Remediation of metals by constructed mixed microbial mats. (Adapted from Goodroad et al. 1994)

Metal	Initial concentration ($mg\ l^{-1}$)	Removal rate ($mg\ metal\ m^{-2}\ mat\ h^{-1}$)
Free floating mats[a]	U^{238}: 0.12	3.19
Mats immobilized on glass[b] wool layered in baffled tanks	Mixture of	
	Cr: 24	10,129
	Co: 24	10,052
Mats immobilized on floaters[c]	Mixture of	
	Zn: 22	313
	Mn: 18	462
Excised mats applied to Iron[d] Mountain mine drainage sample	Mixture of	
	Cu: 284	378
	Zn: 3021	3778
	Cd: 19	356
Acid mine drainage in field[e] ponds	Mn: 3.5–7.6	1.0–4.1 g Mn m^{-2} day^{-1}

[a] Self-buoyant microbial mat was cultured on the surface of laboratory ponds.
[b] Microbial mat, cultured on glass wool, was placed into acrylic tanks constructed with baffles to create a serpentine flow.
[c] Microbial mat was attached to glass wool balls that were floated in metal-contaminated water.
[d] Small sections of microbial mat were excised and applied to a mixed solution of Cu, Zn, Cd and Fe sample from Iron Mountain mine drainage in California, USA. pH was adjusted to 3–4 before adding microbial mat sections.
[e] A floating microbial mat (1 to 2 cm thick), composed of a filamentous green alga and blue-green algae, was developed by enriching with silage and microbial inocula (initially selected from the site) on an approximately 40-m^2 field pond. A second microbial mat formed on the limestone at the pond bottom.

within the mat. The final product is a thick, gelatinous green mat dominated by blue-green algae floating over a clear water column. Under field conditions, these mats are generally modified by the voluntary invasion of local flora, notably filamentous green algae in the field experiments discussed here. This latter assists in the long-term stability of the floating microbial mat.

Field studies, currently in progress, show that the microbial mat is easily generated in the natural environment and exhibits excellent durability in the field. Relevant research and development results from mining pilot field projects using constructed mixed microbial mats are reported here.

19.2 Acid Coal Mine Drainage Treatment

19.2.1 Background

Manganese removal from acid mine drainage represents a unique challenge due to the solubility of manganese sulfide and the alkaline conditions required to precipitate manganese as an oxide or carbonate. Therefore, it is common to find drainage with manganese above US Environmental Protection Agency standards (less than 2 mg l^{-1}). Additionally, in an oxygenated environment, ferric iron precipitates as $Fe(OH)_3$, and the consequent release of hydrogen ions will increase acidity. Thus, the dual goal of simultaneously removing manganese and iron from mine drainage is complicated if the system does not have enough alkalinity and a $pH > 7$.

The Tennessee Valley Authority (TVA) utilizes constructed wetlands technology to treat acid mine drainage. These wetlands have generally been effective in removing Mn (0.15–1.87 g m^{-2} day^{-1}) and Fe (0.4–21.3 g m^{-2} day^{-1}) (Brodie 1993). At one site within the abandoned Fabius coal mine fields in northeast Alabama, USA, the drainage contains total dissolved Mn and Fe at approximately 8 and 6 mg l^{-1} (0.45-μm filtered) after leaving an oxidation pond and before draining toward an extensive constructed wetland.

The research reported here, together with other TVA sponsored projects, have demonstrated the success of a comprehensive solution to this problem. The resolution of the Fe/Mn contamination in an environment of low pH, oxygen and alkalinity requires several components: an anoxic drain, an oxidation pond and a metal-tolerant microbial photosynthetic system.

19.2.2 Methods

Beginning in 1992, at a point downstream from the oxidation pond, our research group conducted a pilot-scale field test to determine if microbial mat would efficiently remove residual manganese in a 44-m^2 pond. Phillips et al. (1994, 1995) and Bender and Phillips (1995) give detailed explanations of the methodology and earlier phases of the project. The acidity in the coal mine drainage was reduced by directing the flow through an anoxic drain, a buried limestone barrier. After the limestone barrier, much of the iron precipitated in an oxidation pond of approxi-

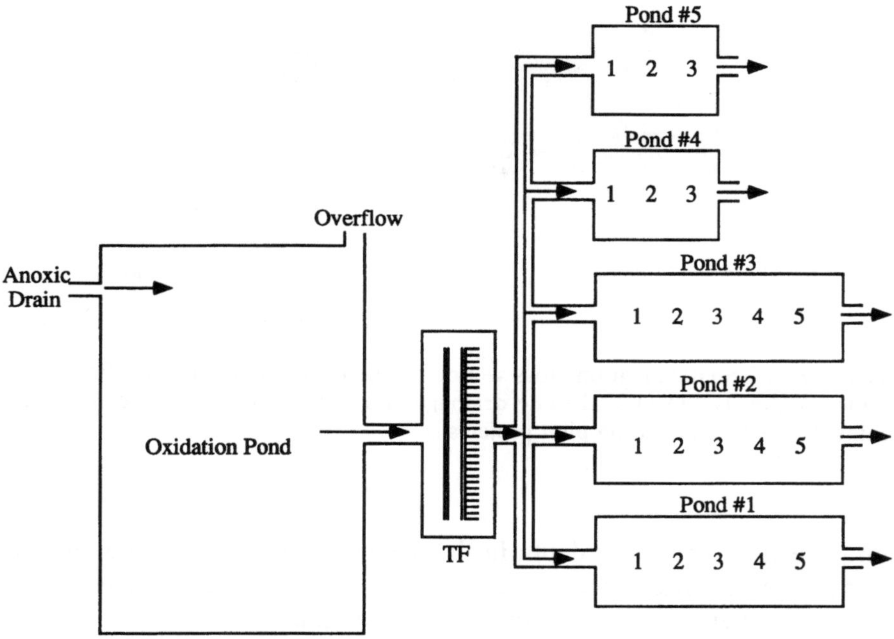

Fig. 19.1. Schematic diagram of the Tennessee Valley Authority (TVA) acid coal mine drainage remediation pilot project, Alabama, USA. Anoxic drain increases pH, and oxidation pond and trickling filter (*TF*) precipitate iron. Pond 1 was inoculated with microbial mat in 1992. Pond 2 was a limestone control pond from 1992 to 1994; then it was inoculated with microbial mat. Pond 3 was a pea gravel control pond. Pond 4 was a second limestone control pond and pond 5 was a soil substrate control pond, both operating from December 1994 to September 1995. *Numbers 1–5* refer to sampling points separated by 1 m each

mately 1 ha. The anoxic drain contributed buffer in anoxic conditions, thereby preventing iron deposit and armoring of the limestone. As the water entered the oxic region (oxidation pond) it precipitated as $Fe(OH)_3$ while maintaining neutral pH. A stream of drainage from the oxidation pond was diverted to a trickling filter system to remove additional iron. The drainage was then flowed through the pilot ponds to determine manganese removal under various treatments. The project began with three ponds (discussed below in periods 1 and 2). Two additional ponds were later added (period 3) (Fig. 19.1). All ponds were lined with 4-mm black plastic and layered to a depth of 10 cm with 2.5-cm-size limestone (ponds 1, 2 and 4), 1-cm-size pea gravel (pond 3) or soil substrate (pond 5) (Fig. 19.2). The bottom was intentionally formed into an undulating configuration to provide variable depth and to prevent vertical stratification

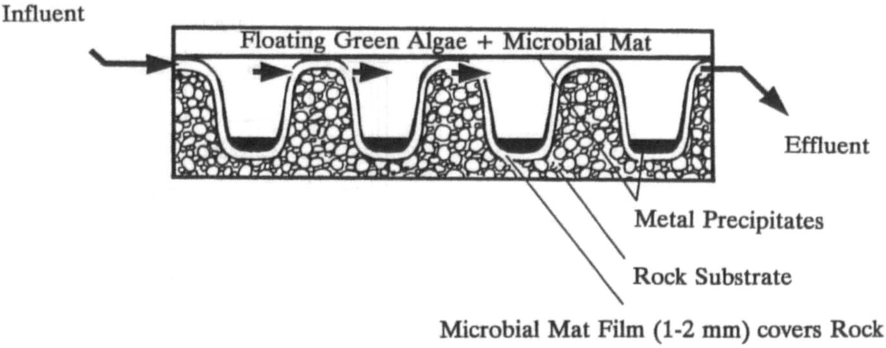

Fig. 19.2. Schematic diagram cross section of microbial mat pond at the Tennessee Valley Authority (TVA) acid mine drainage and Bureau of Mines precious metal mine drainage remediation pilot projects

and sheeting. The water depth in the troughs was 20 cm. The pilot project analysis is divided into three separate periods, which separate distinct events – either climatic or management changes.

Period 1 includes the first 4 months of the experiment from November 1992 to February 1993. Pond 1, 44 m^2 and with a limestone substrate, was inoculated with microbial mat and silage. Pond 2, 44 m^2 and with a limestone substrate, was a control pond for pond 1. Since the carbonate supplied by the limestone was expected to buffer the pond water and, in itself, be responsible for a degree of manganese precipitation, pond 3, 34 m^2 and with a pea gravel substrate, was a control pond for the pond 2 limestone substrate.

Period 2 includes the 4 months of August to November 1994. During period 2, the management strategy for the three ponds was to significantly increase flow through pond 1, inoculate pond 2 with microbial mats and silage and leave pond 3 alone. The objective in pond 1 was to determine if manganese would be released at greater than 2 mg l^{-1} through the effluent (the compliance level) at high flow rates. The objective in pond 2 was to increase the efficiency of manganese removal by direct inoculation.

Period 3 includes the 10 months from December 1994 to September 1995. Two objectives were addressed during this period. First, two small ponds were added to the pilot project. Pond 4, 17 m^2, was a new limestone substrate control to compensate for the loss of the original limestone substrate pond 2 control. Pond 5, 14 m^2, contained a soil substrate and was inoculated with microbial mat and silage. This was a control pond to measure the importance of limestone substrate. Flow rates through these

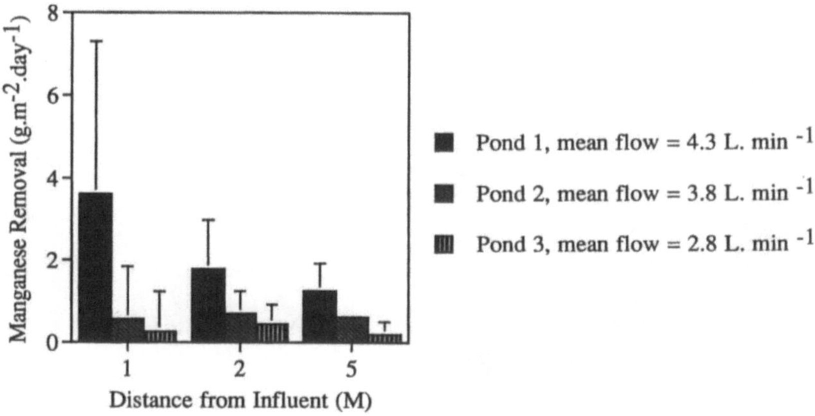

Fig. 19.3. Grams per square meter per day (g m^{-2} day^{-1}) removal of manganese in first 4 months of operation of the Tennessee Valley Authority (TVA) pilot project. The time period extended from November 1992 to February 1993

two new ponds were reduced to reflect their smaller size. Second, all management of the ponds was ended. This meant that after 1 December 1994, no additional inoculation or ensiling took place. Water quality and flow rates were monitored.

19.2.3 Results

During period 1, in the inoculated pond 1, a floating mat (1 to 2 cm thick), composed of microbial mat and a voluntary filamentous green alga, developed. A second microbial mat formed over the rock substrate at the pond bottom (Fig. 19.2). Pond 1 had the greatest manganese removal rate across its length (from influent pipe to 5-m distance). At 1 m, manganese removal was 3.61 g m^{-2} day^{-1}, diminishing to 1.25 g m^{-2} day^{-1} at 5 m (Fig. 19.3). Both control ponds were significantly less efficient. It is important to note that the mean flow rate was greatest in pond 1 (11% greater than pond 2 and 35% greater than pond 3).

During the course of operating the pilot project, microbial mats voluntarily established over the substrate of ponds 2 and 3, similar to the microbial mat substrate cover of pond 1 (Fig. 19.2). Repeated attempts to maintain a biologically nearly sterile rock substrate environment by chlorination of the ponds failed to eliminate the microbial mat cover. Therefore, both ponds 2 and 3 showed increased efficiency of manganese removal as the ponds aged.

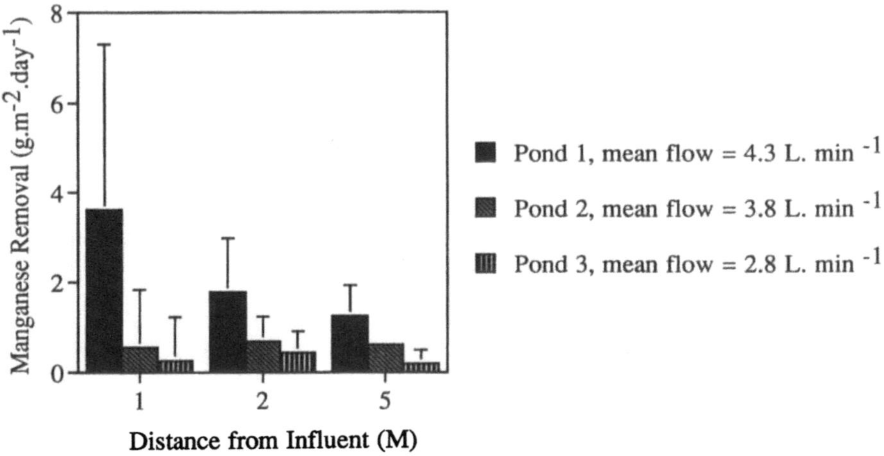

Fig. 19.4. Grams per square meter per day (g m^{-2} day^{-1}) removal of manganese after more renovation (heavy inoculation and ensiling to improve manganese removal efficiency) occurred in ponds 1 and 2. These data are the mean of 4 months from August to November 1994

During period 2, pond 2 operated as a second biological treatment pond. Average pond 1 flow rate was more than twice that of ponds 2 and 3 (Fig. 19.4) and after 2 years of continuous operation, manganese deposits had accumulated to high levels in the sediments of pond 1. The pond 1 high flow and accumulated manganese in the sediments likely resulted in an advancing front of peak manganese removal toward the effluent pipe. For example, manganese was removed at a low rate at 1 m (1.1 g m^{-2} day^{-1}), released at 2 m (−0.3 g m^{-2} day^{-1}), and then removed at a high rate (4.1 g m^{-2} day^{-1}) at 5 m. At flow rates of 6.8 and 5.9 l min^{-1}, respectively, ponds 2 and 3 showed the greatest manganese removal. The relative importance of the effect of the two factors high flow and manganese accumulation in sediments on manganese removal efficiency in pond 1 cannot be determined from this study.

During period 3, the two small control ponds 4 and 5 were operating. The only management included flow rate adjustment and occasional chlorination to maintain low levels of involuntary flora in ponds 3 and 4. The establishment of microbial mat and a green alga was maintained at a low level in pond 4 due to its newness. Pond 3 always had a microbial mat over the substrate.

The flow rates were similar in the first three ponds (Fig. 19.5). The data show that manganese removal in pond 1 is still best at 5-m distance from the influent pipe. Pond 2 manganese removal is best nearer the influent

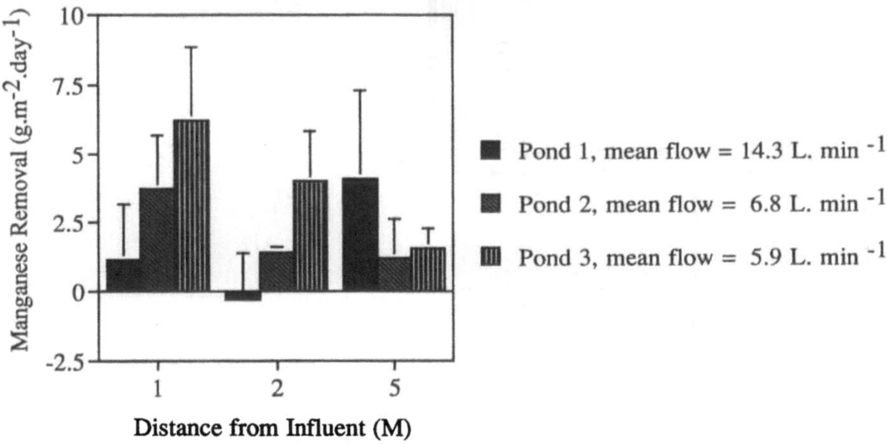

Fig. 19.5. Grams per square meter per day (g m^{-2} day^{-1}) removal of manganese after all active management was suspended (other than repeated attempts to rid pond 3 of microbial mats by chlorination; this was not successful). The intention was to observe ability of system to sustain itself. Pond 4 had a limestone substrate to compensate for loss of pond 2 as a limestone substrate control. Pond 5 had a soil substrate to compensate for the rock substrate in all other ponds. These data are the mean of 10 months from December 1994 to September 1995

pipe. This may likely be due to it being a biologically less mature pond (inoculation with microbial mat occurring in 1994 versus 1992 in pond 1). Superior removal of Mn from pond 2 versus pond 1 was significant ($p<0.05$). Pond 3 microbial mat was not eliminated by chlorination and this may explain the similar rate (grams per meter squared per day) of manganese removal between this pond and ponds 1 and 2 ($p>0.05$). Ponds 4 and 5 (limestone and soil substrate controls, respectively) did not appear to remove as much manganese as the first three ponds. Nevertheless, an analysis of variance revealed no significant difference between pond 1 and ponds 4 or 5 ($p>0.05$). The difference between ponds 2 and 4 was significant ($p<0.05$) and between 2 and 5 highly significant ($p < 0.01$). The difference between the original control ponds 3 and 4 was not significant ($p>0.05$), but pond 3 performed significantly ($p<0.05$) better than pond 5. Of the two newest ponds, the limestone substrate pond 4 removed a greater amount of Mn across its length compared with pond 5, but this difference was not statistically significantly ($p>0.05$). Water quality parameters for the five ponds are shown in Table 19.2. The anoxic drain system raised pH to neutral before entering this pilot project system.

Table 19.2. Water quality parameters for the Tennessee Valley Authority (TVA) Fabius, Alabama, USA, pilot project, 1992–1995

Pond No.	Temperature, °C (range)	Dissolved oxygen, mg l^{-1} (range)	pH (range)	Oxidation-reduction potential, mV (range)	Conductivity, μΩ cm^{-1} (range)	Alkalinity, mg l^{-1} (range)	Manganese at influent, mg l^{-1} (range)	Manganese at final pond sampling point, mg l^{-1} (range)
1	17.2±8.0	6.6±1.7	7.2±0.4	425±69.8	559±128.1	146±59.4	5.0±1.6	1.2±1.6
	(4.0–36)	(2.8–10.5)	(6.4–8.3)	(324–599)	(304–892)	(60–264)	(1.2–7.6)	(0.005–6.5)
2	17.1±8.0	7.1±1.8	7.3±0.4	432±75.6	542±122.4	148±60.0	4.5±2.2	1.3±2.0
	(3.0–33)	(3.0–11.3)	(6.5–8.2)	(328–565)	(301–760)	(72–281)	(0.2–8.3)	(0.02–8.0)
3	16.9±8.3	7.2±2.0	7.2±0.3	425±67.0	526±106.2	139±63.6	5.0±2.0	2.0±2.0
	(3.0–35.5)	(2.8–11.0)	(6.8–8.0)	(336–566)	(306–705)	(41–264)	(0.9–10.3)	(−0.007–7.4)
4	16.4±9.8	7.9±2.1	7.2±0.4	388±51.2	538±51.2	128±41.0	5.2±2.2	1.2±2.2
	(3.5–35.0)	(5.0–11.0)	(6.4–7.6)	(331–488)	(493–618)	(67–190)	(0.07–7.9)	(0.006–6.9)
	17.1±10.2	8.0±2.2	7.1±0.3	396±70.5	319±92.2	116±52.5	5.8±1.5	3.3±3.6
	(2.0–38)	(7.0–12.0)	(6.7–7.6)	(311–567)	(641–505)	(43–194)	(3.4–8.3)	(0.05–8.6)

A comparison of data from pond 1 results with the data shown in Figs. 19.3–19.5 reveals that as manganese fills the sediment zone of the pond, there is an advancing front of peak manganese removal. This would indicate that eventually an individual pond will be saturated with manganese, its presence will interfere with the biologically mediated mechanisms of further manganese removal, and, therefore, manganese concentrations at the effluent pipe will exceed levels permitted by the US Environmental Protection Agency. The pond will then need to be closed and this will be a permanent manganese deposit.

Microbial mat entraps photosynthetic oxygen, generating consistently elevated dissolved oxygen levels (average 7.3 mg l^{-1}). Unlike the control ponds, the microbial mat-containing ponds consistently maintained manganese effluent levels at less than 2 mg l^{-1}. The microbial mat promotes maintenance of high pH, therefore minimizing the possibility of remobilization of manganese.

During period 1, when the pilot project system was new, day/night and winter/summer manganese removal was essentially the same. At this time, control ponds showed manganese breakthrough (manganese at greater than 2 mg l^{-1}) during night-time sampling or when mine drainage flow exceeded 4.5 l min^{-1} (Phillips et al. 1994). Although there was some binding of manganese to the microbial mat, it was primarily deposited as a precipitate at the pond bottom. Samples analysed by x-ray diffraction of floating microbial mat showed that it contained manganous carbonate. In bottom deposits, major minerals were manganese carbonate and calcite. Minor minerals were quartz, gypsum and maghemite. Trace minerals were amorphous iron oxyhydroxides and/or hydroxides.

The problem remains of the validity of comparing results in high flow (10–15 l min^{-1}) versus medium (5–9 l min^{-1}) or slow (1–4 l min^{-1}) flow periods, as the kinetics of manganese removal may be very different at different flow rates. Additionally, pond 1 (containing the heaviest microbial mat) deposited the highest quantity of metal over the duration of the project. This was clearly indicated by the nearly total filling of the sediment troughs with metal. Other ponds showed minimal deposits in their troughs.

Manganese removal rates, calculated under various treatments during this pilot study, have generated the basic information necessary for recommending sizing and flow rates needed for full-scale treatment to compliance levels for manganese removal in coal mine drainage. Each pond creates a sink of precipitated manganese oxides and carbonates, and will eventually require closure.

19.3 Precious Metal Mine Drainage

19.3.1 Background

Microbial mat is being used to treat an abandoned gold and silver mine drainage located at approximately 3100-m elevation in central Colorado, USA. The drainage, of neutral pH, contains manganese and zinc at up to 18 mg l^{-1}. These were the target contaminants for remediation. Additionally, Ag, Cd, Cr, Cu, Pb, Ni and Fe, present in micrograms per liter levels, were also concentrated by microbial mats.

19.3.2 Methods

Microbial mat was applied in two ways: (1) it was previously cultured on coconut mesh, dehydrated and shipped to the site; or (2) dry coconut mesh was applied to the pond surface and microbial mat inocula plus silage were added in the field. Under high altitude conditions, the microbial mat must be shaded to avoid ultraviolet radiation damage.

During summer 1994, microbial mat was applied to three 4×4-m treatment ponds, as well as the larger and deeper retention pond (Fig. 19.6). The pond substrate was similar to the Alabama coal mine drainage ponds. By February 1995, during the harsh winter weather, the treatment ponds had produced a thick green biomass due to the microbial mat inocula, as well as to an invasion of a filamentous green alga.

19.3.3 Results

Historical metal concentrations in the mine drainage are shown in Table 19.3. By contrast, after 2 months (between July and September 1994), the microbial mat biomass had accumulated these same metals to very high mean milligram per kilogram concentrations. In August 1995, manganese and zinc levels entering the three pilot project ponds from the retention pond were significantly reduced in the two biological treatment ponds compared with the control pond which was not inoculated (Table 19.4). In pond 1 (treatment pond), there was a 52% decrease in both manganese and zinc concentrations and in pond 2 (treatment pond), there was a 64% decrease in manganese and a 63% decrease in

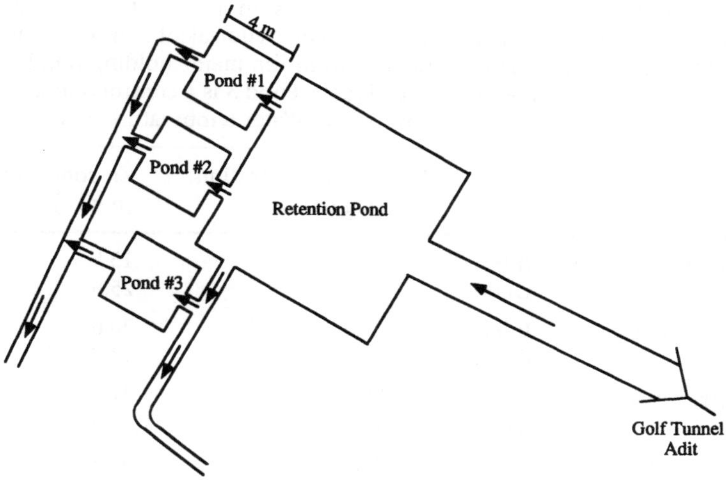

Fig. 19.6. Schematic diagram of the Golf Tunnel mine drainage remediation pilot project, Colorado, USA. The large pond is a retention pond for collecting mine drainage. The three small ponds: *pond 1* with microbial mat; *pond 2* with microbial mat; *pond 3* control, no microbial mat

Table 19.3. Historical Golf Tunnel, Colorado, USA, metal concentrations in mine drainage, and those same metals concentrated by microbial mat

Metal concentration in drainage ($mg\ l^{-1}$):								
Mn	Zn	Ag	Cd	Cu	Cr	Ni	Pb	Fe
3–34	6–43	<0.002	<0.1	<0.3	<0.01	<0.04	<0.04	<3
Metal concentration in microbial mat ($mg\ kg^{-1}$):								
Mn	Zn	Ag	Cd	Cu	Cr	Ni	Pb	Fe
12,050	30,300	14	122	2566	106	32	27,275	118,800

zinc concentrations. In the control pond 3, there was a 4 and 3% decrease in manganese and zinc concentrations, respectively. Table 19.5 summarizes the water quality parameters for the three ponds.

19.4 Mechanisms of Metal Removal

Metals are known to complex with a wide range of organic material, including microorganisms and their organic releases. Dunbabin and Bowmer (1992) identified four dominant binding processes that incorporate metals into organic materials: (1) cation exchange; (2) adsorption; (3)

Table 19.4. Manganese and zinc concentrations in the Golf Tunnel drainage, Colorado, USA, August 1995. Retention pond inlet is equivalent to mine tunnel drainage. Retention pond outlet is the drainage from the major holding pond. Ponds 1 and 2 are those inoculated with microbial mats. Pond 3 is a control pond. All ponds have a limestone substrate and measure 16 m^2. Water temperature was 5 °C

Station		Mn concentration ($mg\ l^{-1}$)	Zn concentration ($mg\ l^{-1}$)
Retention pond	Inlet	13.9	15.5
	Outlet	23.0	25.6
Pond 1 (treatment)	Inlet	25.6	28.6
	Outlet	12.2	13.7
Pond 2 (treatment)	Inlet	28.8	31.8
	Outlet	10.8	11.8
Pond 3 (control)	Inlet	31.8	34.8
	Outlet	30.4	33.6

precipitation and coprecipitation; and (4) complexation or chelation. Although metals that are adsorbed, precipitated or complexed can be released back into solution in an equilibrium response, microbial mat promotes high pH levels to prevent remobilization of metals. Additionally, high dissolved oxygen and redox levels (mediated by the biological component) favour the chemical precipitation of metal oxides and hydroxides. These oxides and hydroxides, in turn, act as reservoirs for additional metal deposit. Other metal removal mechanisms available with microbial mat include flocculation, cell sorption and autocatalysis. For example, laboratory research has shown that specific bioflocculants are released by microbial mat in response to the presence of a positively charged metal ion (Bender and Phillips 1994). These materials carry surface charges ranging from −58.8 to −65.7 mV. The charges changed to +1.8 in the presence of divalent metal, indicating metal binding to the bioflocculant. This provides initial protection to the microbial community contacting the toxic metal. Another example would be that at the community level, the anaerobic zones harbour sulfur-reducing bacteria, which generate hydrogen sulfide in the anoxic zones. Thus, sulfide is available for metal precipitation in the interstitial spaces of the mat. A third example is the mechanism of metal transport through the water. Scanning electron microscopy/microanalysis research, correlated with chemotaxis studies of the motile bacteria in microbial mat, suggests that these microbes become bonded to the metals and migrate to the mat by responding chemotactically to the blue-green algae and silage (Bender et al. 1989).

Table 19.5. Water quality parameters for the Bureau of Mines, Colorado, USA, pilot project, 1994–1995

Pond No.		Temperature, °C (range)	Dissolved oxygen, mg l^{-1} (range)	pH (range)	Oxidation-reduction potential, mV (range)	Alkalinity, mg l^{-1} (range)	Manganese mg l^{-1} (range)	Zinc, mg l^{-1} (range)
1	Inlet	3.0±2.0 (0.9–5.0)	10.4±2.9 (8.4–13.8)	7.3±0.4 (6.7–7.9)	359±156.6 (235–584)	48±4.6 (40–51)	12.1±7.9 (4.0–25.6)	13.2±8.5 (5.9–28.6)
	Outlet	3.5±2.7 (0.6–6.5)	9.8±2.4 (8.1–11.5)	7.3±0.5 (6.8–7.9)	394±168.4 (235–582)	47±4.8 (40–50)	10.3±4.3 (6.1–16.1)	10.9±4.2 (6.4–16.5)
2	Inlet	3.0±2.0 (1.0–5.1)	9.0±0.2 (8.8–9.1)	7.4±0.3 (7.0–7.9)	359±168.9 (215–592)	52±8.3 (40–60)	11.4±8.9 (3.8–28.8)	12.6±9.3 (5.6–31.8)
	Outlet	3.6±3.6 (0.0–9.0)	8.6±2.2 (7.2–10.3)	7.4±0.3 (7.1–7.9)	359±156.5 (235–586)	51±4.1 (45–56)	8.4±4.2 (3.3–13.4)	9.3±3.1 (5.3–13.3)
3	Inlet	3.0±1.9 (1.1–5.0)	9.0±0.4 (8.7–9.2)	7.4±0.4 (7.1–8.1)	353±161.6 (220–590)	45±8.1 (40–58)	12.9±10.3 (3.9–31.8)	14.0±10.9 (5.8–34.8)
	Outlet	3.0±2.3 (0.7–5.5)	8.0±1.3 (7.0–8.9)	7.4±0.4 (7.0–7.9)	353±155.2 (235–590)	45±7.4 (40–56)	12.6±9.7 (4.1–30.4)	14.1±10.2 (6.8–33.6)

19.5
Microbial Mat as an Ideal Bioremediation System

Microbial mat has broad potential in the treatment of mine wastes. Because microbial mat is photosynthetic and nitrogen-fixing, it is generally self-maintained and, after establishment, does not require outside supplies of nutrients. A central issue in microbial biotechnology is retention of the integrity of the biological system by maintaining the optimum populations of inoculated microbes. The mixed microbial consortium of microbial mat possesses a natural system of checks and balances and it should be less difficult to achieve this integrity in a complex microbial mat system than it is in technologies employing single species. Another distinct advantage of a mixed microbial remediation system is that specific detoxification mechanisms unique to constituent strains of microbial mat are accessible to all members of the consortium. Thus, a broader variety of cellular releases (enzymes, bioflocculants) are available within a microbial mat consortium compared with that offered by a single microorganism treatment system. Challenging the complex system of microorganisms in microbial mat will likely result in population shifts favouring the microorganisms most efficient in that particular remediation. These properties of self-maintenance, resiliency and efficiency under fluctuating environmental conditions may resolve a number of maintenance problems often associated with bioremediation technologies. These properties suggest that microbial mats have excellent potential for mine drainage remediation.

Acknowledgements. Support of the United States Environmental Protection Agency Assistance ID CR-818689, the Tennessee Valley Authority Contract TV-89721V and the United States Department of the Interior Bureau of Mines Cooperative Agreement 1432-C0240006 is gratefully acknowledged. J. Neil, United States Geological Survey, conducted the X-ray diffraction analysis of metal precipitates.

References

Bender J, Phillips P (1994) Implementation of microbial mats for bioremediation. In: JL Means, Hinchee RE (eds) Emerging technology for bioremediation of metals. Lewis, Boca Raton, pp 85–98

Bender J, Phillips P (1995) Biotreatment of mine drainage. Mining Environ Manag (UK) 3(3):25–27

Bender J, Graves B, Wright W (1995) Evaluation of the removal of BTEX from groundwater using a microbial mat. Air and Waste Management Association, 88th Annu Meeting and Exhibition, San Antonio, Texas, 18–23 June 1995, 15 pp

Bender JA, Archibold ER, Ibeanusi V, Gould JP (1989) Lead removal from contaminated water by a mixed microbial ecosystem. Water Sci Technol 21:1661–1664
Brodie G (1993) Aerobic constructed wetlands and anoxic limestone drains to treat acid drainage. Constructed wetlands workshop for electric power utilities, Tennessee Valley Authority, Chattanooga, Tennessee, 24–26 Aug 1993, 10 pp
Caumette P (1989) Ecology and general physiology of anoxygenic phototrophic bacteria in benthic environments. In: Cohen Y, Rosenberg E (eds) Microbial mats, physiological ecology of benthic microbial communities. American Society for Microbiology, Washington, DC, pp 283–304
Dunbabin J S, Bowmer KH (1992) Potential use of constructed wetlands for treatment of industrial wastewaters containing metals. Sci Total Environ 111:151–168
Goodroad L, Bender J, Phillips P, Gould J, Saha G, Rodríguez-Eaton S, Vatcharapijarn Y, Lee R, Word J (1994) Potential for bioremediation using constructed mixed microbial mats. Hazardous Materials Control Resources Institute, Washington, DC, 29 Nov–1 Dec 1994
Goodroad L, Bender J, Phillips P, Gould J, Hater G, Burrow B (1995) Use of constructed mixed microbial mats for landfill leachate treatment. 18th Int Madison Waste Conf, Department of Engineering Professional Development, University of Wisconsin, Madison, 20–21 Sept 1995, 13 p
Phillips P, Bender J, Simms R, Rodríguez-Eaton S (1994) Use of microbial mat and green algae for manganese and iron removal from coal mine drainage. 1994 Int Land reclamation and mine drainage Conf and the 3rd Int Conf on Abatement of acidic drainage, Pittsburgh, Pennsylvania, 24–29 April 1994, vol 1, pp 99–108
Phillips P, Bender J, Simms R, Rodríguez-Eaton S (1995) Manganese removal from acid coal mine drainage by a pond containing green algae and microbial mat. Water Sci Technol 31:161–170
Shilo M (1989) The unique characteristics of benthic cyanobacteria. In: Cohen Y, Rosenberg E (eds) Microbial mats, physiological ecology of benthic microbial communities. American Society for Microbiology, Washington, DC, pp 207–213
Stal LJ, Heike H, Bekker S, Villbrandt M, Krumbein WE (1989) Aerobic–anaerobic metabolism in the cyanobacterium *Oscillatoria limosa*. In: Cohen Y, Rosenberg E (eds) Microbial mats: physiological ecology of benthic microbial communities. American Society for Microbiology, Washington, DC, pp 255–276

20 Managing the pH of an Acid Lake by Adding Phosphate Fertiliser

D. G. George[1] *and W. Davison*[2]

[1] Institute of Freshwater Ecology, Windermere, Cumbria, UK
[2] Institute of Environmental and Biological Sciences, Lancaster University, Lancaster, UK

20.1 Introduction

In recent years, it has become clear that a large proportion of the acid neutralising capacity of softwater lakes is generated within such lakes by biological processes (Davison 1986, 1987; Schindler et al. 1986). A number of biological processes can alter the acid–base balance of a lake, but the most important are those associated with the synthesis and decomposition of organic matter. The organic matter produced during photosynthesis is not a simple carbohydrate but includes nitrogen, phosphorus and other elements which are assimilated as nutrients and released during decomposition. Changes in the chemical form of these nutrients affect the acid–base balance and may, under certain circumstances, bring about a sustained increase in the pH. Since most freshwater systems are phosphorus-limited, the most effective way of generating base is to add phosphorus to stimulate the uptake of nitrate by primary producers. The efficiency with which additions of phosphorus generate base depends on a number of factors, such as the flushing rate of the basin, the quantity of nitrate available and the ultimate fate of the organic material produced. The treatment is likely to prove most effective in lakes with a long residence time and relatively high loadings of nitrate nitrogen. Most acidified lakes contain relatively high concentrations of nitrate and only small concentrations of phosphorus are required to counter the effects of the episodic input of acid from the surrounding catchment.

In this chapter we describe the results of a large-scale field experiment designed to test the feasibility of increasing the pH of an acid lake by fertilising with phosphorus. The central objective of the experiment was to develop a system of internal base generation that could be used to buffer the pH of any moderately acid lake. The results presented here cover the critical period of predominantly assimilative base generation where the quantity of base produced can be predicted from mass-balance calculations and the known additions of phosphate. The chapter summarises the most important physical, chemical and biological effects of adding fertiliser and compares the quantity of base generated with the quantity produced by more conventional neutralization strategies.

20.2 Theoretical Background

The long-term acid–base balance of a lake (as expressed by the mean annual alkalinity) is quite distinct from the instantaneously measured pH which is influenced by the rate at which CO_2 is removed by photosynthesis. A number of biologically mediated reactions can consume or release hydrogen ions and so affect the alkalinity of a water body. Davison (1987) has reviewed the chemical processes responsible for this internal buffering and demonstrated that the most important transformations are those associated with the assimilation and dissimilation of nitrate nitrogen. Figure 20.1 shows the two base-generating mechanisms that are stimulated by the regular additions of phosphate fertiliser. In the initial stages of treatment, the key base-generating process is the assimilative uptake of nitrate. When nitrate is assimilated it generates base (consumes acid) according to Eq. (1):

$$106CO_2 + 138H_2O + 16NO_3^- = (CH_2O)_{106}(NH_3)_{16} + 16OH^- + 138O_2 . \quad (1)$$

If the organic matter produced during this process decays in the water column (i.e. in the presence of oxygen), there is no overall change in the acid–base balance. In contrast, if most of the additional production sinks to the bottom and is buried in the sediment, more base will be generated during its anoxic decomposition. The quantity of base generated by these assimilative and dissimilative processes depends on a number of factors, including the flushing rate of the basin, the rate of incorporation into the sediment and gaseous transfers to the atmosphere. In the initial stages of fertilisation, the only process of any significance is the assimilative uptake of nitrate, but dissimilative processes become increasingly important as the decomposition of organic matter outstrips the supply of oxygen to the surface sediments.

20.3 Description of Site

Seathwaite Tarn, a deep, unproductive upland lake, was chosen as the site for the experiment because it is one of a small number of moderately acid lakes in the English Lake District. Figure 20.2 shows the acid–base status

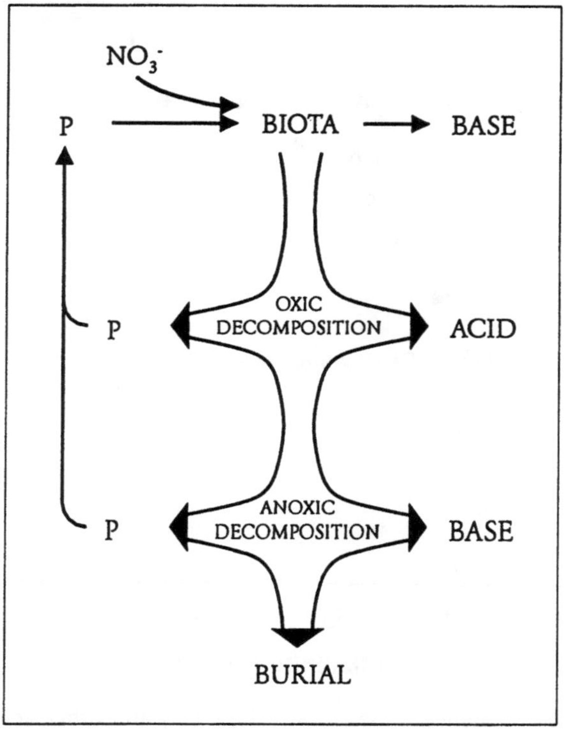

Fig. 20.1. Schematic diagram showing the two base-generating mechanisms that are stimulated by regular additions of phosphorus

of the lake when compared with other acid and circumneutral waters in the English Lake District. These data are taken from a survey of more than 200 lakes and tarns (Carrick and Sutcliffe 1982) and demonstrate that a relatively small amount of base generated within the lake would have a major effect on the measured pH. Seathwaite Tarn (Fig. 20.3) has a surface area of 27 ha, a mean depth of 12 m, a maximum depth of 26 m and typically contains over 3,000,000 m^3 water. The lake was formed by impounding the outlet from a natural tarn and is now managed by North West Water PLC as a source of compensation water for the River Duddon. Water level fluctuations are generally low, but reductions of 2–6 m are occasionally recorded during dry periods in summer. The lake has one main inflow (Tarn Head Beck) which drains 64% of the surrounding catchment and a number of minor inflows that are difficult to monitor. The land surrounding the lake is predominantly rough pasture and moorland with isolated areas of scree and bare rock. There are local deposits of sphagnum peat in the valley above the lake and a thin layer of

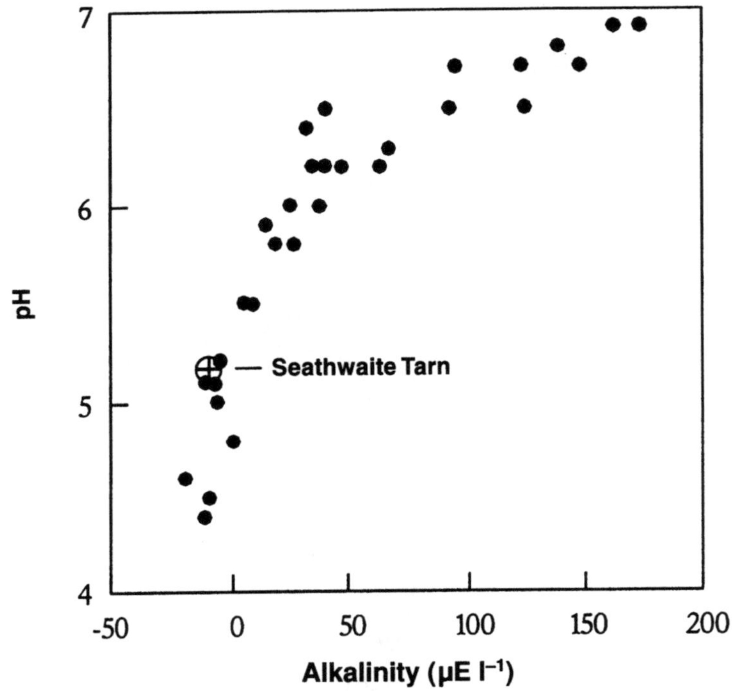

Fig. 20.2. Acid-base balance of Seathwaite Tarn (⊕) and other acid and circum-neutral lakes in the English Lake District

blanket peat covers a large proportion of the catchment area. The lake is consequently very unproductive and supports a very low biomass of phytoplankton and zooplankton. The maximum summer biomass of phytoplankton seldom exceeds 2 μg l^{-1} and the growing season is short, typically extending from the first week in June to the end of October.

20.4 Methods

20.4.1 Design of the Experiment

The first phase of the experiment was designed to cover a 3-year period with one pre-treatment year followed by 2 years of regular fertilisation. Intensive monitoring started in January 1991 and the first dose of fertiliser was added in March 1992. The fertiliser used was a concentrated solution of sodium phosphate (50 g P l^{-1}) specially formulated by

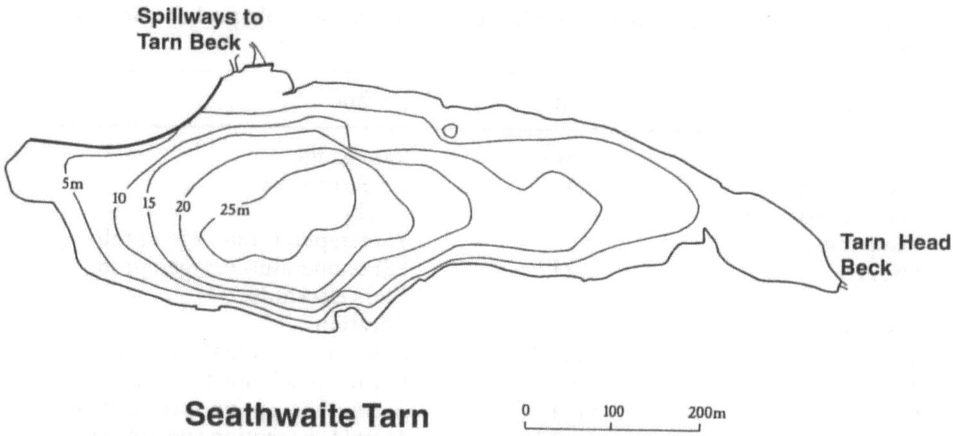

Fig. 20.3. Bathymetric map of Seathwaite Tarn

Albright and Wilson PLC, an international chemical company. The fertiliser solution was introduced directly into the lake using a simple gravity feed system installed in a fast-moving boat. Approximately 75 l was added every fortnight in 1992 and the dosing rate increased by 50% in the summer of 1993. This treatment was estimated to produce an average concentration of phosphate of 20 $\mu g\ l^{-1}$ in 1992 and 28 $\mu g\ l^{-1}$ in 1993, assuming no uptake by biota. The experiment, one of the most ambitious field studies ever undertaken by the Institute of Freshwater Ecology, was designed as a multidisciplinary venture with teams of specialists working on the plant and animal communities in the littoral zone as well as the open water. In this chapter we describe the general pattern of change in the water chemistry and the open water biota. Other papers in press (Edwards et al. 1993; May 1995) provide more detailed accounts of the flux of heavy metals and the seasonal succession of the planktonic rotifers.

20.4.2
The Monitoring Programme

Daily meteorological measurements were obtained from North West Water PLC who maintain an automatic weather station at a treatment plant 3 km below the lake. Automatic recording instruments in the main inflow monitored the hourly variations in stream flow, air temperature, water temperature and pH, and a level recorder on the dam measured the hourly change in lake volume. Paired combination pH electrodes (Davison and Harbison 1988) were used at each location and calibrated at fortnightly intervals using buffer solutions equilibrated in the field to

Table 20.1. Methods used to record physical, chemical and biological characteristics of the lake

Variable	Units	Method
Rainfall	mm	Rain gauge
Lake volume	m3	Level recorder
Water transparency	m	Secchi disc
Temperature	°C	YSI temperature/oxygen probe
Oxygen	mg l^{-1}	YSI temperature/oxygen probe
pH		Glass electrode
Alkalinity	µE l^{-1}	Titration
Total CO_2	µM l^{-1}	Gas-liquid chromatography
Al	µM	Pyrocatechol violet
Ca	mg l^{-1}	Atomic absorption spectroscopy
Mg	mg l^{-1}	Atomic absorption spectroscopy
Na	mg l^{-1}	Atomic absorption spectroscopy
K	mg l^{-1}	Atomic absorption spectroscopy
SO_4	µg l^{-1}	Ion chromatography
Cl	µE l^{-1}	Ion chromatography
NO_3	µE l^{-1}	Ion chromatography
DRP	µg l^{-1}	Molybdenum blue
Total P	µg l^{-1}	Molybdenum blue
DOC	mg l^{-1}	Carlo Erba
Chlorophyll *a*	µg l^{-1}	Methanol extraction/665 nm
Phytoplankton counts	Cells ml^{-1}	7-m tube
Zooplankton counts	Numbers l^{-1}	Conical net (120-mm mesh)

YSI, Yellow Springs Instruments; DRP, dissolved reactive phosphorus; DOC, dissolved organic carbon.

match the stream temperature. Integrated water samples were collected from the main inflow and outflow at fortnightly intervals using flow-proportional samplers mounted on the stream bed. The physical, chemical and biological characteristics of the lake were monitored at 1- to 2-week intervals throughout the experiment. Chemical samples were collected at fortnightly intervals throughout the year and biological samples at fortnightly intervals from November to May and weekly intervals during the summer. Table 20.1 lists the key variables covered by this monitoring programme and notes some of the methods used to analyse the samples. In-situ measurements of water transparency were recorded using a white Secchi disc mounted on a graduated line. Depth-specific measurements of water temperature and dissolved oxygen were recorded using a Yellow Springs Instruments oxygen probe fitted with a magnetically driven stirrer. Samples of water for chemical analyses and phytoplankton counts were collected from the deepest point using a 7-m length of weighted plastic tubing (Lund and Talling 1957). Zooplankton samples were collected by hauling a conical net through the water column at a series of fixed stations. In winter, two replicate samples were collected at a central site. At other times, samples were collected from five or more stations positioned so that the average length of haul was approximately equal to the mean depth of the lake. The net was made of stainless steel mesh (120-μ aperture) and was fitted with a mouth-reducing cone to increase the efficiency of filtration. The animals collected were killed in the field by adding a few drops of alcohol and more alcohol was added before the samples were stored for counting. The samples were counted in a circular Perspex trough fitted to the stage of a stereo-zoom microscope. The numbers of attached and free eggs were also counted and the body lengths of the dominant species measured with an eyepiece micrometer. These body length measurements were later used to calculate the daily grazing rate of the dominant herbivore using depth-specific measurements of water temperature and published estimates of individual clearance rates (Knoechel and Holtby 1986).

20.4.3 Budget Calculations

The significance of the observed changes in lake chemistry were assessed by constructing mass balance budgets for the lake from samples collected from the main inflow, the lake and the outflow. The samples from the lake were collected with the 7-m-long tube and those from the inflow and outflow with the flow-proportional sampler. The main inflow supplies 64% of the water reaching the lake. The contribution from the other inflows was estimated on a pro rata basis on the assumption that their chemical composition was the same as the main inflow. Periodic checks demonstrated that the chemistry of these minor inflows was very similar to the main stream in the early part of the year, but became more variable during the growing season. Chemical budgets were constructed for successive 2-week periods by assuming that the lake was instantaneously mixed:

$$M_{end} = M_{st} + M_{in} + M_{pn} - M_{out},$$

where:

- M_{end} is the mass in the lake at the end of the period,
- M_{st} is the mass in the lake at the beginning of the period,
- M_{in} is the mass entering the lake through the inflows,
- M_{pn} is the mass entering the lake through deposition,
- M_{out} is the mass leaving the lake through the outflow.

Since the discharge from the lake could not be monitored accurately, the outflow volume was calculated from the inflow and the lake level measurements taken at the beginning and end of the period. The predicted lakewater concentration at the end of each fortnight was then used as the initial concentration in the calculation for the following period. This procedure allowed the concentration of any component entering and leaving the lake to be predicted for the full 3-year period from the hydrological data and the concentrations in the inflow.

20.4.4
Modelling Changes in pH

Modelling the changes in pH is not as straightforward as for the other elements in the mass balance calculation since the pH electrode provides an estimate of the H^+ ion which interacts with both the aluminium and carbon species in the water. Changes in the acidity status of the lake can therefore only be properly assessed when variations in the concentration of Al, organic acid and dissolved CO_2 are considered along with the hydrogen ion activity. In the modelling part of this study, the catchment-induced variations in the pH of the tarn have been predicted using an acidity function, A_f, that represents the total concentration of acid species:

$$A_f = [H^+] + [Al(OH)^{2+}] + 2[Al^{3+}] - [Al(OH)_3] - [Al(OH)^{4-}] \qquad (2)$$
$$- [HCO^{3-}] - 2[CO^{2-}] - [OH^-] + 3[DOC],$$

where DOC stands for dissolved organic carbon.

The acidity function was calculated from the measured temperature and pH, and the concentrations of aluminium, carbon dioxide and dissolved organic carbon. The function was then used in the mass balance calculations in the same way as other components and the resulting acidity converted to a predicted pH using the measured values of total CO_2 and Al at the end of each fortnight.

20.5
Results

20.5.1
Physical Characteristics of the Lake

Seathwaite Tarn is a deep valley lake that usually remains thermally stratified from the beginning of June to the end of September. Year-to-

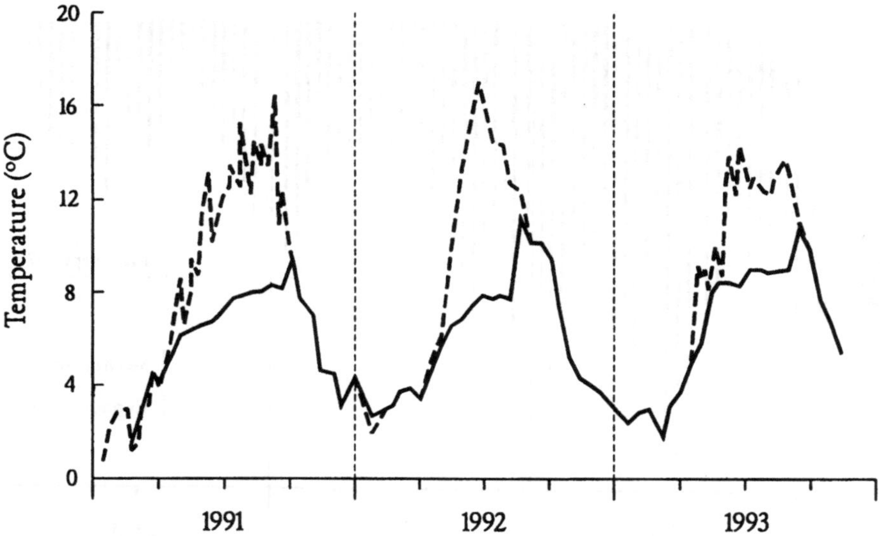

Fig. 20.4. Seasonal variation in surface (*broken lines*) and bottom (*solid lines*) temperatures recorded in the lake in 1991, 1992 and 1993

year variations in the timing and intensity of thermal stratification do, however, have a major effect on the seasonal succession of plankton and can influence the rate at which this organic production accumulates in the deep sediment. The time series graphs in Fig. 20.4 show the surface and bottom temperatures recorded in the lake in 1991, 1992 and 1993. In 1991, the weather in early summer was unsettled, but the lake warmed up rapidly in July and remained thermally stratified until the end of September. In 1992, the lake stratified rather earlier, but strong winds in mid-summer depressed the seasonal thermocline and the water column was completely mixed by the end of August. In 1993, the lake became thermally stratified in early May and remained stably stratified until the end of September.

The time series graphs in Fig. 20.5 show the seasonal variation in the transparency of the water column in 1991, 1992 and 1993. The water in the untreated lake was very clear, but the growth of phytoplankton had a pronounced effect on the vertical penetration of light in the fertilised lake. The solid lines in Fig. 20.5 show the Secchi disc depths recorded in 1991, 1992 and 1993. In 1991, the average Secchi depth was 12.1 m, but this decreased to 7.5 m in 1992 and 4.6 m in 1993. The broken lines in Fig. 20.5 show the depth at which the underwater light intensity was attenuated to 1% of its surface value. This depth is commonly referred to as the euphotic depth and provides a better measure of the area of lake bed that

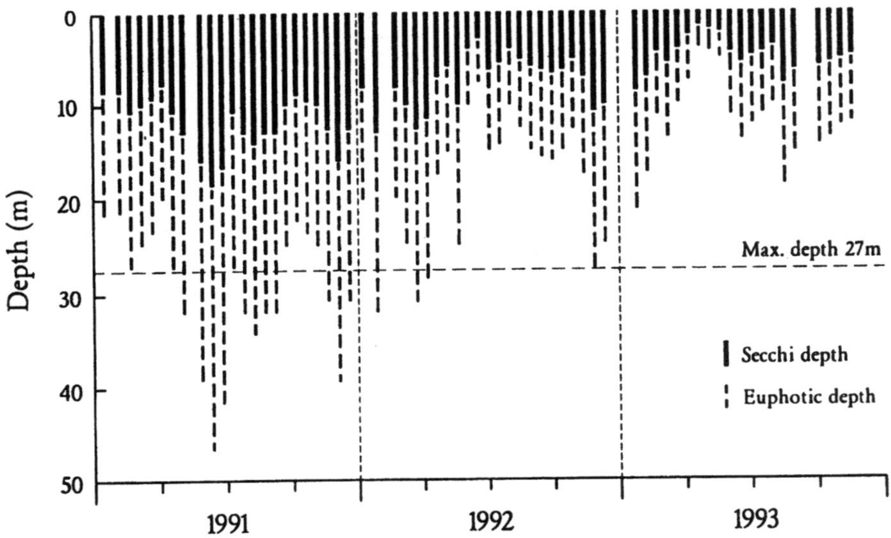

Fig. 20.5. Seasonal variation in Secchi disc depth and euphotic depth in 1991, 1992 and 1993

could be colonised by photosynthetic organisms. The euphotic depths were calculated by multiplying the individual Secchi disc measurements by a factor of 2.5 (Talling, pers. comm.). In 1991, the mean euphotic depth was 29 m, but this fell to 18 m in 1992 and 11 m in 1993.

20.5.2
Chemical Characteristics of the Lake

Figure 20.6a shows the seasonal variation in the concentration of dissolved reactive phosphorus (DRP) recorded in the lake in 1991, 1992 and 1993. In the untreated lake, there was no measurable DRP, but concentrations of 1–2 µg DRP l^{-1} were periodically recorded in 1992 and relatively large amounts of DRP were recorded in September 1993 when the growth of the phytoplankton was temporarily limited by the supply of nitrate nitrogen. Figure 20.6b compares the actual nitrate concentrations recorded in the lake in 1991, 1992 and 1993 with the concentrations predicted by the mass balance calculations. In 1991, the measured and predicted concentrations of nitrate were very similar for most of the year, but the measured concentrations were rather higher than the predicted concentrations in late summer and early autumn. In 1992 and 1993, the measured concentration of nitrate in the lake was consistently lower than that predicted from the composition of the main inflow. The 'missing' nitrate was,

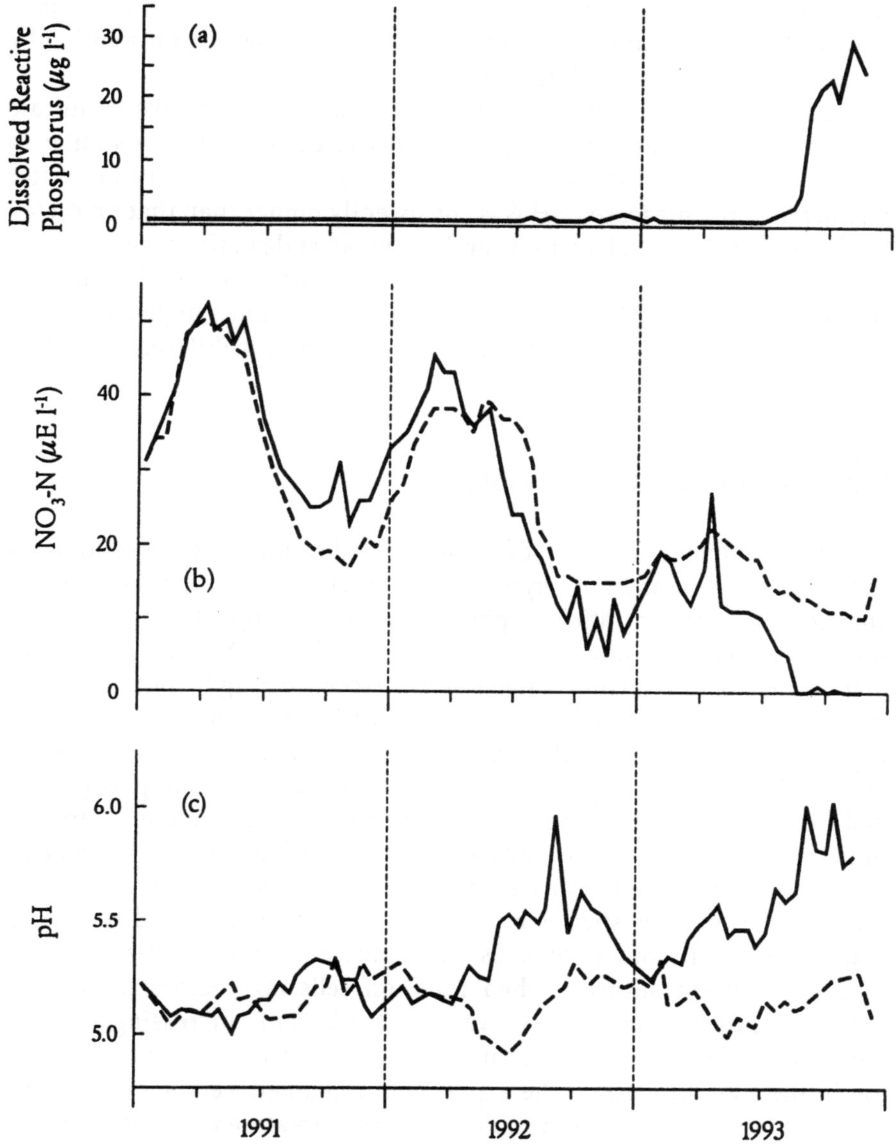

Fig. 20.6. **a** Seasonal variation in the concentration of dissolved reactive phosphorus in 1991, 1992 and 1993. **b** Seasonal variation in the actual (*solid lines*) and 'baseline' (*broken lines*) concentration of nitrate nitrogen in 1991, 1992 and 1993. Baseline concentrations have been estimated by mass balance calculations on inflow and outflow and assume no assimilative uptake of nitrate nitrogen. **c** Seasonal variation in the actual (*solid lines*) and 'baseline' (*broken lines*) pH in 1991, 1992 and 1993. Baseline pH values have been estimated by mass balance and assume that no base is being generated within the lake by biological processes

however, stoichiometrically consistent with the amount of phosphate added, clearly indicating that nitrate was being removed from solution by assimilative uptake according to Eq. (1).

Figure 20.6c shows the effect that this assimilative uptake of nitrate had on the pH of the lake. In 1991, the measured pH was very similar to that predicted from the mass balance of inflow concentrations. After fertilisation, the measured pH was consistently higher than that predicted by the mass balance calculations and the greatest deviations were recorded in summer following periods of rapid phytoplankton growth. In 1992, the average summer (July to October) pH was 0.31 units higher than that predicted from the mass balance calculations, but this difference increased to 0.49 units in 1993.

20.5.3 Biological Characteristics of the Lake

The most immediate effect of regular fertilisation was a pronounced increase in the biomass of phytoplankton. Figure 20.7 shows the seasonal change in the concentration of phytoplankton chlorophyll *a* measured in the lake in 1991, 1992 and 1993. In the untreated lake, the maximum summer concentration of chlorophyll seldom exceeded 2 $\mu g\ l^{-1}$, but this maximum reached 9 $\mu g\ l^{-1}$ in 1992 and 36 $\mu g\ l^{-1}$ in 1993. The annotations in Fig. 20.7 show the seasonal succession of dominant genera. Despite the pronounced increase in productivity, the regular additions of phosphorus had little effect on the qualitative composition of the phytoplankton. In 1991, the phytoplankton crop was dominated by *Chlorella* (cf. *ellipsoidea*), which reached a maximum density of 2200 cells ml^{-1} at the end of the growing season. This species remained dominant in 1992 and reached a maximum density of 43,000 cells ml^{-1} at the end of September. In 1993, *Chlorella* concentrations reached 1 million cells ml^{-1} before the end of May, but another green alga (*Cryptomonas*) became increasingly dominant towards the end of the summer. The concentrations of phytoplankton in the lake throughout the experiment remained well below the concentrations judged to cause problems in water treatment plants. The first summer maximum recorded in 1993 did, however, exceed the maximum summer concentrations agreed by the water company and the late summer crop could well have reached similar densities had it not been grazed by the large numbers of microcrustacea that appeared in July.

The crustacean zooplankton that live in the open water form an important link in the food chain between the phytoplankton and the fish. In oligotrophic lakes, the microcrustacea also consume a significant proportion of the phytoplankton and play a key role in the recycling of

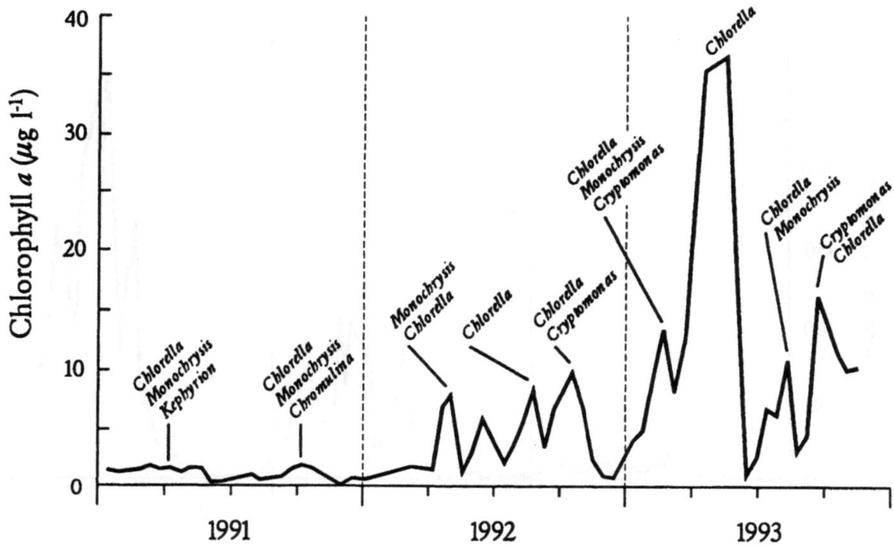

Fig. 20.7. Seasonal variation of phytoplankton chlorophyll *a* in 1991, 1992, and 1993. *Annotations* show seasonal succession of dominant genera

nutrients. Two species of microcrustacea (*Bosmina coregoni* and *Eudiaptomus gracilis*) dominated the zooplankton community in Seathwaite Tarn throughout the period of the experiment. The most abundant species was *Bosmina coregoni*, a cladoceran species that is widely distributed in the less productive lakes in the English Lake District. Figure 20.8a shows the seasonal change in the numbers of *Bosmina* in 1991, 1992 and 1993. In 1991, the numbers of *Bosmina* present in the open water ranged between 2 and 15 individuals l^{-1}. In 1992, densities of 20–100 individuals l^{-1} were frequently recorded in summer and this range increased to 50–280 individuals l^{-1} in 1993. The number of eggs carried by a female *Bosmina* is strongly influenced by the quality as well as the quantity of food consumed. The small flagellates that dominated the phytoplankton in Seathwaite are, however, easy to ingest and form an almost ideal food source for the filter-feeding *Bosmina*. The time series plots in Fig. 20.8b show that there has been a significant increase in the number of eggs laid by the *Bosmina* since the lake was fertilised. In 1991, the average number of eggs laid by a gravid female was 1.1, but this increased to 1.2 eggs per female in 1992 and 1.4 eggs per female in 1993.

Secondary production increases of such magnitude clearly have a pronounced effect on the rate at which organic material is lost from the water column and accumulates in the deep sediment. Figure 20.9 shows the seasonal variation in the quantity of water filtered by the *Bosmina* in

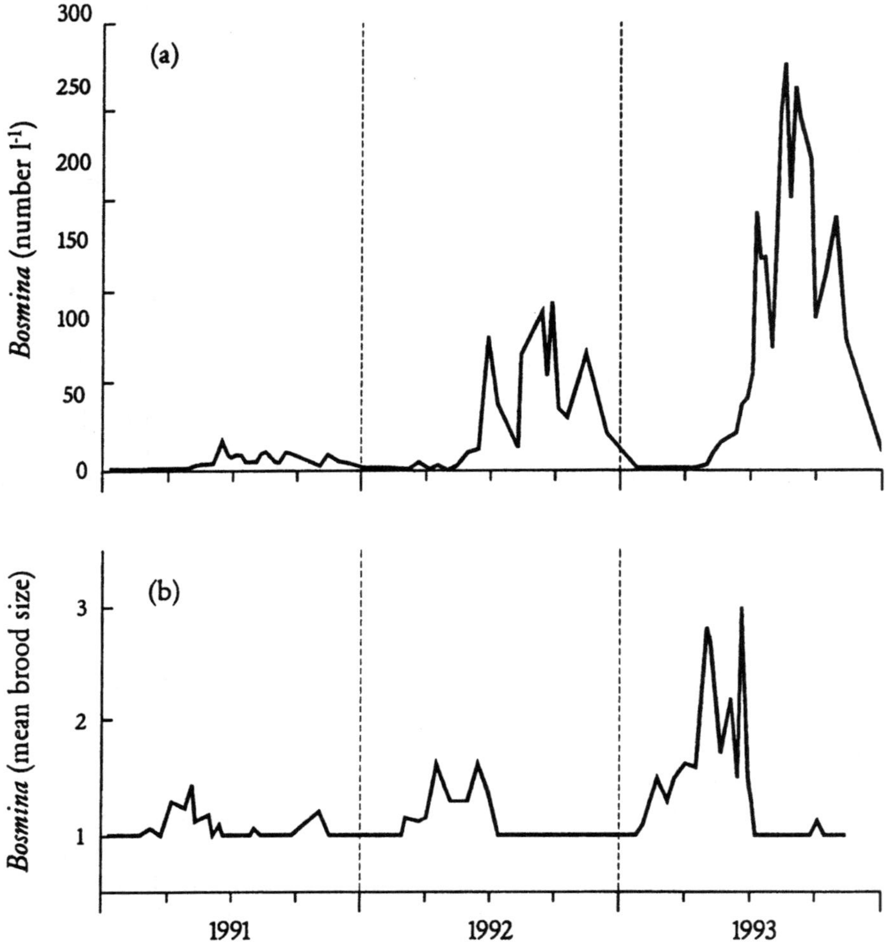

Fig. 20.8. **a** Seasonal variation in the numbers of *Bosmina* in 1991, 1992 and 1993. **b** Seasonal variation in the mean brood size of *Bosmina* in 19912, 1992 and 1993

1991, 1992 and 1993. The filtration rates in the figure have been derived by combining the temperature-specific observations of Knoechel and Holtby (1986) with the temperature-dependent filtration curves derived by Jones et al. (1979). The plotted rates are population averages calculated for a notional 1-l sample of lake water, i.e. a filtration rate of 250 ml day^{-1} implies that the animals were notionally able to filter 25% of the lake volume in 24 h. These average filtration rates are not particularly high, but the local impact of grazing could be much higher in late summer when the *Bosmina* and the flagellates are both concentrated in the top 10–15 m of water.

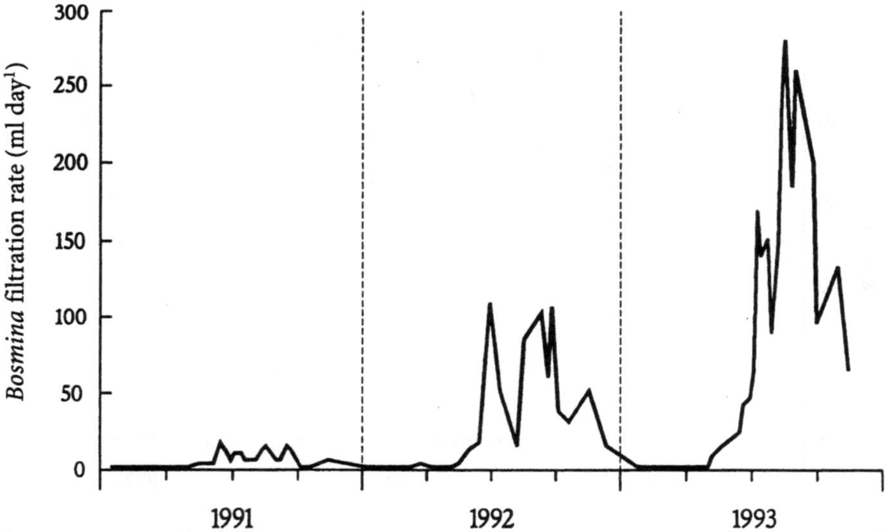

Fig. 20.9. Seasonal variation in the estimated filtration rate of *Bosmina* population in 1991, 1992, and 1993

20.5.4
Dissimilative Production of Base in Deep Sediment

In the first phase of this study, we have only considered the generation of base by the assimilation of nitrate nitrogen from the water column. Our calculations of base generation have not taken into account the fact that a significant proportion of the organic matter produced during the summer sinks into the hypolimnion and will ultimately decompose anoxically in the sediment. In the longer term, we expect the supply of O_2 to the sediment to become increasingly limiting, stimulating the dissimilative production of additional quantities of base. The time series plots in Fig. 20.10 show that the regular additions of phosphorus have already had a marked effect on the amount of oxygen consumed in deep water, i.e. the summer oxygen deficit. The quantity of oxygen consumed in the hypolimnion during the period of thermal stratification is a good measure of the rate at which organic matter accumulates in the deep sediment. In 1991, the late summer oxygen deficit was about 45% of the early summer maximum, but this increased to 52% in 1992 and 63% in 1993.

In order to estimate the net amount of base produced by a combination of assimilative and dissimilative processes, it is necessary to consider

Fig. 20.10. Seasonal variation in the concentration of oxygen measured approximately 1 m above the deep sediment in 1991, 1992 and 1993

the ultimate fate of the organic matter synthesised (Davison 1987, 1990). There are few quantitative studies of the pathways of organic matter in lakes, but in one productive system it was found that 25% was washed from the lake, 25% was buried in sediments, 32% decomposed anaerobically, 13% was dissimilatively reduced to N_2 and 5% was reduced to form insoluble iron sulphide (Jones and Simon 1980). In Fig. 20.11 these estimates have been used to predict how much base might be produced in a system where phosphate has been added for some time and resulted in significant anoxic decomposition. The figure shows how much base would be generated when 1 µmol DRP l^{-1} (31 µg DRP l^{-1}) is added to a deep lake where a high proportion of the organic matter produced decomposes anoxically in the deep sediment. We suspect that about 30% of the base produced by the assimilative uptake of nitrate is lost when the phytoplankton sink and decompose in the aerobic part of the water column. Some organic matter is also lost through the outflow, but a significant amount is buried in the deep sediment and generates additional amounts of base by dissimilative reduction. The estimates in Fig. 20.10 are relatively conservative and assume that only 50% of the added P would be released from decomposing organic material. If the recycling process is 100% efficient, 1 µmol P l^{-1} would generate 42 µmol base l^{-1} according to the same scheme.

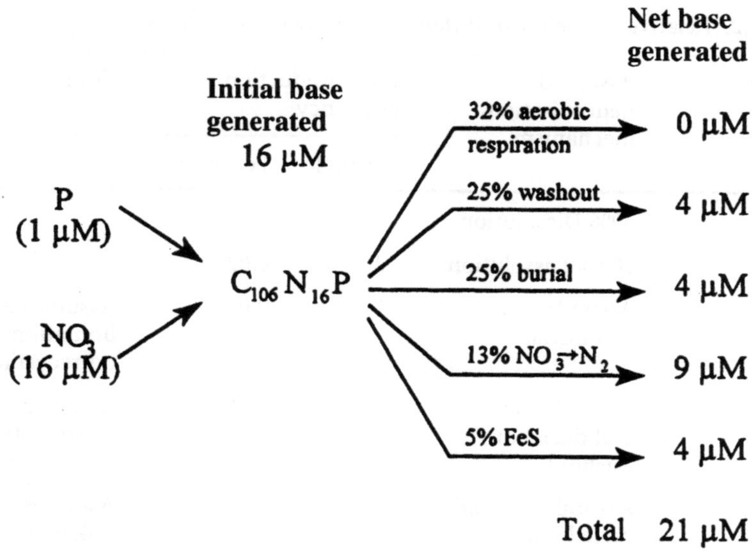

Fig. 20.11. Schematic diagram showing amount of base generated by different physical and biological processes following the addition of 1 µM P l^{-1}

20.6 Discussion

In recent years, considerable effort has been directed to the treatment of acid waters to improve water quality for fisheries and habitat conservation. In most instances this has involved the application of limestone in some form, either directly to the lake (Dickson 1988; Alenas et al. 1990) or to the soils of the catchment area (Rosseland and Hindar 1988; Howells and Dalziel 1992). Liming is not, however, a permanent solution to the restoration of acid waters since the treatment has to be repeated at regular intervals and produces calcium-rich waters that support biological communities that are unlike those found in natural softwater lakes. In the Seathwaite experiment we have demonstrated that it is possible to generate sufficient base by adding phosphate to raise the pH of an acid lake without drastically altering the community structure. In the first phase of this experiment, the principal mechanism of base generation was the assimilative uptake of nitrate nitrogen. In the second phase of the experiment (now in progress), we expect additional quantities of base to be generated by the dissimilative reduction of organic material in the sediments. We have also demonstrated that the phytoplankton growth required to bring about these changes was modest and readily controlled by the

Table 20.2. Relative efficiency of different neutralising strategies

Treatment	Assumed neutralisation mechanism	Base production efficiency		Comments
		Per mole	Per gram	
$CaCO_3$	50% Dissolution	1	0.01	
$NaCO_3$	100% Dissolution	2	0.02	
$NaNO_3$	Assimilation and burial	1	0.01	Assumes N limitation and no recycling
Na_2HPO_4	Assimilation and burial or washout	16	0.11	Assumes no dissimilative base generation
Na_2HPO_4	Assimilation and varied fate of organic material	21	0.15	Assumes assimilation of nitrate
Na_2HPO_4	As above, but with recycling of P	42	0.30	Assumes all P released by decomposition
Na_2HPO_4	Assimilation and anoxic decomposition to form FeS	94	0.66	Assumes no oxic decomposition or washout of organic material

very high concentrations of filter-feeding crustacea that appeared in the fertilised lake. Several authors have already demonstrated that it is possible to combat the effects of acidity by adding nutrients (Yan and Lafrance 1984; Schindler et al. 1985), but they used nitrogen compounds or combinations of nitrogen and phosphorus at dosages that gave rise to unacceptable biological changes. Table 20.2 compares the base-generating capacity of these contrasting fertiliser treatments with the quantity of base produced by treating a lake with lime. These calculations suggest that treating a lake with nitrate produces much the same quantity of base as that generated by conventional liming. In contrast, adding phosphate can generate between 11 and 66 times more base than adding the same weight of calcium carbonate. The quantity of base generated by the regular addition of phosphate is, however, critically dependent on the ultimate fate of the organic matter produced. If all the phosphate added is locked into organic production that sinks to the bottom and decomposes anoxically, 94 mol base could be generated from each mole of phosphorus added. If a significant proportion of the organic matter produced decomposes aerobically, less base would be generated, but most lakes should still be able to produce between 21 and 42 mol base for each mole of phosphorus added. A major advantage of the treatment is the very small quantity of fertiliser required and the ease with which the phosphate solution can be transported to remote locations. For the treatment described here, a total of 5.9 m^3 phosphate solution was applied over a 2-year period. To bring about the same effect using lime, more than 34 t calcium carbonate would have had to be added.

Acknowledgements. A number of staff and students from the Institute of Freshwater Ecology have contributed to the success of this project. Special thanks are due to Dr. C.S. Reynolds, Dr. E.W. Tipping, Ms. D.P. Hewitt, Mr. G.H.M. Jaworski, Mr. B. James, Mr. B.M. Simon and Mr. N.J.A. Edwards (University of Lancaster). The project was supported by the Natural Environment Research Council, the National Rivers Authority, Albright and Wilson, the Department of the Environment and National Power/Powergen.

References

Alenas I, Anderson IB, Hultberg H, Romarin A (1990) Liming and reacidification reactions of a forest lake ecosystem, Lake Lysevatten, in SW Sweden. Water Air Soil Pollut 59:55–57

Carrick TR, Sutcliffe DW (1982) Concentration of major ions in lakes and tarns of the English Lake District (1953–1978). Occas Publ Freshwat Biol Assoc 16:168

Davison W (1986) Sewage sludge as an acidity filter for groundwater-fed lakes. Nature 322:820–822

Davison W (1987) Internal element cycles affecting the long-term alkalinity status of lakes: implications for lake restoration. Schweiz Z Hydrol 49:186–201

Davison W (1990) Treatment of acid waters by inorganic bases, fertilizers and organic material. Trans Inst Min Metall 99:A153–A157

Davison W, Harbison TR (1988). Performance testing of pH electrodes suitable for low ionic strength solutions. Analyst 113:709–713

Dickson W (ed) (1988) Liming of Lake Gardsjon, an acidified lake in SW Sweden. Rep no 3426. National Swedish Environmental Protection Board, 327 pp

Edwards NJA, Davison W, George DG, Hamilton-Taylor J (1993) Trace metal dynamics and fluxes in an acid lake and stream. In: Allen RJ, Nriagu JO (eds) Heavy metals in the environment, vol 2. CEP Consultants Ltd, Edinburgh, pp 363–366

Howells G, Dalziel TRK (1992) Restoring acid waters: Loch Fleet 1984–1990. Elsevier, Amsterdam, 421 pp

Jones HR, Lack TJ, Jones CS (1979) Population dynamics and production of *Daphnia hyalina* var. *lacustris* in Farmoor I, a shallow eutrophic reservoir. J Plankton Res 1:45–65

Jones JG, Simon BM (1980) Decomposition processes in the profundal region of Blelham Tarn and the Lund Tubes. J Ecol 68:493–512
Knoechel R, Holtby LB (1986) Construction and validation of a body-length based model for the prediction of cladoceran community filtration rates. Limnol Oceanog 31:1–16
Lund JWG, Talling JF (1957) Botanical limnological methods with special reference to the algae. Botan Rev 23:489–583
May L (1995) The effect of lake fertilisation on the rotifers of SeathwaiteTarn, an acidified lake in the English Lake District. Hydrobiologia 313/314:333–340
Rosseland BO, Hindar A (1988) Liming of lakes, rivers and catchments in Norway. Water Air Soil Pollut 41:165–188
Schindler DW (1985) The coupling of elemental cycles by organisms: evidence from whole-lake chemical perturbations. In: Stumm W (ed) Chemical processes in lakes. Wiley, New York
Schindler DW, Turner MA, Hesslein RH (1985) Acidification and alkalinization of lakes by experimental addition of nitrogen compounds. Biogeochemistry 1:117–133
Schindler DW, Turner MA, Stainton MP, Lindsey GA (1986) Natural sources of acid neutralizing capacity in low alkalinity lakes of the Precambrian Shield. Science 232:844–847
Yan ND, Lafrance C (1984) Responses of acidic and neutralized lakes near Sudbury, Ontario, to nutrient enrichment. In: Nriagu J (ed) Environmental impacts of smelters. Wiley, New York

21 Model Calculations for Active Treatment of Acidic, Iron-Containing Lake Water Assuming Mining Lakes as Chemical Reactors

R. Fischer[1], *T. Guderitz*[2] *and H. Reißig*[3]

[1] Dresden University of Technology, Institute of Water Chemistry and Chemical Water Technology, Mommsenstraße 13, 01062 Dresden, Germany
[2] IDUS GmbH, Biological Analytical Environmental Laboratory, Dresdner Straße 43, 01458 Ottendorf-Okrilla, Germany
[3] Arltstraße 10, 01189 Dresden, Germany

21.1 Introduction

In the Lusatian area of former East Germany, lignite surface mining left numerous mining lakes, the utilization of which for recreation, fishing and drinking water supply or for nature conservation is considerably impaired and does not fulfil the threshold values of the EG (Europäischen Gemeinschaften) guidelines for water quality with respect to ammonium, pH value and heavy metals (Cu, Zn) (EG 76/160; EG 78/659). The retention time of the water in the lakes is generally greater than 5 years. We intend to develop and to test ecologically and economically justifiable techniques for the restoration of acidic mining lakes which are highly loaded with iron. The pH level of these waters has to be increased to 5.5–7 and the iron has to be removed to a great extent. Thus, it is necessary to develop control mechanisms and to substantiate them by laboratory tests in order to obtain practicable restoration methods.

21.2 Direct Neutralization by the Addition of Alkaline

21.2.1 Chemical Fundamentals

The acidity of the mining lake water has three main components:

- The mineral acidity.
- The latent acid fraction, generated by hydroxo complexation or Al precipitation as gibbsite.

- The acid fraction, generated by hydrolysis of iron(III)-species and the precipitation as ferric hydroxide.

As a result, the base neutralizing capacity (BNC) of the solution is defined, taking into consideration the reference states HCO_3^-, $Al(OH)_3$ and $Fe(OH)_3$, according to Eq. (1):

$$[BNC] = [H^+] + [CO_2H_2O] - [CO_3^{2-}] + 3[Al^{3+}] + 2[AlOH^{2+}] + [Al(OH)_2^+] + 3[Fe^{3+}] + 2[FeOH^{2+}] + [Fe(OH)_2^+]. \tag{1}$$

This equation is only valid if iron(III) is in excess in the mining lake water. However, because iron(II) is present due to the continual weathering of pyrite and marcasite, the restoration of the natural groundwater level after mining and the thermal stratification of the water body, the BNC calculation has to include the alkalized fraction.

$$4Fe^{2+} + O_2 + 4H^+ \rightarrow 4Fe^{3+} + 2H_2O. \tag{2}$$

$$[BNC]^* = [BNC] + 2[Fe^{2+}]. \tag{3}$$

The acidic portion of the silicic acid need not be taken into consideration because the silicic acid as H_4SiO_4 or $Si(OH)_4$ in the mining lake water remains unchanged during neutralization up to pH 7 [Eqs. (4) and (5)]:

$$Si(OH)_4 + H_2O \rightleftharpoons SiO(OH)_3^- + H_3O^+ \qquad pKs_1 = 9.5; \tag{4}$$

$$SiO(OH)_3^- + H_2O \rightleftharpoons SiO_2(OH)_2^{2-} + H_3O^+ \qquad pKs_2 = 12.6, \tag{5}$$

where pKs_1 and pKs_2 stand for the first and second acidity constant, respectively.

Knowledge about the BNC of the mining lake is of great importance in determining the alkali demand of the mining lake water. The buffer intensity of the mining lake, as the gradient at each point of the titration curve, is described as follows:

$$\beta = -\frac{d\,[BNC]}{d\,pH} \tag{6}$$

The BNC of the mining water is the integral of the buffer intensity over the pH range 1–2 and represents a conservative parameter (Stumm and Morgan 1981) that is not affected by temperature and pressure.

$$[BNC] = -\int_{pH_1}^{pH_2} \beta\, dpH \tag{7}$$

21.2.2 Internal Treatment

If neutralizing agents (lime hydrate, soda lye) have to be added to the already filled mining lake, the following ways of complete dissolution and mixing of the chemicals with the water of the mining lake should be considered. The objective is to design the neutralization, ferrous ion oxidation, hydrolysis and precipitation so that: (1) the neutralization capacity of the alkalization agents is completely used; and (2) the precipitation products are deposited in the deepest parts of the mining lake.

For a complete utilization of the alkali potential of the $Ca(OH)_2$, considering Eqs. (1) and (2), the following reactions are important:

1. Neutralization of the mineral acid

   $H^+ + OH^- \Leftrightarrow H_2O$

pH value of the mining lake water:	pH 3
Goal of the treatment:	pH 7
Alkali demand:	17 mg OH^-/l
$Ca(OH)_2$ demand:	37 mg $Ca(OH)_2$/l

2. Neutralization of the acid portion, caused by iron hydrolysis

   $Fe^{3+} + 3OH^- \Leftrightarrow Fe(OH)_3$

Ferric content of the mining lake:	26.1 mg/l
Ferrous content of the mining lake:	0.9 mg/l
Goal of the treatment:	more or less iron-free
Alkali demand for iron(III):	23.83 mg OH^-/l
$Ca(OH)_2$ demand for iron(III):	51.87 mg $Ca(OH)_2$/l
Alkali demand for iron(II):	0.548 mg OH^-/l
$Ca(OH)_2$ demand for iron(II):	1.192 mg $Ca(OH)_2$/l

3. Neutralization of the acid portion, caused by aluminium hydrolysis

   $Al^{3+} + 3OH^- \Leftrightarrow Al(OH)_3$

Aluminium content of the lake water:	12 mg/l
Goal of the treatment:	more or less aluminium-free
Alkali demand caused by the aluminium content:	22.67 mg OH^-/l
$Ca(OH)_2$ demand:	49.34 mg $Ca(OH)_2$/l

For the neutralization of 1 l mining lake water with the composition of 26.1 mg Fe^{3+}/l, 0.9 mg Fe^{2+}/l and 12 mg Al^{3+}/l at pH 3. The alkali demand amounts to 64.05 mg OH^-/l and thus the $Ca(OH)_2$ demand is 139.4 mg $Ca(OH)_2$/l.

An extrapolation to a mining lake volume (V) of 10×10^6 m^3 results in 0.6405×10^3 t OH^-, which corresponds to 1.394×10^3 t $Ca(OH)_2$. At a price of $65/t $Ca(OH)_2$, the cost of the alkali amounts to about $90,610 with this method.

21.3
Percolation Through Buffering Soil Layers

21.3.1
Chemical Processes

During percolation of mining lake water through the soil, the proton-binding components of the soil interact with the mining water. As an additional phase, besides the liquid phase, three solid phases occur in this system: soil material with cation-exchange properties, the soluble natural or artificially added calcite mineral and the silicate mineral matrix (primary silicates). As the latter components are only important when minor quantities of lime are used and in case of considerations over a long time period, the weathering of these components need not be taken into account here after large doses of lime have been applied. Calcite in soils reacts with dissolved species of the soil solution, producing 2 mol alkalinity and 1 mol Ca^{2+} ions for each mole of $Ca(OH)_2$, which can react further with exchangeable cations in the soil phase:

$$Ca(OH)_2(s) + 2H^+ \rightarrow Ca^{2+} + 2H_2O. \quad (8)$$

$$R\text{-}H_2 + Ca^{2+} \rightarrow R\text{-}Ca + 2H^+. \quad (9)$$

During the percolation of acidic, ferrous and ferric mining waters through buffering soil layers, ferric hydroxides are precipitated and the protons produced are neutralized by lime:

$$Fe^{2+} + H^+ + 1/4O_2 \rightarrow Fe^{3+} + 1/2H_2O. \quad (10)$$

$$Fe^{3+} + 3H_2O \rightarrow Fe(OH)_3 + 3H^+. \quad (11)$$

$$R\text{-}Ca_{1/2} + H^+ \rightarrow R\text{-}H + 1/2Ca^{2+}. \quad (12)$$

Using this method, attention should be paid to a careful analysis and control of the chemical and soil physical conditions, in order to avoid an immediate blockage of the soil and to ensure sufficient water drainage.

21.3.2
Treatment by Infiltration

A possibility for the treatment of acidic, iron-containing mining lakes is the infiltration of water near the mining lake through soil material or through buffering and neutralizing material (mixtures of clay, sand, gravel, ash from power stations and lime) analogous to the artificial enrichment of groundwater, if the hydrogeologic conditions permit the flow of the infiltrated and neutralized water into the mining lake. After neutralization, the precipitated iron (II)- or iron (III)-hydroxide predominantly remains in the aquifer when this procedure is applied. Therefore, conditions have to be created such that neither iron clogging nor iron condensation areas arise after the iron elimination in the aquifer (see Sects. 21.3.2.2 and 21.3.2.3).

For the study and determination of important parameters, dynamic tests were carried out at room temperature in 42-cm-long and 5.2-cm-diameter soil columns, which were filled with 1200 g natural soil material. By infiltration of the original mining lake water against gravity, we prevented an air pad from forming and so the actual filtration parameters changed considerably compared with the natural conditions. Using an appliance for the measurement of the pressure difference at the column input and column output, it was possible to selectively calculate the k_f value (Darcy coefficient) during the filtration process in order to furnish proof of possible clogging effects. Original mining lake water was used to simulate the natural conditions (ion strength, organic and inorganic xenobiotics, trace elements) as far as possible. The following three tests were carried out with the experimental arrangement described above.

21.3.2.1
Filtration Through Natural Soil Layers at the Precalculated Alkalization

The iron (II) oxidation kinetics is a strongly pH-dependent process. At a pH range of 1–5, the oxidation rate at $pO_2 = 0.2$ atm and temperature 25 °C can be described by the following equation:

$$-d \lg [Fe(II)]/dt = k'', \qquad (13)$$

where d is the differential, lg the logarithm, dt the differential time, and k″ the zero-order rate constant.

At pH values greater than 5, the iron (II) oxidation follows a first-order reaction regarding the oxygen and iron (II) concentration and a second-order reaction regarding the hydroxide concentration [Eqs. (14) and (15)]:

$$-d[Fe(II)]/dt = k^* [Fe(II)] (pO_2)(OH^-)^2 \text{ and} \quad (14)$$

$$-d[Fe(II)]/dt = k^{**} [Fe(II)] [O_2(aq)]/(H^+)^2, \quad (15)$$

where k* and k** are formally fourth-order rate constants. Thus, at 20 °C, $k^* = 8 \times 10^{13}$ min^{-1} atm^{-1} mol^{-2} l^2 and $k^{**} = 3 \times 10^{-12}$ min^{-1} mol^1 l^{-1}.

Based on the rate constants determined in experiments (Singer and Stumm 1968, 1970), the following iron(II) oxidation kinetics results (Fig. 21.1). From these theoretical considerations, it should be possible to achieve a controlled iron(II) oxidation and hydrolysis in the soil with a calculated preliminary increase in the pH value of the mining lake water after pumping the water from the mining lake and before the infiltration through natural soil material. The result of this procedure is a regular deposition of the $Fe(OH)_3$ material in the soil matrix.

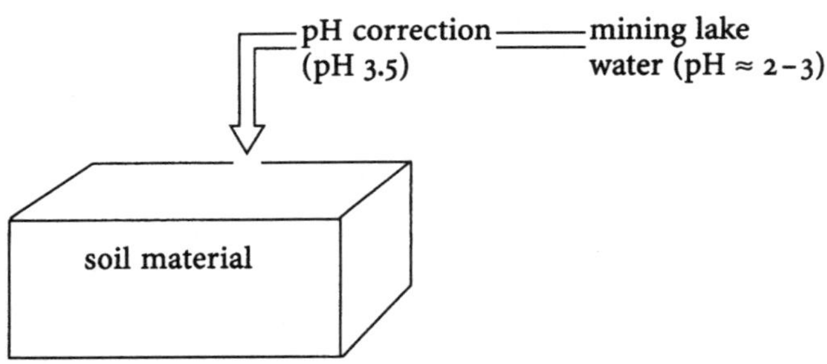

The goal of the test was to neutralize the greatest part of the mineral acid before percolation of the mining lake water through natural soil material to make use of the exchange capacity of the soil in the following step (soil percolation) of iron elimination. Variance of the intended pH correction to a value of 3.5 with subsequent soil percolation was tested. The measured data are represented in Figs. 21.2 and 21.3. The results clearly show that in this case (pH correction to 3.5, followed by percolation through a soil column) of more than 52-l filtration volume, the total iron concentration measured in the outflow was always lower than 1.5 mg/l, at a concentration at the column inflow of 26.1 mg Fe(III)/l. It was also observed, however, that iron hydroxide precipitated after a longer residence time in the supply bottle and in the tubes; this made the iron balance complicated. For this reason, only fresh mining lake water was used for the per-

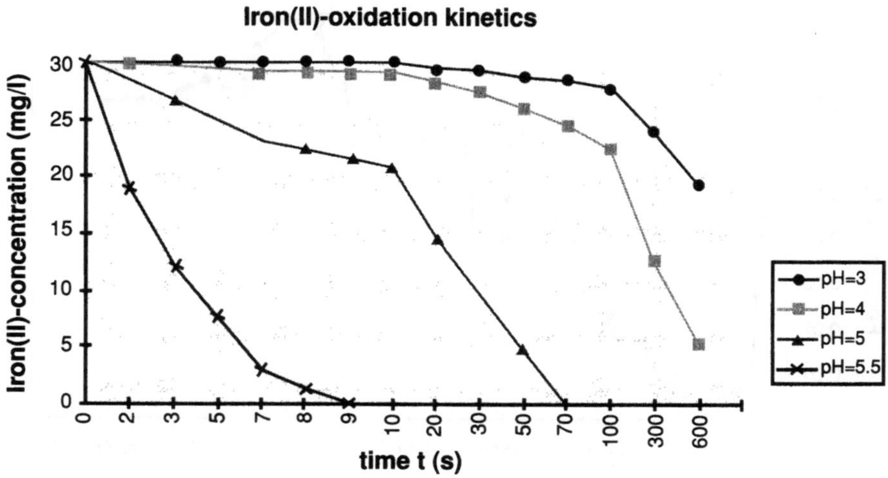

Fig. 21.1. Iron(II) oxidation kinetics at 30 mg Fe(II)/l, pO_2 0.2 atm and temperature T = 25 °C [calculated with Eq. (21.15)]

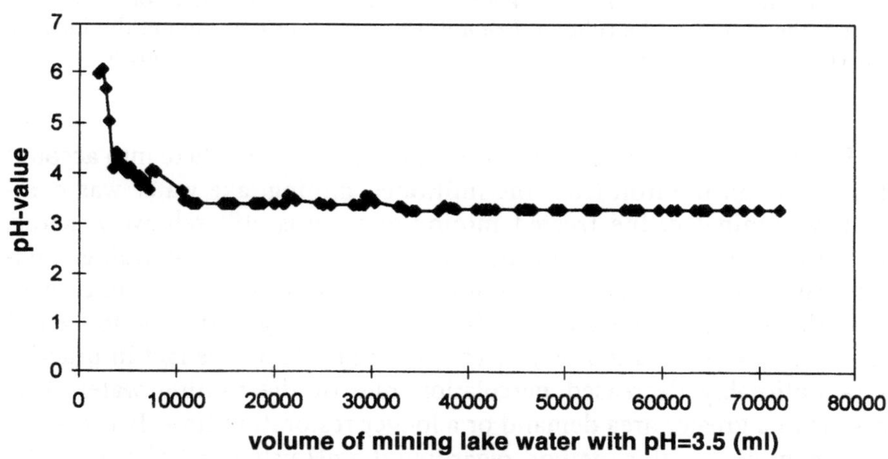

Fig. 21.2. The course of pH during percolation of acidic, iron-containing mining lake water after a previous artificial pH increase and subsequent filtration through natural soil material in a laboratory columm

colation. In contrast to iron, the protons percolated considerably faster through the soil material, so that already after the percolation of 10 l water, a pH value of 3.5 at the column exit was reached. The further pH decrease below the value of the infiltrate (pH < 3.5) can be explained by the release of protons during the hydrolysis of iron in the soil body, which was in accordance with expectations.

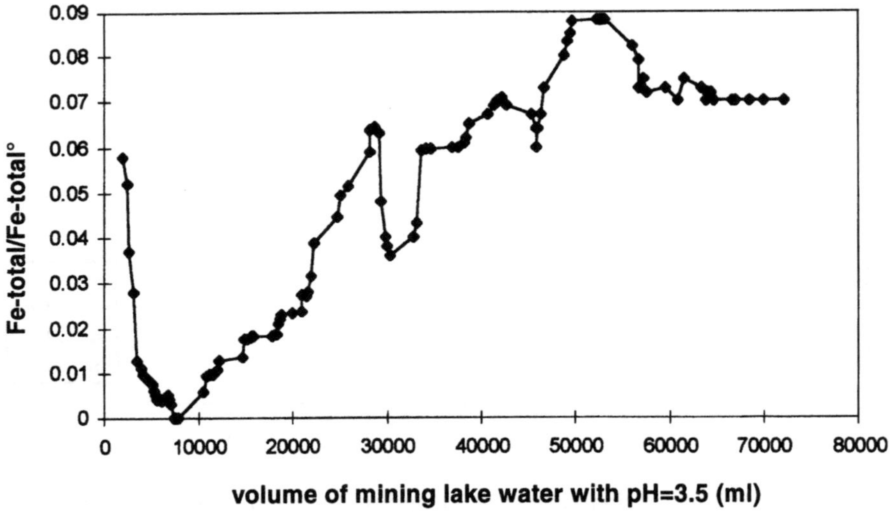

Fig. 21.3. The course of total iron during percolation of acidic, iron-containing mining lake water after a previous artificial pH increase and subsequent filtration through natural soil material in a laboratory column. *Fe-total* Total iron concentration at the output of the column: *Fe-total°* total iron concentration at the input of the column; *Fe-total/Fe-total°* relative value without unit of measurement

For the practical application of this method, we must take into account that although the iron from the infiltrated mining lake water was completely eliminated, the treated mining water was still relatively acidic, because the natural soil material has no sufficient proton-binding capacity. An additional disadvantage of this method is that the k_f value declines relatively fast from 5×10^{-4} m/s at the beginning of the test to 5×10^{-5} m/s after the percolation of 52 l. The consequence of this fact in practice is a noticeably decreased percolation rate of the mining water. This results in a greater area demand or a longer restoration time. Because the treatment goal of this method regarding the pH value (pH 6) will not be reached and therefore the procedure has to be ruled out for the treatment of the mining lake water, the area and alkali demand was not calculated.

21.3.2.2
Alternating Percolation and Saturated $Ca(OH)_2$ Solution Through Soil Material

The aim of this experiment was to test, by alternating percolation of mining water and saturated $Ca(OH)_2$ solution, to what extent the incorporation capacity of the soil for iron and H^+ ions is dependent on the pH of the outflow if a measurable clogging of the soil occurs and how the ferric hydroxide is distributed in the soil. For this purpose, the soil column was filled with 1200 g natural sandy soil through which the mining lake water [pH 3, 26.1 mg Fe(III)/l, 0.9 mg Fe(II)/l] was filtrated. During the filtration at a temperature of 25 °C, protons and iron are bound until a dynamic exchange equilibrium at the soil is reached. During the following "regeneration" of the ion-exchange complex of the soil with a $Ca(OH)2$ solution, the iron (II and III) is dissolved out from the exchange complex and precipitates as hydroxide.

Now the ion-exchange complex is saturated with calcium, and the protons are neutralized at the exchange complex. This process is repeated with the infiltration of acidic, iron-containing mining lake water, as described by the following equations:

Interactions of mining lake water and soil

1. *For iron(II):*

$$Fe^{2+} + 1/4O_2 + H^+ \rightarrow Fe^{3+} + 1/2H_2O \qquad \text{Iron (II)-oxidation}$$

$$Fe^{3+} + 3H_2O \rightarrow Fe(OH)_3 + 3H^+ \qquad \text{Iron (III)-precipitation}$$

$$\begin{array}{l} =S=Ca \\ \vert \\ =S=Ca \end{array} + Fe^{2+} + 2H^+ \rightarrow \begin{array}{l} =S-H \\ \vert \diagdown H \\ =S=Fe \end{array} + 2Ca^{2+} \qquad \text{Ion-exchange process}$$

$$\begin{array}{l} =S-H \\ \vert \diagdown H \\ =S=Fe \end{array} + 2Ca^{2+} + 4OH^- + 1/4O_2 \rightarrow \begin{array}{l} =S=Ca \\ \vert \\ =S=Ca \end{array} + 1.5H_2O + Fe(OH)_3$$

"Regeneration" with a $Ca(OH)_2$ solution

2. *For iron(III)*

$$\begin{array}{l} =S=Ca \\ \vert \\ =S=Ca \end{array} + FeOH^{2+} + 2H^+ \rightarrow \begin{array}{l} =S-H \\ \vert \diagdown H \\ =S=FeOH \end{array} + 2Ca^{2+} \qquad \text{Ion-exchange process}$$

$$\begin{array}{l} =S-H \\ \vert \diagdown H \\ =S=FeOH \end{array} + 2Ca^{2+} + 4OH^- \rightarrow \begin{array}{l} =S=Ca \\ \vert \\ =S=Ca \end{array} + Fe(OH)3 + 2H2O$$

"Regeneration" with a $Ca(OH)_2$ solution

From the practical test concerning this problem, whose experimental results are documented in Figs. 21.4 and 21.5, the following important findings are derived. The natural soil material has a considerably lower incorporation capacity for protons and iron (Table 21.1, first cycle) com-

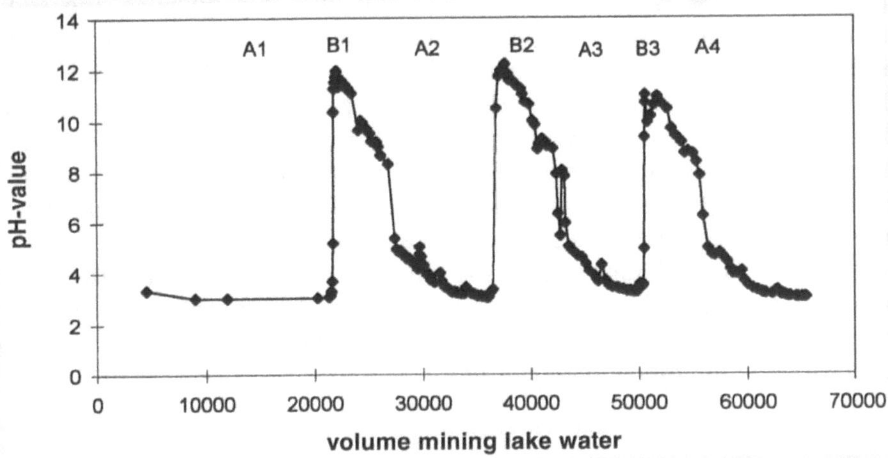

Fig. 21.4. The course of pH during alternating percolation of acidic, iron-containing mining lake water and of saturated $Ca(OH)_2$ solution through natural soil material in a laboratory column. *A* Filtration of mining lake water; *B* filtration of saturated $Ca(OH)_2$ solution. *Subscripts 1–4* Number of the cycle

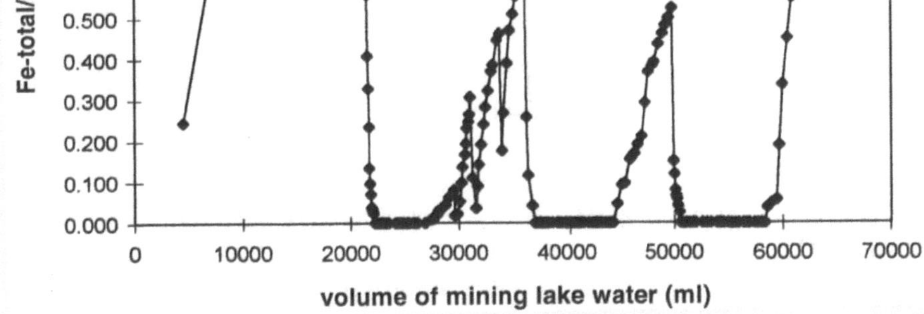

Fig. 21.5. The course of total iron during alternating percolation of acidic, iron-containing mining lake water and of satured $Ca(OH)_2$ solution through natural soil material in a laboratory column. *A* Filtration of mining lake water; *B* filtration of saturated $Ca(OH)_2$ solution. *Subscripts 1–4* Number of the cycle

Table 21.1. Exchange capacities for protons and iron depending on pH value and regeneration cycle

Cycle	Percolated volume of mining lake water	Cumulative H^+ input		Cumulative iron-total input
	(l)	(mmol/1200 g soil)	(mmol/1200 g soil)	(mmol/1200 g soil)
I	4.5	4.5	Up to pH 3.33	2.1
II	5.35	5.35	Up to pH 5.4	2.44
	2.05	8.05	Up to pH 4.04	3.68
	12.8	13.9	Up to pH 3.03	6.35
III	6.03	5.03	Up to pH 5.43	2.11
	8.7	7.7	Up to pH 4.03	3.24
	13.15	12.15	Up to pH 3.22	5.11
IV	5.36	6.0	Up to pH 6.2	2.45
	8.86	9.5	Up to pH 4.03	3.88
	13.12	13.76	Up to pH 3.03	5.62

pared with the soil that had already been brought into contact with the $Ca(OH)_2$ regeneration solution. This result is understandable owing to the fact that due to the application of a concentrated $Ca(OH)_2$ solution, all ion-exchange places available in the soil are occupied by calcium, so that we can compare this process with a distinct, fixed exchange capacity regarding H^+ and Fe ions. If we interrupt the percolation of mining lake water, the pH will be raised and the iron content in the eluate will decrease simultaneously. This phenomenon indicates a non-equilibrium of the participating reactions. For the regeneration of the soil material, constantly 31.1 mmol $Ca(OH)_2$ in the form of a saturated solution were used.

A very important observation that was made during this experiment is that the parameter k_f only slightly decreased during the filtration. At the beginning of the test, k_f amounted to 5×10^{-4} m/s, and after filtration of more than 65 l mining lake water including $Ca(OH)_2$ solution it was measured as 3×10^{-4} m/s. An analysis of the ferric hydroxides in the soil showed a relative regularity. Although at the column-input slightly higher values were measured which indicate a faster iron precipitation in the soil, we can start from a nearly homogeneous longitudinal distribution of the ferric hydroxide, which facilitates the practical application of this variant.

21.3.2.2.1
Soil and Lime Demand for Realization of the Procedure

Proceeding from a mining water quality that was also the basis of the laboratory filtration tests: [pH 3, 26.1 mg Fe(III)/l, and 0.9 mg Fe(II)/l]; we aimed at the following treatment goal: pH 6, 0.2 mg Fe-total/l, and 0 mg Fe(II)/l. On the basis of the laboratory filtration tests, we carried out calculations for the determination of the necessary soil volume and the quantity of lime to be employed. In laboratory experiments, the following data were found:

- Used soil mass (m): 1200 g
- Used soil volume (V): 0.800 l
- Cumulative H^+ input: 5.5 mmol/1200 g soil
- Cumulative total iron input: 2.3 mmol/1200 g soil
- Percolated volume of mining lake water: 5.6 l
- Necessary volume of saturated $Ca(OH)_2$ solution: 0.760 l

Depending on the number of treatment cycles and the soil depth, the areas shown in Table 21.2 are needed to treat 1 m^3 mining lake water. With this method, the necessary mass of Ca(OH)2 for the treatment of a mining lake with a volume of 10×10^6 m^3 amounts to 2.239×10^3 t $Ca(OH)_2$. This quantity corresponds to about 1.6 times the quantity of $Ca(OH)_2$ of about 1.394×10^3 t which is necessary for direct lake restoration.

21.3.2.3
Percolation Through Natural Soil Material by Use of Soil–Ca(OH)$_2$ Mixtures

Percolation of mining lake water through natural soil material by use of soil–$Ca(OH)_2$ mixtures is based on the fact that a layer of a mixture of sand and alkalizing substances [preferably $Ca(OH)_2$, possibly also power-station ash] makes it possible to achieve a fast alkalization of the mining lake water as well as iron precipitation. The iron (III)-hydroxide deposit is

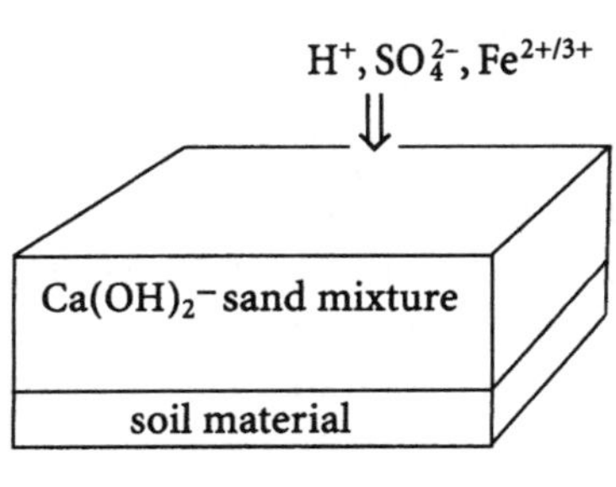

Table 21.2. Soil demand at alternating percolation of mining lake water and a saturated $Ca(OH)_2$ solution

Treatment cycle	Depth of soil reactor (m)	Necessary area/volume lake water of mining (m^2/m^3)
1	1	0.1428
	10	0.0143
	30	0.00476
10	1	0.0143
	10	0.00143
	30	0.000476
100	1	0.000143
	10	0.0000143
	30	0.00000476

concentrated in the pore space of the mixture and in the adjacent soil layer which, after a certain time, become clogged. Using a device, the highest layer is cleared away and the filling of the mixture of sand and alkalizing substances is repeated.

In this experiment, the soil column was filled according to Fig. 21.6. The $Ca(OH)_2$-sand mixture in the middle of the column properly acts as the neutralizing layer. With the stated sequence of layers (see Fig. 21.6), we aimed to prevent the finely dispersed lime hydrate from being washed out either upwards or downwards. As already stated in the test in Section 21.3.2.2, also here the protons are less retained than iron by the soil. Whereas after the filtration of 9 l mining lake water through the soil column a pH value of 3.8 was measured at the outlet (pH_{in} 3), at that time iron was not yet detected (Figs. 21.7 and 21.8).

After an ion breakthrough of about 50% (V = 14.57 l), the filtration was interrupted, and the mining lake water was allowed to stand in the column for about 7 days. After the resumption of the filtration, higher pH values (see Fig. 21.7) and much lower iron values were measured at the column exit (see Fig. 21.8). As mentioned in Section 21.3.2.2, also there was evidence of a slow kinetics of dissolution of Ca(OH)2 which can be directly proved. Another important finding derived from this experiment is that at a Ca(OH)2-sand mass ratio of 1:100 during the whole percolation phase, no reduction of the k_f value could be measured. After a filtration volume of almost 24 l mining lake water in the outflow, k_f value amounted to about 4.5×10^{-4} m/s. If the theoretical buffering capacity of the $Ca(OH)_2$ used in the sand stratum is compared with the real con-

Fig. 21.6. Column contents

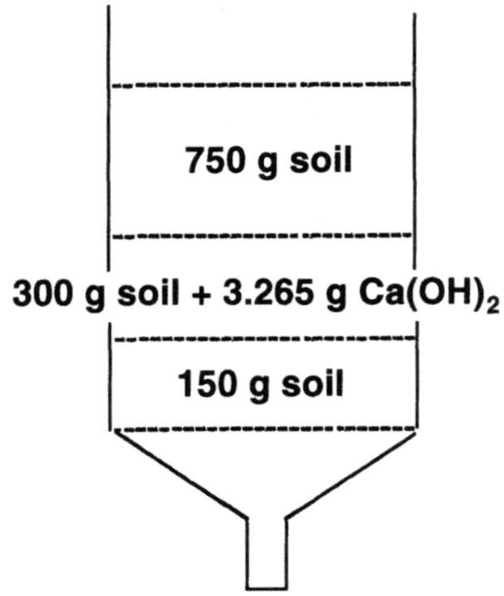

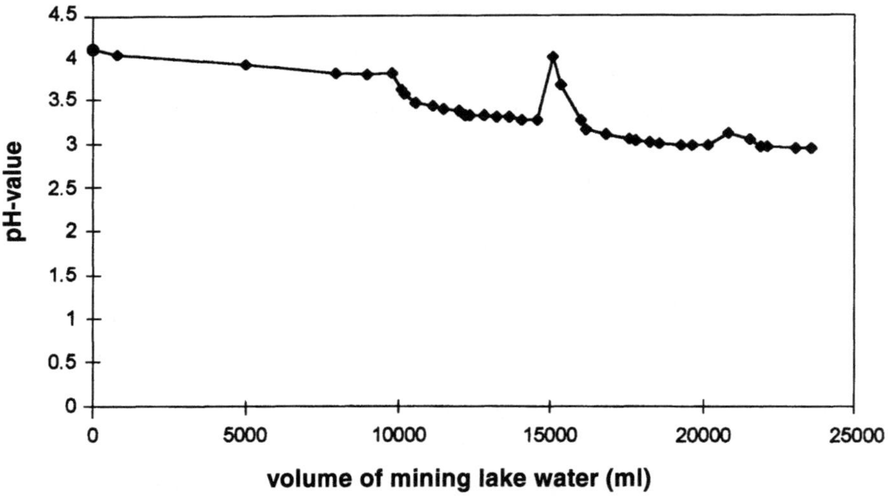

Fig. 21.7. The course of pH during percolation of acidic, iron-containing mining lake water through a soil column with a $Ca(OH)_2$-sand layer

sumption used, a discrepancy results. Theoretically, 3.265 g $Ca(OH)_2$ (which corresponds to 1.5 g OH^-) is sufficient for the treatment of 23.4 l mining lake water having the following quality: pH 3, 26.1 mg Fe(III)/l, 0.9 mg Fe(II)/l and 12.0 mg Al^{3+}/l. The breakthrough of iron occurred after a passage of about 9 l.

This great difference becomes understandable due to the fact that the rate-determining step is the solution reaction of the powdered lime

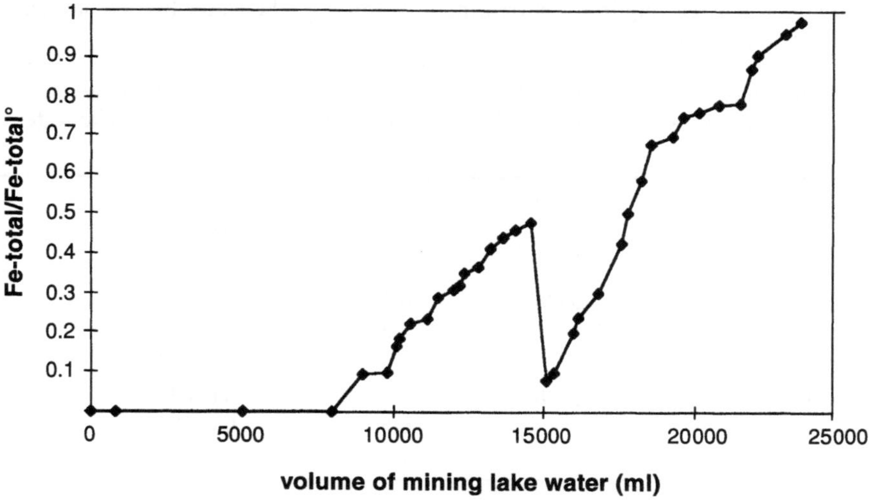

Fig. 21.8. The course of total iron during percolation of acidic, iron-containing mining lake water through a soil column with a $Ca(OH)_2$–sand layer

hydrate and that at a great mixing mass ratio of $Ca(OH)_2$ and sand the alkalinity will be washed out from the layer. At a lower rate of filtration these values would become very similar if it is assumed that there is no additional acidity in the mining lake water (increased concentrations of humic acids, fulvic acids and others).

For the treatment of 10×10^6 m^3 mining lake water, 3628 t $Ca(OH)_2$ is required when this method is applied. This is an additional demand of about 2234 t compared with the direct addition of the alkalizing agent to the mining lake. The required area of land depends on the $Ca(OH)_2$–soil mixing ratio. In our first orientational laboratory experiment, a mixing ratio of 1:100 was chosen. This ratio has to be increased by further optimization experiments in order to reduce the area required. In Table 21.3, the required area is given for the application of the method "Percolation of the mining lake water over an alkalizing layer", depending on the layer thickness of the $Ca(OH)_2$–soil mixture and on the mass ratio of the $Ca(OH)_2$–soil mixture.

Under the assumption that in all cases the $Ca(OH)_2$ demand is constantly 3628 t, chemical costs of $ 235,820 result for the restoration of the mining lake.

Table 21.3. Area required for treatment of mining lake water by percolation over alkalizing layers

Mixing mass ratio of $Ca(OH)_2$: sand	$Ca(OH)_2$ demand (t)	Layer thickness (m)	Required area/volume of mining lake water (m^2/m^3)
1:100	3628	0.3	0.08143
		0.5	0.04889
		1.0	0.02443
1:50	3628	0.3	0.04111
		0.5	0.02467
		1.0	0.01234
1:10	3628	0.3	0.00887
		0.5	0.00532
		1.0	0.00266

21.4 Conclusions

As possible restoration methods for an acidic, iron-containing mining lake, we have tested and discussed direct neutralization with alkalizing agents and soil treatment. In a compared assessment of the restoration of a fictitious mining lake with a volume of 10×10^6 m^3, it is sensible to apply the direct neutralization with lime hydrate method without risk, with acceptable costs and restoration time.

References

EG 78/659 (1978) Richtlinie des Rates vom 18.7.1978 über die Qualität von Süßwasser, das schutz- oder verbesserungsbedürftig ist, um das Leben von Fischen zu erhalten. Amtsblatt der Europäischen Gemeinschaften, no L 222/1

EG 76/160 (1975) Richtlinie des Rates vom 8.12.1975 über die Qualität der Badegewässer. Amtsblatt der Europäischen Gemeinschaften, no L 31/1

Singer PC, Stumm W (1968) Kinetics of the oxidation of ferrous iron. Presented at the 2nd Symp on Coal mine drainage research. Mellon Institute, Pittsburgh

Singer PC, Stumm W (1970) Acidic mine drainage: the rate-determining step. Science 167:3921

Stumm W, Morgan JJ (1981) Aquatic chemistry. Wiley Interscience, New York

22 Ways of Controlling Acid by Ecotechnology

H. Klapper, K. Friese, B. Scharf, M. Schimmele and M. Schultze

UFZ Centre for Environmental Research, Inland Water Research, Magdeburg, Germany

22.1 Introduction

Geogenic acidification results from the oxidation of sulfidic minerals that had been stable for millennia because of anaerobic conditions in the underground. The geochemical process is microbially intensified and therefore a natural process. However, it is also man-made, because the sulfidic minerals are oxidized as a consequence of aeration due to mining activities. Open-cast lignite mining starts with the dewatering of the overburden, of the lignite and of the uppermost layers below the coal. Pyrite and marcasite are then in contact with atmospheric oxygen instead of anaerobic groundwater. The acidity results from the oxidation of sulfur and iron and from the hydrolysis of iron (see also Evangelou, this Vol.). Bodies of water become acidified when the sulfuric acid and iron(II)-sulfate are leached and transported into the lake by the groundwater. In the case of refilling with groundwater, the resulting mining lakes are acidic, with pH between 2 and 3. Their water is brown because of the high content of dissolved iron hydroxide. The low pH is strongly buffered by iron. Other heavy metals formerly present in the overburden are dissolved and contaminate the lake water. Living conditions differ widely from those in natural lakes in Germany.

The general question of whether to do something against this acidity or not is a matter of water policy. One has to take into account:

- Demands of the society (local and governmental)
- Options for lake utilization
- Quality demands of the users in question
- Protection of the natural environment
- Structure and function of the ecosystems

A number of relatively small acidic lakes have existed for many decades. They have been and will be excellent sites for field research on the very extreme habitats and natural successions as well as natural neutralization processes. For large mining lakes – say 10 km^2 in area or more – the aim only should be a usable lake. In acidic lakes water-oriented recreational activities are restricted to water sports not involving body contact with the water. Fish do not survive at $pH < 5$, so fishing of any kind is excluded.

As mentioned above, the acidic water is strongly buffered and neutralization is complicated. Nevertheless, with occasional exceptions, in general the aim is a neutral lake where fish life is possible. The following recommendations are derived from experiences with land and water reclamation in the brown-coal region of East Germany, from limnological observations of existing mining lakes and from international literature on acid mine drainage. In this state-of-the-art report further research fields are defined.

22.2 Prophylactic Measures During the Mining Process

To minimize acid formation at the outset, some inhibitory measures should be taken during the mining process:

- Minimization of dewatering before excavation in volume and time.
- Deposition of sulfur-rich overburden material in the deepest part of the mining hole with little water exchange.
- Inhibition of the acid-forming metabolism of the *Thiobacilli* by treating the refuse piles with bactericides (Kleinmann et al. 1981; Onysko et al. 1984).
- Refilling of the dewatered subsurface and heaps after coal extraction as soon as possible.
- The shape of the lakes and their orientation with respect to the groundwater flow are important factors for the water quality. Elongation of mining lakes in the groundwater flow direction should be avoided. Such lakes are steps in the slope of the groundwater table and pull groundwater from above together with the acidity (Luckner et al. 1995).
- The final state of the mining activity and therefore the lake should preferably be such that the inflowing groundwater stems from the undisturbed rocks and not from the oxidized heaps.

22.3 Measures to Counteract Acidity in the Drainage Basin

The recultivation of the landscape should include measures to stop the acidification process at underground level and the migration of acid together with dissolved heavy metals. Iron, stemming from pyrite (FeS_2), is most important. It buffers the water at low pH between 2 and 3. Rules for good farming practice, forestry and fishery are known from acid rain research and experience. The landscape and its groundwater hydrology should look and function like or ecologically better than that before mining:

- Neutralizing fertilization with lime, dolomite powder (forestry), alkaline ashes, etc. to enhance the revegetation is also helpful in improving water quality (see also Katzur and Liebner, this Vol.). Liming dose is usually 2–4 t ha^{-1} (carbonate content 95–98%) and thus has an acid neutralization capacity (ANC) of about 40–80 kmol ha^{-1} (Kreutzer 1994).
- The groundwater table has to be kept high; fluctuations should be avoided. The result is to minimize aeration, acidification and heavy-metal migration.
- Evaporation and transpiration losses should be high. This reduces formation and throughflow of the groundwater.
- On the other hand, a higher air moisture in wetlands allows water condensation and thus a stabilization of the water in the subsoil.
- The vegetation cover should be close, if possible continuously during the whole year.
- Infiltration of appreciable amounts of carbon from organic fertilizer, such as dung, liquid manure and sewage sludge, and from harvest residues is useful to consume oxygen in the subsurface.
- Because nitrification produces acidity, ammonia fertilizer should be avoided and all N fertilizer has to be applied in nitrate form (Kelly 1994).
- The monocultural pine forests predominant before mining should be replaced by mixed forests with a large proportion of deciduous trees, creating a humus-rich topsoil.
- Long growth periods and restriction of wood-harvesting decrease the acid-forming cation export from the area.
- Clear cutting of large stands of trees must be avoided (Kreutzer 1994).
- One option for using the recultivated land is to build fish ponds, especially those for intensive fish cultivation with pellet feeding of carp. The sealing of the surface against atmosphere and oxygen is beneficial, as is the trickling of organic compounds into the ground, where they stimulate oxygen consumption.
- Another option for neutralizing land use is the establishment of wetlands. Wetlands are good habitats for sulfur-reducing bacteria. The product of desulfurication in an iron-containing environment is black

iron sulfide. This may be found in anaerobic parts of wet topsoils and especially between the rhizomes of the reed. In the USA, constructed wetlands have been proved to be service ecosystems for abatement of acidity in drainage waters from mining heaps (Hedin 1989; Kleinmann et al., this Vol.). Fish ponds and wetlands have had a long tradition in the Lusatian lignite district, East Germany, which should be continued now after mining.

All the measures recommended for drainage basins are at the same time part of the concept of a sustainable development of the landscape (Ripl et al. 1992):

1. Minimizing the groundwater throughflow.
2. Decreasing the coupled transport of substances.
3. Enhancing evaporation from the vegetation cover.
4. Enrichment of soils with organic substances to increase the binding capacity for water and other materials.

The post-mining landscape is much higher in relief than before and the groundwater table cannot reach the surface of higher heaps. These oxidized parts are sources for reacidification. Harmful, because intensifying the acidification, are fluctuations of groundwater and lake water tables. By these means new acid is produced and transported, together with acid-soluble heavy metals. The use of mining lakes for water storage therefore may be endangered by the acidity of the surroundings. Water table changes in this case are the operational normality. Evidently, the only way to avoid pyrite oxidation is to refill the overburden with water. The diffusion coefficient of oxygen in a water-filled heap is about 1/10,000 of that for an aerated one.

22.4 Neutralizing In-Lake Quality Management

22.4.1 Abatement of Acidification During Filling

Where there is danger of geogenic acidification, the possibility of filling with surface water has to be tested at an early stage. Criteria include the availability of water in quantity and quality as well as the costs of necessary pipelines, pumping stations, inlet and outlet buildings and sometimes surface water treatment plants. Natural elimination mechanisms such as incorporation, degradation, sedimentation, etc. may achieve the same quality targets in a longer time but without costs. There is a risk

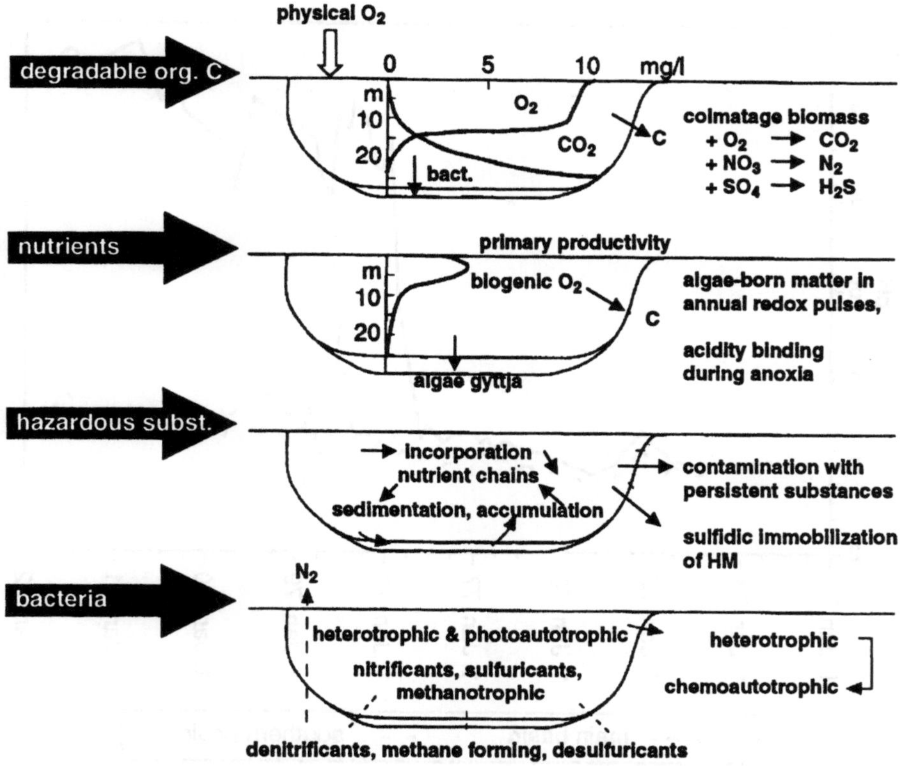

Fig. 22.1. Loading of a mining lake with contaminated surface waters and the most important consequences. *HM* Heavy metals. (After Glässer and Klapper 1992)

because of unwanted contaminants in the rivers in question, e.g. oxygen-consuming organics, plant nutrients, hazardous substances and unwanted bacteria including pathogens. Behaviour, metabolisms and pathways of matter have to be investigated in advance (see Fig. 22.1; Glässer and Klapper 1992).

The large running waters in and around the mining districts in East Germany contain neutral water buffered by hydrogen carbonate. The salt content is generally lower than that of groundwaters in the coal region. With surface water added, the water table in the mining hole may be kept higher than the groundwater level in the surroundings. The flow direction will be from the lake into the dewatered underground spaces. The infiltration of degradable substances during this first filling is advantageous with respect to the abatement of acidification. Under anoxic conditions, the acidification process may be stopped and eventually turned into the contrary – iron immobilized in sulfidic form. The third

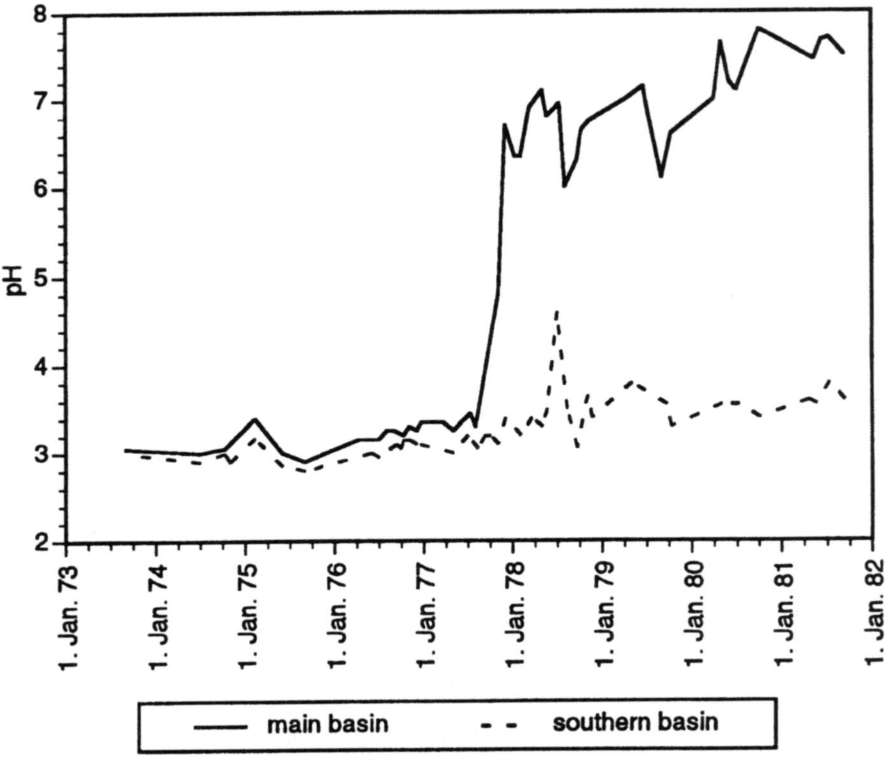

Fig. 22.2. Spatial distribution of pH values in Senftenberger See 1973–1981. A throughflow of Schwarze Elster River water exists since 1976. The water in the main basin is effectively exchanged, in contrast to the southern partial basin, being separated by an island and macrophyte stands. (After Puetz et al. 1991, cited in Benndorf 1994)

acidification step, the hydrolysis of iron sulfate to hydroxide, may be suppressed in this way.

The addition or throughflow of surface water in an already filled acidic lake causes two processes that counteract acidification. Firstly, a part of the acidic lake water is replaced by neutral river water. Secondly, a distinct neutralization results from carbonate hardness. The ANC of the surface water is consumed by the iron (and aluminium) buffer of the acid lake water. When the iron buffer is overcome, the pH shifts to the bicarbonate buffering system, i.e. into the pH range from 6 to 8. An impressive example was given at the mining lake Senftenberger See. For neutralizing purposes, 1.2 times the volume was replaced by river water from the river Schwarze Elster. The change in the pH value of the lake water looks like a titration curve (see Fig. 22.2). To avoid unwanted loading and decay of water quality, the addition of surface water should be limited to the amount necessary for neutralizing. ANC should be thoroughly monitored. Reacidification may occur when the buffering capacity is low.

In the Senftenberger See, in 1995 the water level was lowered for operational purposes. The pH decreased immediately, because of groundwater inflow, with the consequence of a fish-kill. The lake should be in a bypass to a stream. Acidity then may be controlled by the carbonate hardness and amount of the surface water added. Another option for control during the filling process is the utilization of treated groundwater from the active mine industry. For example, water treatment plants exist in the large power plants Schwarze Pumpe, Vetschau and Lübbenau in Lusatia. Mining water treatment plants are situated in Burgneudorf, Rainitza and Lichterfeld; these employ liming and separation of iron hydroxide. It is planned to treat two-thirds of the filling water in these facilities (Luckner et al. 1995). In cases with low acidification potential in the surrounding rock layers, a decision to do nothing and to wait the years necessary until all acidity is removed might also be a rational option.

22.4.2
Abatement of Acidification by In-Lake Measures

In-lake liming has been used in many soft water lakes impaired by acid rain. Throughout 1988, about 5000 surface waters were treated in the operational liming program conducted by the Swedish government (Nyberg 1989, cited in Olem 1991). Successful lake liming by different application technologies has been reported in Canada, USA and Norway, among others, as a suitable neutralizing measure for lakes with low ionic content. These lakes are quite different from the mining lakes with hard waters. They are buffered by iron in a lower pH range. A rough estimate of the limestone demand for the neutralization of the acidic mining lakes of the Lusatian district comes to about 235,000 t limestone powder. This would have to be distributed over a large number of lakes by a potential liming campaign (Schultze and Geller 1996). The neutralizing chemistry is shown in Chapter 3. For mining lakes, chemical alkalization is generally possible but expensive because of the high alkalinity demand. Today, worldwide attention is focused on the biological production of alkalinity – desulfurication and pyrite formation – that is, a reversal of the process of pyrite oxidation. Pyrite oxidation and reduction of sulfate to pyrite are shown in the following equation:

$$\begin{aligned} 2FeS_2 + 7O_2 + 2H_2O &\rightarrow 2Fe^{2+} + 4SO_4^{2-} + 4H^+ + 7C \\ &\rightarrow 2FeS_2 + 7CO_2 + 2H_2O . \end{aligned}$$

To remove sulfuric acid by desulfurication is a relatively cheap method, but only applicable under anaerobic conditions. Denitrification of nitrate also produces alkalinity, but it is not as important because the nitrate concentrations are low. Such an anaerobic environment is found in the deep water of stratified eutrophic lakes. In oligotrophic or mesotrophic lakes, anaerobiosis occurs only in the monimolimnia of the meromictic types. In these deep water bodies, which do not take part in any mixing for many years, the oxygen demand accumulates.

After depletion of dissolved oxygen, the heterotrophic degradation of organic matter is continued with the help of nitrate- and sulfate-oxygen. This sulfate reduction is a desirable process as it eliminates sulfuric acid. Pyrite oxidation produces a high concentration of iron. Therefore the end product is not hydrogen sulfide but iron sulfide. This is insoluble and settles as black mud. Other (toxic) metals are also transferred in their insoluble sulfidic form. This immobilization seems to be an important step from the extreme environment of acidic to more nearly natural neutral lakes.

The living conditions required by the heterotrophic sulfate-respiring bacteria must be investigated in order to design an ecotechnology (Wendt-Potthoff and Neu, this Vol.). At first, observations from nature should be evaluated. Further ideas may be gained from technological solutions for denitrification and desulfurication in drinking-water treatment (Fichtner 1983; Klapper 1991; Brettschneider and Pöpel 1992). Environmental conditions which have to be realized for sulfate respiration within an ecotechnology are:

- Exclusion of dissolved (and nitrate-) oxygen
- Presence of degradable organic substrate
- Biologically inert or organic materials as supporting structures or biofilm carriers
- High content of sulfate (and iron)
- Initial microhabitats with $pH > 4$

The neutralized water needs reaeration after desulfurication to become a fish habitat.

Most acidic mining lakes are oligoproductive in the early stages of their development. The oxygen demand of sedimenting and degrading algae is not high enough for oxygen depletion (see for example Fig. 22.3, mining lake Koschen). A similar productivity but very small hypolimnion volume is producing oxygen depletion and the first signs of alkalization near the bottom in mining lake 117 near Lauchhammer. The lake treatment plant Laubusch is heavily loaded with domestic sewage. Deep water is strongly anaerobic and neutralized up to the hydrogen carbonate range.

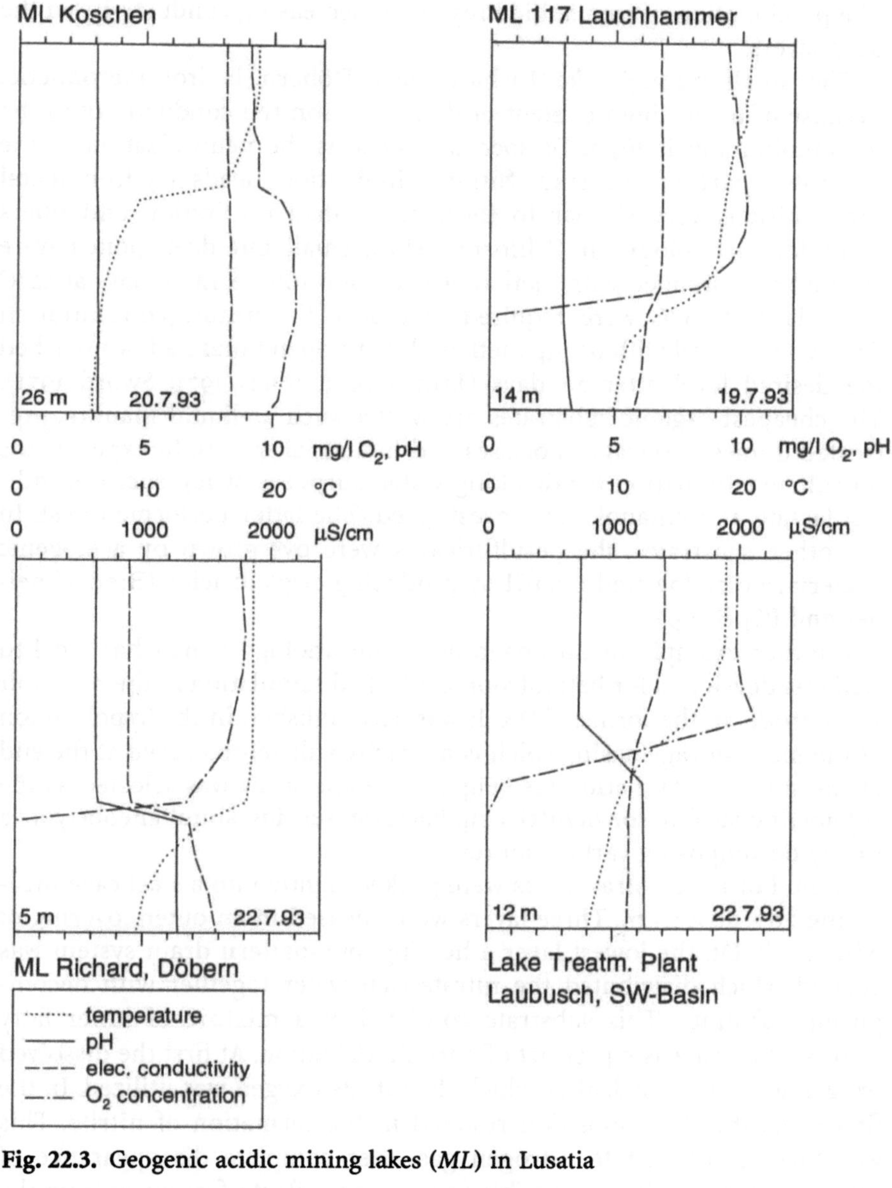

Fig. 22.3. Geogenic acidic mining lakes (*ML*) in Lusatia

The precipitation of iron sulfide results in decreasing conductivity in the deep layers.

The small mining lake Richard near Döbern is iron-meromictic. Because of a very high content of dissolved iron the conductivity in the monimolimnion is high. Of special interest is the neutralization in the anaerobic part of the lake. Nitrate elimination needs environmental preconditions very similar to those for sulfate respiration and offers ideas for technology. In California, USA, small but deep ponds were covered and supplied with methanol as carbon source. Eight days at 22 °C or 15 days at 16 °C were required to reduce the nitrate concentration from 20 to 2 mg/l with 65 mg methanol /l. An open pond had not reached the desired level after 20 days (Brown 1971; Jones 1971; Sword 1971). The cheapest organic substrates are wastes such as liquid manure, pretreated domestic sewage, molasses and industrial wastes. In experiments on sulfate elimination for drinking-water purposes, whey, sucrose, sodium lactate and ethanol were investigated. The latter performed best. In the other substrates, the desulfuricants were overgrown by acetogenic bacteria, which lowered the pH by producing organic acids (Brettschneider and Pöpel 1992).

Another example of an anaerobic ecotechnology which has worked well was developed for heterotrophic nitrate dissimilation in the reservoir Zeulenroda, in the former GDR. It was accomplished in the hypolimnion of the main storage basin, which is normally still aerobic, even at the end of the summer stagnation (see Fig. 22.4). Rape straw was selected as the supporting surface for denitrifying bacteria, serving simultaneously as a slowly decomposing carbon source.

A total of 13,000 straw bales were packed tightly into a steel cage measuring $20 \times 60 \times 1.5$ m. Three layers were encased in an outer covering of wire mesh. On the lowest layer a herringbone-pattern drain system was created which distributed the nitrate-rich water together with decomposing substrate. This substrate consisted of a mixture of lower fatty acids and was a waste product of paraffin oxidation. At first the dissolved oxygen was consumed, after which the nitrate oxygen was utilized. In the first stage the decomposition resulted in the formation of nitrite. This was further reduced to nitrogen gas after complete disappearance of nitrate. In this way it was possible to discharge nitrate-free water from the hypolimnion into a non-impounded river segment. There it can become saturated with oxygen before it reaches the terminal barrage with its raw water intake (Klapper 1991).

Natural processes of acidification and alkalinization are quite different in different lake environments. Polymictic lakes and those filled mainly with groundwater may remain acidic and oligoproductive for decades. In

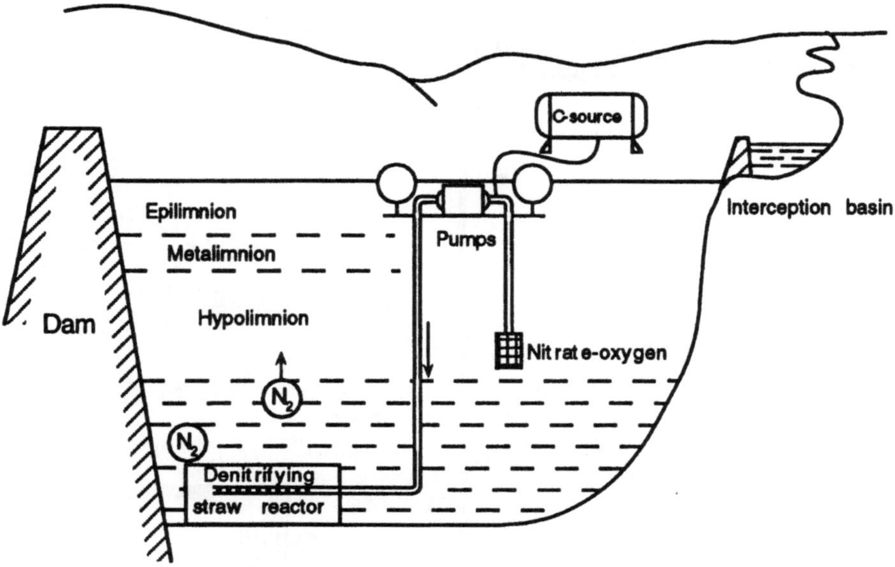

$$C_{12}H_{12}O_{12} + 8\ NO_3^- \longrightarrow 4\ N_2\uparrow + 12\ CO_2 + 11\ H_2O$$

Fig. 22.4. Schematic diagram of heterotrophic nitrate dissimilation in the hypolimnion of a reservoir. (After Klapper 1991)

deeper, stratified lakes and those filled with nutrient-rich surface water, alkalinization can take place in the deep water and in the surrounding groundwater as well (see Fig. 22.5).

Many of the mining lakes have critical depths and the neutralizing processes are interrupted when stratification breaks down. In this case it is useful to shorten the fetch of the wind by artificial barriers, which divide the surface into some shorter segments; by this means the mixing depth is decreased. From an ecological point of view natural materials should be preferred, e.g. floating reed installed across the lake (see Fig. 22.6).

For research purposes, lake volumes can be partitioned by introducing enclosures or limnocorrals. Standing waters generally are protected by law against nutrient input. A controlled addition of nutrients or organic substrates for controlling acidity goes against the commonly accepted water policy and should be used only temporarily to stimulate alkalinization processes. As a long-term goal, the water quality in a nearly natural mining lake should correspond to its hydrography. Deep lakes will be kept oligotrophic or mesotrophic, shallow ones more or less eutrophic. The finished neutralization will be accompanied by flocculation of iron hydroxide together with phosphorus. Macrophyte stands will assist this

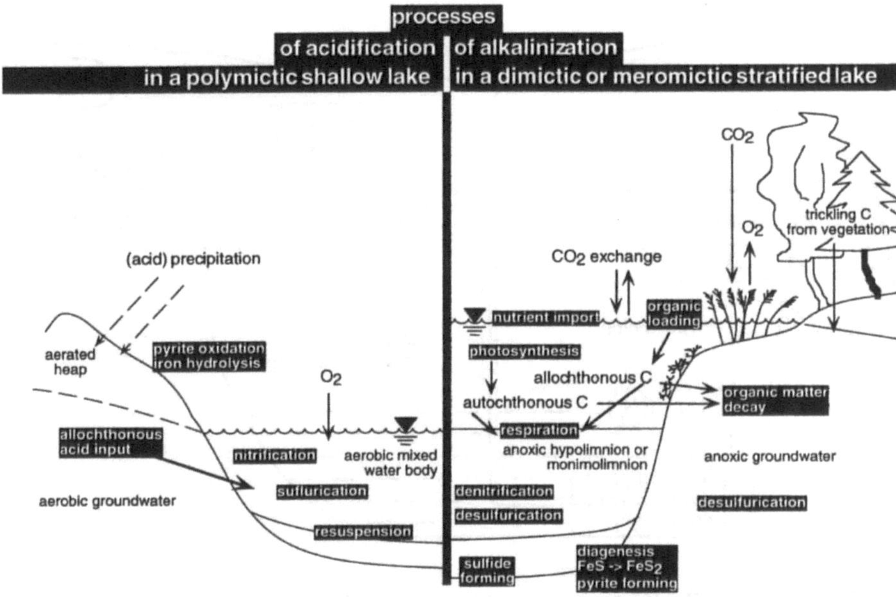

Fig. 22.5. Processes affecting acidity in shallow and in stratified water bodies

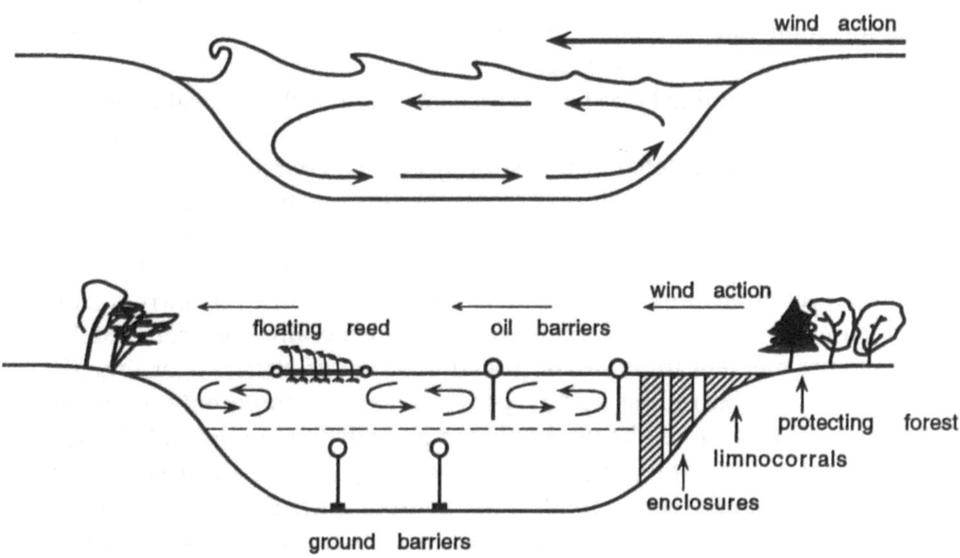

Fig. 22.6. Circulation and stability without and with barriers

Table 22.1. Ecotechnologies for recovery of acidic mining lakes

In situ (within the mining lake)	Whole lake anaerobic	Only very small bodies of water with high organic load; covered surface
	Anaerobic parts	Use of deep water for desulfurication Use of enclosures for research Stabilization of stratification Sediments as traps for sulfur and metals Addition of organic substrate Addition of nutrients and production of organics by controlled eutrophication
	In-lake placement of a throughflow reactor	Throughflow straw reactor positioned in the deep water and addition of organic substrate
Ex situ (before and between mining lakes)	Anaerobic treatment	Anoxic straw-filled trenches Anoxic limestone drains Closed subsoil reactor, filled with inert or degradable biofilm carriers; addition of C-substrate Infiltration ponds with compost bottom and drainage
	Aerobic treatment	Reed-bed treatment plants for combined purification of tailings and sewage Reaeration channels with limestone overflow barriers

process. They are suitable in different service ecosystems for polishing the water.

Some of these "constructed wetlands", "float reed" and „shore bioplateaus" may be called well-tried and successful ecotechnologies. The recommended combinations of different ecotechnologies are summarized in Table 22.1. Further intensification is possible by a partial chemical alkalinization with lime, ashes, etc. The different conditions at each individual mining lake have to be checked by monitoring and limnological expertise in advance of filling. Additional investigations should be carried out as the lake fills and during the first succession stages of the young lake, until the water quality has stabilized. Ecological engineering may help to develop mining lakes as new sustainable ecosystems that have high human and ecological value. Sustainability has to function indefinitely through the design of the lake itself, with only a modest amount of human intervention (Mitsch 1993).

22.5 Conclusions

Acidity of geogenic origin released by mining activities is the most severe water quality problem in the lignite mining lakes in the former GDR. Only a few smaller examples of these extreme habitats should remain unaltered and be preserved for natural succession, in the interest of con-

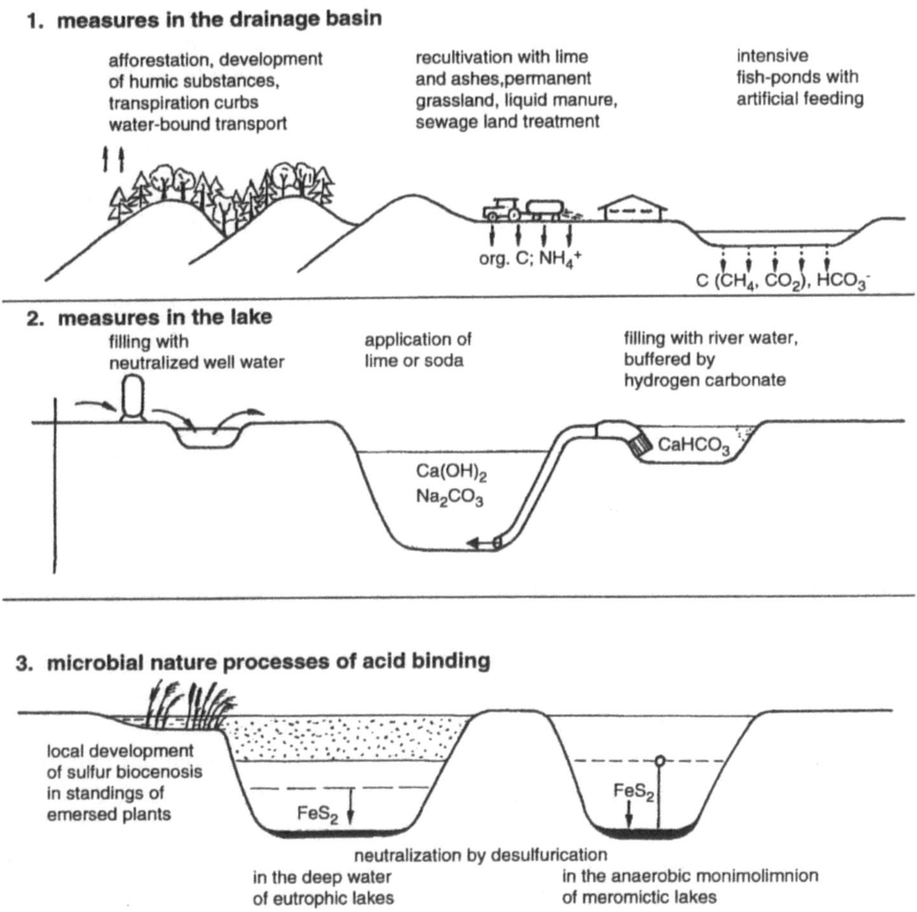

Fig. 22.7. Neutralization of geogenically acidified mining lakes. (Klapper and Schultze 1995)

servation and limnological research. In the majority of cases, water quality management has to be directed towards creating more nearly natural environments. Abatement of the acidification starts from sites of pyrite oxidation, includes the groundwater and acidity transport and concentrates on the mining lakes, where in-lake measures must be taken. Chemical neutralization is often not feasible because such large amounts of alkalizing agents are required. The large lakes are preferably filled with surface waters containing bicarbonate. The disadvantage of a higher trophic level has to be tolerated temporarily. Another neutralization alternative is the microbial process of acid binding desulfurication. There is an urgent need for research into ways to stimulate this strong anaerobic process. Further ecotechnological methods with limestone barriers, constructed wetlands, etc. are suitable for water polishing by aerobic flocculation of iron hydroxide (see Fig. 22.7).

References

Benndorf J (1994) Sanierungsmaβnahmen in Binnengewässern: Auswirkungen auf die trophische Struktur. Limnologica 24:121–135

Brettschneider U, Pöpel HJ (1992) Sulfatentfernung aus Wasser. Wasser Abwasser Praxis 1:298–304

Brown RL (1971) Removal of nitrogen of tile drainage – a summary report. EPA. Bio-engineering aspects of agricultural drainage, rep no. 13030 ELY 7/71, West Raleigh, North Carolina

Fichtner N (1983) Verfahren zur Nitrateliminierung im Gewässer. Acta Hydrochim Hydrobiol 11:339–345

Glässer W, Klapper H (1992) Stoffflüsse beim Füllprozeß von Bergbaurestseen – Entscheidungsvorbereitung für die Sanierung von Tagebaulandschaften. In: Boden, Wasser und Luft. Umweltvorsorge in der AGF. Arbeitsgemeinschaft der Großforschungseinrichtungen, Bonn, pp 19–23

Hedin RS (1989) Treatment of coal mine drainage with constructed wetlands. In: Majumbar SK et al. (eds) Wetland. Ecology and conservation: emphasis in Pennsylvania. The Pennsylvanian Academy of Science, pp 349–362

Jones JR (1971) Denitrification by anaerobic filters and ponds – phase II. EPA. Bio-engineering aspects of agricultural drainage, rep no 13030 ELY 06/71 – 74, West Raleigh, North Carolina

Kelly C (1994) Biological processes that affect water chemistry. In: Steinberg CE, Wright RF (eds) Acidification of freshwater ecosystems: implications for the future. Wiley, Chichester, pp 201–215

Klapper H (1991) Control of eutrophication in inland waters. Ellis Horwood, Chichester

Klapper H, Schultze M (1993) Das Füllen von Braunkohlerestseen. Wasserwirtschaft Wassertechnik 43:34–36

Klapper H, Schultze M (1995) Geogenically acidified mining lakes – living conditions and possibilities of restoration. Int Revue Ges Hydrobiol 80:639–653

Kleinmann RLP, Crerar DA, Pacilli RR (1981) Biogeochemistry of acid mine drainage and a method to control acid formation. Mining Eng 33:300–304

Kreutzer K (1994) The influence of catchment management processes in forests on the recovery in fresh waters. In: Steinberg CE, Wright RF (eds) Acidification of freshwater ecosystems: implications for the future. Wiley, Chichester, pp 325–344
Luckner L, Eichhorn D, Gockel G, Seidel K-H (1995) Durchführbarkeitsstudie zur Rehabilitation des Wasserhaushaltes der Niederlausitz auf der Grundlage vorhandener Lösungsansätze. Dresdner Grundwasserforschungszentrum, Dresden
Mitsch WJ (1993) Ecological engineering; a cooperative role with the planetary life-support system. Environ Sci Technol 27:438–445
Olem H (1991) Liming acidic surface waters. Lewis, Chelsea
Onysko SJ, Kleinmann RLP, Erickson PM (1984) Ferrous iron oxidation by *Thiobacillus ferrooxidans*: inhibition with benzoic acid, sorbic acid and sodium lauryl sulfate. Appl Environ Microbiol 48:229–231
Ripl W, Gerlach-Koppelmeyer I, Wolter K-D (1992) Steuerung des Wasser- und Stoffhaushaltes in einer durch Braunkohlentagebau geschädigten Landschaft für die Wiederherstellung ihrer nachhaltigen Nutzbarkeit. Bericht TU Berlin und Gesellschaft für Gewässerbewirtschaftung GmbH (unpublished report)
Schultze M, Geller W (1996) The acid lakes of the East-German lignite mining district. In: Reuter R (ed) Geochemical approaches for environmental engineering of metals. Environmental science series. Springer, Berlin Heidelberg New York, pp 89–105
Sword BR (1971) Denitrification by anaerobic filters and ponds. EPA. Bio-engineering aspects of agricultural drainage, rep no 13030 ELY 04/71–8, West Raleigh, North Carolina

Part 5
Summary of Group Discussions and Conclusions

In three groups and in a final assembly of all participants the topics listed below were addressed which previously were identified during the workshop as basically unclear points and unsolved problems. The results from the three groups were reported by V.P. Evangelou, M. Kalin and W. Geller, and R.L.P. Kleinmann. Additional short texts on details were given by M. Maiss and B. Scharf. A discussion of the options for chemically classifying the types of acidic lakes emphasized the problem of the appropriate definition of alkalinity and acidity for weakly and strongly acidic waters. The discussion was initiated during the meeting in Magdeburg, Germany, and is still in progress among some of the participants (S. Peiffer, M. Schultze and C. Steinberg). The resulting papers are not included in the present volume (see citations in Chap. 24).

The following topics were discussed:

- Hydrological, geochemical and biological processes in lake and watershed, especially pyrite oxidation and its control.
- Typology and chemical characteristics of acidic lakes.
- Succession from young, acidic lakes to old, neutral lakes, assuming a natural development through time.
- Modelling of processes in lakes, groundwater and watersheds.
- Approaches to the restoration of strongly acidic mining lakes and watersheds.

Further Reading

Geller W, Klapper H, Schultze M (1997) Typologie der sauren Seen. Deutsche Gesellschaft für Limnologie (DGL) Tagungsbericht 1996 (Schwedt), pp 577–581

Peiffer S, Lindemann J (1996) Kritische Anmerkungen zur Verwendung des Alkalinitätsbegriffs in der Limnologie. Deutsche Gesellschaft für Limnologie (DGL) Tagungsbericht 1995 (Berlin), pp 389–393

Reuss JO, Johnson DW (1985) Effect of soil processes on the acidification of water by acid deposition. J Environ Qual 14:26–31

Schultze M, Geller W (1996) The acid lakes of the lignite mining district of the former German Democratic Republic. In: Reuther R (ed) Geochemical approaches to environmental engineering of metals. Environmental science series. Springer, Berlin Heidelberg New York, pp 89–105

Sigg L, Stumm W (1994) Aquatische Chemie. Verlag der Fachvereine, Zürich

Stumm W, Morgan JJ (1981) Aquatic chemistry. 2nd edn. Wiley, New York

Stumm W, Morgan JJ, Schnoor JI (1983) Saurer Regen, eine Folge der Störung hydrogeochemischer Kreisläufe. Naturwissenschaften 70:216–223

23 Pyrite Oxidation Control

V. P. Evangelou (Rapporteur)

R. Chmielewski, K. Friese, H. Klapper, H. Lamb, M. Maiss, J. Meier, T. Neu, B. Nixdorf, T. F. Pedersen, M. Schultze, F. Wisotzky

23.1 Introduction

In general, the consensus of the group was that the physical properties play a very important role in the oxidation of pyrite mainly because of the following two reasons: (1) physical properties are regulating mass/diffusion transport of atmospheric oxygen into the overburden, and (2) physical properties include particle size which affects specific surface (metres squared per gram). The latter influences fate of pyrite oxidation products, including Fe^{3+} which is the main pyrite electron acceptor.

23.2 Composition, Compaction and Porosity of the Overburden

In addition to the above, oxygen diffusion may also be influenced by overburden composition, e.g. sandstone vs. shale or organic matter content. This influence can be dependent on microporosity and/or reactivity of the surface with respect to acting as a chemical sink for O_2. For example, organic matter may consume O_2 due its microbial decomposition, or the presence of Fe^{2+} and Mn^{2+} on the surfaces of overburden, where during metal transformation to higher oxidation states O_2 may be consumed.

Overburden compaction would have a large influence on mass transfer of atmospheric O_2 as well as on water movement. The biggest factor for the influence of compaction is the type of heavy equipment used during reclamation and the mineral composition. For example, it is known that heavy use of scraper pans leads to heavy compaction. Furthermore, the composition of the overburden may have a great deal to do with degree of compaction. For example, sandstone may disintegrate to sand particles, but their compaction potential is limited. On the other hand, the solution composition of shale stones composed of 2:1 clay minerals, as they disintegrate under mechanical compression, may induce dispersion which may bring about maximum compaction. This would limit O_2 and water movement in the overburden.

Physical size of overburden plays a major role in oxidation as well as mass transport of the oxidation products. There are two basic schools of thought on the subject of porosity and pyrite oxidation. The presence of macropores limits the availability of surfaces exposed to oxidation. Furthermore, large quantities of water going through such overburden introduce a dilution effect on the pyrite oxidation products and, thus, the impact of acid mine drainage (AMD) on the environment would be limited. Micropore flow, on the other hand, may increase the production of AMD because of large contact time between overburden solution and large oxidizable pyrite surfaces. However, depending on the degree of microporosity, O_2 and water transfer processes may be impacted and thus pyrite AMD production would be limited. Unfortunately, there are not much data verifying or disputing such micropore/macropore phenomena in overburden. However, there are computer models available that one may use to simulate such processes on a watershed basis.

23.3 Temperature and Availability of Water

Rainfall may influence AMD in several ways. One way is through intensity of mass-transfer as affected by macropore flow. A second way is through rainfall distribution. Distribution could affect AMD production through water availability which could be a pyrite-oxidation-limiting factor (recall that the oxygen atoms making up SO_4 have as source H_2O). A third factor could influence AMD production by considering rainfall distribution relative to temperature distribution. Note that the maximum rate of pyrite oxidation is observed at around 30 °C.

Pyrite oxidation proceeds under low water availability. The presence of free water but not necessarily excess water is necessary for AMD production. Availability of free water in the overburden, especially overburden exposed to the atmosphere, appears to be regulated by two factors: (1) low temperature/freezing conditions, and (2) humidity. Under freezing conditions, availability of water as a pyrite reactant is very low, and microbial activity is negligible. Drought and low humidity would also limit AMD production. Not much is known about the relationship between humidity and pyrite oxidation. One possibility is that under conditions of low quantities of free water, pyrite oxidation, for instance, is driven by water condensing on the pyrite surfaces, and the pH of pyrite surfaces is controlled by hydrolysis of Fe^{3+}. This maintains the pH at the pyrite surfaces in the ideal range (pH 2–3) of microbial pyrite oxidation. The water condensation process on the pyrite surfaces would be the rate-limiting factor. On the other hand, the impact of such AMD on the environment would depend on the nature of mass-transfer processes (macropore flow vs. micropore flow).

23.4 Erosion and Vegetational Succession

Oxidation of pyrite in overburden is highly dependent on availability of oxygen. Because of this limiting factor, pyrite oxidation mostly takes place in the first metre of the overburden. Processes such as erosion would have a large impact on AMD production because of the continuous exposure of pyrite to atmospheric air.

There are various ways of controlling erosion. They vary from heavy mulching to temporary vegetative cover [annual species to permanent green cover (perennial species)]. However, little is known about how vegetation affects pyrite oxidation. One hypothesis is that upon establishing long-term vegetation, O_2 depletion takes place at lower depths due to root respiration. Another hypothesis is that organic matter built up in the overburden over time may be complexing with the surfaces of pyrite, and thus limiting oxidation. There is little if any experimental support for these hypotheses. One may also present arguments that the type of vegetation may play an important role in accelerating pyrite oxidation.

Vegetation does deplete O_2 in the soil. However, it is known that every autumn, when a fraction of the vegetative cover decays, large quantities of nitrate are produced. Leguminous plants, e.g. *Lespedeza*, produce much more nitrate because of the high nitrogen content in their tissue. NO_3 contents as high as 100 mg/l were observed in the water at 2-m depth in overburden covered with *Lespedeza*. Nitrate serves as a pyrite-oxidizing agent. This protects the groundwater from NO (converts to N_2) and acid (the reaction consumes H^+), but the groundwater is enriched with sulfate.

Based on the above, vegetation succession would have an impact on pyrite oxidation as nitrogen-producing plants are replaced by nitrogen-consuming plants or vice versa, or shallow-rooted plants are replaced by deep-rooted plants. The latter would impact on mass-transfer processes.

24 Limnological Fundamentals of Acid Mining Lakes

M. Kalin and W. Geller (Rapporteurs)

Working group: A. Chabbi, U. Grunewald, M. Maiss, A. Peine, S. Peiffer, W. Pietsch, H. Sahin, B. Scharf, C. Steinberg

24.1 Typology and Chemical Characteristics of Acidic Lakes

Acidic mining lakes are clearly different from softwater lakes that have been acidified by aerial deposition, but despite their unusual ecosystems, they have attracted little scientific interest in the past. Their only natural counterparts appear to be acid crater lakes which, similarly, have been poorly studied. With acid mining lakes, the acids (mainly sulphuric acid), result from drastic man-made changes in the subsoil redox system, i.e., by aeration and the weathering of sulfuric minerals. Any discussion of the chemical characteristics of different types of mining lakes, therefore, is not based on the chemical composition of the lake water alone. Equally important is the hydrogeological environment of the lake, in that it functions as a source of, or sink for, acids, iron and dissolved metals.

A survey of the data available on mining and aerially-acidified lakes showed that the pH values are stabilized in three distinct ranges. The three buffering systems are potentially those of CO_2, Al and Fe to which these pH ranges can be attributed (Schultze and Geller (1996); Geller, Klapper and Schultze (1997)). Triangular distribution diagrams of elements, used in mineralogy and soil science, were proposed to be used in order relate the molar ratios among Fe^{3+}, Ca^{+} and Mg^{2+}, and other graphic modes were presented to elucidate the relations between pH, Al_{tot}, Fe^{3+}, SO_4^{2-}, conductivity, acidity and alkalinity.

The debate among participants in the group revealed the importance of relevant definitions, starting with those of acidity and alkalinity. Both criteria may be measured directly, by titration, as base or acid neutralization capacities, but they can also be calculated by summing ions and their species. These two processes lead to different results (see Stumm and Morgan (1981), Stumm, Morgan and Schnoor (1983), Reuss and Johnson (1985), or Sigg and Stumm (1994) for definitions). Peiffer and Lindemann tentatively proposed that the term alkalinity be replaced by a balance of protons for aluminum, iron, and ammonium-rich waters. The definition of relevant criteria is not only a question of terminology and academic interest. Since the two methods can lead to severe discrepancies in the calculated annual loads of alkalinity/acidity of the lake catchment area the resulting estimation of fluxes of acidity/alkalinity. This is a crucial point in any analysis of the acidification potential and forms the whole basis of a scientific forecast for predicting the future development of an existing mining lake or for a planned flooding programm of mine workings or Tagebaurestseen. The entire restauration effort can be called into question.

24.2 Succession from Young, Acidic Lakes to Old, Neutral Lakes

Limnologists often presume that an autochthonous succession from young to mature lakes exists, resulting in lakes with accumulated sediments and a well-developed ecosystem. New mining lakes are considered to be empty habitats which initially have to be colonized by plankton and macrophytes, thereby being altered in a stepwise progression. This view is also supported by authors in this volume, as presented in Part 2 (Chapters 4, 8 and 9). The participants in the group discussions, however, noted that no long-term investigation (covering decades) is known for any mining lake, and so no real data on the question exist. In fact, whether young, acid lakes become neutral by internal processes is still an open question. While there are a number of lakes where such a process has clearly taken place, there are also many examples of 40-year-old acid lakes where no change has occurred. It was proposed (B. Scharf) that a palaeolimnological approach be used to clarify the question. The investigation of a series of lakes which were, presumably, equally acidic in the past, but different today would be useful. The early diagenetic processes in sediments should reflect the chemical successions by accumulation of reduced sulfur compounds (i.e., net alkalinity gain).

The natural succession may be seen on three levels. For the hydro geochemistry of lakes, the large-scale level, the hydro geochemical sources and sinks of the catchment and drainage areas are the driving forces. A lack of information related to case studies about this level makes modelling an urgent need. Hierarchically linked minimum models were recommended. The extension and community structure of the next level, the macrophyte belt, are determined by the physical and chemical characteristics of the littoral zone. While a set of data is available on this smaller scale, evaluation of these data is necessary with focus on the biogeochemical aspects. The driving force for the smallest-scale level, plankton succession and sedimentation, is the lake chemistry, possibly the least complex of the three levels. Although investigations of plankton micro and pico) are in progress, there is still a large deficit of knowledge.

24.3 Modelling of Processes in Lakes and the Hydrological Environment

In order to test new approaches for restoring existing mining lakes and to plan for future restoration efforts, management strategies need the support of prognostic models. Specific physical and biogeochemical inlake processes need to be considered, the results of which would be of benefit to ecological engineering techniques. For example, the pH development of a mining lake may be dominated, but not exclusively controlled by, the surrounding hydro geochemical constraints. An anaerobic layer at the bottom of a lake which is undisturbed by wind because of a stable stratification (meromixis), might, for instance, act as a sink for sulfate, metal and nutrient loadings, as sulfate will be reduced leading to biomineralisation processes.

Since experimental examination of a large number of options is not a possibility, numerical simulations are a promising alternative. To date, models exist for physical and geochemical processes in lakes, and for the ground water. However models integrating both components are not easily available or not existing. While work in this area should be a priority, care must be taken not to incorporate details that would result in excessive program compexity complexity. The group participants recommended starting with hierarchical linked minimum models.

25 Approaches for the Restoration of Strongly Acidic Mining Lakes

R.L.P. Kleinmann (Rapporteur)

H.-D. Babenzien, S. Clasen, M. Hupfer, B. Johnson, Y. T. Kwong, D. Leßmann, G. Packroff, P. Phillips, W. Salomons, K. Wendt-Potthoff, J. Wurl

25.1 The Problem of Acid Lakes in Germany

The acid lakes in Germany are the final result of lignite mining that lasted for decades. The overburden:lignite ratio ranges from 3:1 to 6:1; the resulting material deficit accumulates as mining progresses. In addition, at some sites, the overburden removed during the initial excavation was piled up and revegetated; had this material been replaced in the final cut, the lake volume would be significantly reduced. Also, at some sites, some of the overburden was removed for use as sand and gravel, adding again to the material deficits. For all of these reasons, some of the lakes are very large (over 9 km^2).

During mining, the water table was kept artificially low by pumping. After mining ceased, the water table was allowed to return to pre-mining levels. Average precipitation in this area is only about 0.5 m/year. Therefore, at most sites, the water table is largely being recharged by groundwater. In turn, the contaminated groundwater is affecting lower aquifers. The primary problem is acidity from overburden rock that oxidized while exposed to the atmosphere. With the recovery of the water table, the stored acidity is being released. Thus, acid production peaks some time after water table recovery and then gradually decreases. Ongoing pyrite oxidation is viewed as a secondary problem.

Various ways to modify the shape of the peak were discussed, including compaction, infiltration control and grouting. We eventually concluded that the scale of the problem means that such approaches are probably not appropriate for these old sites because they would negatively affect revegetation, eventual land use, regional groundwater recharge, etc. An alternative treatment discussed was the idea of increasing the initial acidity and the rate of acid dissolution to decrease the long-term acid problem. One possibility that was discussed was pumping the lake water onto the land surface above the overburden rock that recharges the lake.

At an active operation south of Berlin, a mining company is doing something slightly similar, applying neutralized water to the slope, attempting to create an infiltration barrier. We discussed a different approach – pumping the lake water to the recharge area. This could be accompanied by neutralizing chemicals, phosphate addition, etc. to immobilize some of the metals in place. The water could also be treated biologically by constructing wetlands on top, which would then recharge the lakes with clean water.

Potential problems with this approach are decreased slope stability and the fact that much of the land has already been reclaimed for agriculture, limiting the available area of application. Thus, the technique is not applicable to all sites. Another consideration is the nature of the stored acidity, e.g. melanterite, which is very soluble, or, at the other end, jarosite, which will not dissolve if the material is even partially neutralized in place. Geochemical evaluation of the acid salts remaining in the overburden is obviously necessary.

We then talked about treating the lake and discussed establishing an anaerobic zone on the bottom of the lake using organic waste (ideally composted) to treat the acidic groundwater by sulfate reduction. Cost considerations require the use of locally available waste organics (e.g. brewery waste, dairy wastes, sugar beet processing wastes, etc.). Because ferric iron precipitates phosphate, lakes with a pH less than 3 should not shift to eutrophy. At higher pH, eutrophication can also be avoided as long as care is taken in balancing the waste additions with the acid load. The idea is to continue to add organics only until the acidity peak is passed, allowing lake productivity to sustain the process after that point.

We discussed whether a stable stratification would develop, whether the waste would be resuspended and if this would be a problem. We decided that the lakes vary tremendously, but that this approach would be appropriate at some sites where stratification is present or easily induced. Seasonal turnover would periodically mix the acid and alkaline water. This approach has been successfully tested, on a pilot scale, in the USA and UK, and on a large scale in a quarry pit in England. The English test, which consisted of a single application of sewage sludge, was very successful, but showed that periodic addition of additional organic waste is needed to maintain neutralization.

25.2 Conclusions and Research Needed

Studies on available organics in the laboratory, followed by mesocosm experiments and studies on the hydrologic cycles in lakes, i.e. stratification, recharge rates, acid input, etc., are needed. Regarding treatment by stimulating sulfate reduction, the following questions should be considered:

- Which process of adding organic waste should be used – pipeline to bottom of lake?
- How much material needs to be added and at what rate?
- Should limestone be added to encourage initial sulfate reduction?

At sites where the first technique (pumping the lake water to a constructed wetland) has been used, the eventual wetland material can be pumped to the bottom of the lake after it is no longer needed, thereby insuring that the mineral sulfides do not reoxidize.

Diversion of river water to flush out acid lakes has been very successful at several sites and should be utilized wherever possible. However, this diversion should be viewed as a long-term commitment. Short-term diversions will not work at sites where acid generation is a long-term problem. Future lakes may benefit from rapid establishment of non-acidic water body, to inundate the mine water as rapidly as possible.

Liming was viewed as too expensive for most sites. The acid load is several orders of magnitude higher than that found in lakes acidified by rainfall. Neutralization would require the use of $Ca(OH)_2$ rather than limestone. However, liming may be useful as a final polishing step, after most of the acidity has been otherwise neutralized.

Finally, applicability of Canadian and US research is viewed as limited because, by and large, the problems are different. North Americans have generally treated flowing water. The Americans are, however, beginning to deal with the problem of mining lakes at some old abandoned metal mines which have filled up with very acidic waters high in toxic metals. Perhaps German technology, once it has been developed and evaluated, can be exported to North America for use at such sites.

Subject Index